CALCULUS

Louis J. DeLuca
DEPARTMENT OF MATHEMATICS
UNIVERSITY OF CONNECTICUT

James T. Sedlock
DEPARTMENT OF MATHEMATICS
RHODE ISLAND COLLEGE

CALCULUS

A FIRST COURSE

Prentice-Hall, Inc., Englewood Cliffs, New Jersey

© 1973 by Prentice-Hall, Inc., Englewood Cliffs, New Jersey.
All rights reserved.
No part of this book may be reproduced
in any form or by any means
without permission in writing
from the publisher.

10 9 8 7 6 5 4 3 2 1

Library of Congress Catalog Card No. 72-13444

ISBN 0-13-110270-2

Printed in the United States of America

PRENTICE-HALL INTERNATIONAL, INC., *London*
PRENTICE-HALL OF AUSTRALIA, PTY. LTD., *Sydney*
PRENTICE-HALL OF CANADA, LTD., *Toronto*
PRENTICE-HALL OF INDIA PRIVATE LIMITED, *New Delhi*
PRENTICE-HALL OF JAPAN, INC., *Tokyo*

CONTENTS

PREFACE

As the title indicates, this book is intended for use by anyone taking a first course in calculus. Some of the noteworthy features of the text are as follows.

1. The book strikes a balance between concept and computation.
2. The text contains 308 examples, 268 diagrams, and 499 exercises.
3. Applications of the calculus to physical science, social science, and business are included.
4. Historical comments are inserted at appropriate places.
5. To facilitate the computational aspects of the calculus, the Introduction reviews certain topics from high school algebra.
6. No knowledge of trigonometry is required.
7. Certain sections, section supplements, and examples are specifically marked "optional." This material may be omitted in whole or part without handicap. It is intended primarily for those interested in more theoretical aspects of the calculus.

As a result the book is appropriate for a variety of audiences, with its contents easily adapted to one-quarter or one-semester courses of varying credit hours. It is especially appropriate for those having a minimal mathematics background who seek an initial exposure to the concepts, techniques, and applications of the calculus.

The authors wish to thank Professor Kenneth Hoffman of M.I.T. both for his critical analysis of the original manuscript and, subsequently, for his many insightful comments and suggestions. We are grateful to Art Wester of Prentice-Hall for his constant support and assistance, and to Bruce Williams for the support of the Production Department. Another vote of thanks goes to Lauren Gertner for her efficient and speedy typing of the manuscript. We also want to express our thanks to those people who were helpful with their reviews of earlier versions of the manuscript: Gerald Bradley, Claremont Men's College; Robert Mosher, California State College at Long Beach; Richard L. Mentzer, Central Connecticut State College; Dorothy Schrader, Southern Connecticut State College; and Frank Lether, University of Georgia.

Louis J. DeLuca
James T. Sedlock

CALCULUS

INTRODUCTION

I.1 HISTORICAL BACKGROUND

In this short book we shall study a subject which took thousands of years to evolve and crystallize into its present form.

As far back as the fifth century B.C., Greek thinkers sought to determine the areas of certain regions bounded by curves other than straight lines. For instance, Antiphon the Sophist (circa 430 B.C.), Eudoxus (circa 370 B.C.), and Archimedes (287–212 B.C.) attacked area problems by using various *methods of exhaustion*. One such method was to successively increase the number of sides of a polygon inscribed in a given region (see Figure I.1) and

Figure I.1

compute the areas of the resulting polygons. The works of these men—particularly Archimedes—marked the beginning of a segment of the calculus that we now call integration, or **integral calculus**.

It was not until the sixteenth century A.D. that the groundwork laid by the Greeks received further development. Such men as the Flemish engineer

1

Simon Stevin (1548–1620), the Italian mathematicians Luca Valerio (circa 1552–1618) and Bonaventura Cavalieri (1598–1647), and the astronomer Johannes Kepler (1571–1630) made contributions to the development of the theory of integration.

Another geometric problem, that of finding tangents to curves (see Figure I.2), was also investigated by the ancient Greeks. But it was the work of Pierre de Fermat (circa 1601–1665) on this problem which marked the beginning of what we now call differentiation, or **differential calculus**.

Figure I.2

The shining lights in the history of the calculus are, however, Isaac Newton (1642–1727) and Gottfried von Leibniz (1646–1716). Working independently, each created a workable combination that is called the calculus. Two prominent English mathematicians immediately preceding Newton were John Wallis (1616–1703) and Isaac Barrow (1630–1677); Wallis had made key contributions to the development of integration theory, and Barrow's chief contributions were in differentiation theory. It is Barrow who is generally credited with being the first to fully realize the natural inverse relationship between integration and differentiation—a relationship that mathematicians call the fundamental theorem of the calculus.

Newton was only 26 years old when Barrow, recognizing his pupil's great mathematical ability, graciously resigned the Lucasian chair of mathematics at Cambridge in favor of Newton. Stung early in life by criticism of a paper he published dealing with optics, Newton hesitated to publish any of his subsequent findings in science and mathematics. Pressure from his friends ultimately prevailed, but usually long after Newton's original discovery. For instance, his *Method of Fluxions*, dealing with differential calculus, was written in 1671 but was not published until 1736. By 1676, Leibniz had already formulated his version of the calculus, but his first published paper on the subject did not appear until 1684. His use of well-devised notation in treating the calculus initially gave the mathematicians on the European continent an advantage over their counterparts in England. Much of the notation we employ in the calculus today originated with Leibniz.

As indicated above, it is generally held that Newton and Leibniz discovered the calculus independently—Newton being the first to discover it, while Leibniz was the first to publish it. This codiscovery led to a dispute between the mathematicians in England (pro-Newton) and those on the continent (pro-Leibniz), each group claiming priority to the discovery. This unfortunate controversy tainted the later years of both these great men, and it led to the temporary estrangement of English mathematicians from mathematicians on the continent and their developments.

The utility of the calculus lies in the fact that it allows us to analyze *continuous phenomena* such as motion, change, and time. Newton's *Principia* (three books, the first appearing in 1685) was perhaps the most influential and revolutionary scientific work of all time. In it Newton gave for the first time a mathematical description of terrestrial and celestial motion, thus laying a foundation for much of our present-day engineering physics. As we shall see, these basic principles can be used to investigate problems not only in physical science but also in business, economics, and the social sciences.

I.2 MATHEMATICAL LANGUAGE

In the study of mathematics we are often called upon to determine the truth or falsity of certain types of mathematical statements by means of deductive reasoning—the process of drawing inescapable conclusions from given hypotheses.

Consider the following statements:

1. "If $x = 3$, then $x^2 = 9$."
2. "If $x^2 = 9$, then $x = 3$."
3. "If $x^2 \neq 9$, then $x \neq 3$."
4. "If $x^2 = 9$, then $x = 3$ or $x = -3$."
5. "If $x^2 = 9$, then $x = 3$ or $x = -3$; and if $x = 3$ or $x = -3$, then $x^2 = 9$."

Let us denote

$$\text{"}x = 3\text{"}$$

by the letter p, and

$$\text{"}x^2 = 9\text{"}$$

by the letter q. Then Statement 1 can be written as

$$\text{"If }p\text{, then }q\text{,"}$$

which is abbreviated

$$\text{"}p \Rightarrow q\text{"}$$

and is read

$$\text{"}p\text{ implies }q\text{."}$$

Also, Statement 2 can be written as

$$\text{"}q \Rightarrow p\text{,"}$$

which is called the **converse of p $\Rightarrow$ q**.

In mathematics, the statement

$$"p \Rightarrow q"$$

is said to be **true** if q follows inescapably from p. It should be noted that the mathematical use of the word *implies* is different from common usage, where *implies* might mean *suggests* or *indicates*.

EXAMPLE 1   Let p be "$x = 3$" and q be "$x^2 = 9$." Then "$p \Rightarrow q$" is a true statement since the square of 3 is inescapably 9.

EXAMPLE 2   Let p be "$x = 3$" and q be "$x^2 = 9$." Then "$q \Rightarrow p$" is not a true statement since $x = 3$ does not follow inescapably from $x^2 = 9$. ($x = -3$ is another possibility.)

The first two examples show that the converse of a true statement need not be true.

EXAMPLE 3   "If $x^2 \neq 9$, then $x \neq 3$" is a true statement since a number is not equal to 3 if its square is not equal to 9.

EXAMPLE 4   "If $x^2 = 9$, then $x = 3$ or $x = -3$" is a true statement since "$x = 3$ or $x = -3$" is an inescapable conclusion if $x^2 = 9$.

The statement

$$"p \Rightarrow q \text{ and } q \Rightarrow p"$$

is abbreviated as

$$"p \Leftrightarrow q,"$$

and is read, "p if and only if q." We say that

$$"p \Leftrightarrow q"$$

is **true** if both "$p \Rightarrow q$" and "$q \Rightarrow p$" are true.

EXAMPLE 5   Let p be "$x^2 = 9$" and let q be "$x = 3$ or $x = -3$." Then

$$p \Leftrightarrow q$$

is a true statement.

It should be observed that the mathematical usage of the word *or* is different from common usage. In common usage we say

$$"p \text{ or } q" \text{ is true}$$

if p is true or q is true but not both are true, whereas in mathematics we say

$$\text{"}p \text{ or } q\text{" is true}$$

if p is true or q is true or both are true.

EXAMPLE 6 Each of the statements

$$\text{"}1 + 2 = 3 \text{ or } 2 + 3 = 5\text{"}$$

and

$$\text{"}1 + 2 = 3 \text{ or } 2 + 3 = 4\text{"}$$

is true, whereas the statement

$$\text{"}1 + 2 = 4 \text{ or } 2 + 3 = 6\text{"}$$

is false.

EXERCISES

1. In each of the following cases determine whether or not the mathematical statement $p \Rightarrow q$ is true.

	p	q
(a)	$x = 5$	$x^2 = 25$
(b)	$x = 0$	$x^3 = 0$
(c)	$x^2 = 16$	$x = 4$
(d)	$x^2 = 16$	$x = 4$ or $x = -4$
(e)	$2x = 14$	$x = 7$
(f)	$6x - 1 = 2$	$x = \frac{1}{3}$
(g)	$3x + 2 = 11$	$x = 3$ or $x = 1$
(h)	$2xy = 8$	$x = y = 2$

2. For each of the cases in Exercise 1, write the converse of the mathematical statement $p \Rightarrow q$. Which of these are true?

3. For each of the cases in Exercise 1, write the mathematical sentence $p \Leftrightarrow q$. Which of these are true?

4. Which of the following mathematical statements are true?

   (a) $3 + 2 = 5$ or $4 + 1 = 6$.   (b) $1 + 4 = 4$ or $1 + 1 = 2$.

   (c) $6 + 2 = 8$ or $2 + 5 = 7$.   (d) $0 + 3 = 4$ or $2 - 1 = 3$.

   (e) $7 - 3 = 4$ or $8 - 0 = 8$.

5. Recall that an **even integer** is an integer which is divisible by 2 (such as 76) and an **odd integer** is an integer which is not divisible by 2 (such as -23).

With these definitions in mind, determine which of the following mathematical statements are true.

(a) If x is an even integer, then x^2 is an even integer.

(b) If x is an odd integer, then $x + 1$ is an even integer.

(c) If x is an even integer, then $2x + 1$ is an even integer.

(d) If x and y are even integers, then $x + y$ is an even integer.

(e) If x and y are odd integers, then $x + y$ is an odd integer.

(f) If x and y are even integers, then xy is an even integer.

(g) If x and y are odd integers, then xy is an odd integer.

6. For each of the parts of Exercise 5, write the converse of the given implication. Which of these statements are true?

7. Which of the following implications are true?

(a) Triangles A and B are congruent $\Rightarrow$ Triangles A and B are similar.

(b) Trianges A and B are equilateral $\Rightarrow$ Triangles A and B are isosceles.

(c) Triangles A and B are equiangular $\Rightarrow$ Triangles A and B are equilateral.

(d) Triangles A and B are similar $\Rightarrow$ Triangles A and B have equal areas.

(e) Triangles A and B are congruent $\Rightarrow$ Triangles A and B have equal areas.

(f) Lines L_1, L_2 are parallel $\Rightarrow$ Lines L_1, L_2 intersect in a unique point.

(g) Lines L_1, L_2 are perpendicular $\Rightarrow$ Lines L_1, L_2 intersect in a unique point.

8. Write the converse of each of the statements in Exercise 7. Which of these are true?

9. In each of the parts of Exercise 7, replace the implication symbol $\Rightarrow$ by the equivalence symbol $\Leftrightarrow$. Which of the resulting mathematical statements are true?

10. Which of the following mathematical statements are true?

(a) $x = \frac{1}{3} \cdot \frac{1}{4} \Rightarrow x = \frac{1}{12}$. (b) $x = \frac{1}{3} + \frac{1}{4} \Rightarrow x = \frac{2}{7}$.

(c) $x = 2/(2 + 5) \Rightarrow x = \frac{1}{5}$. (d) $a/8 = b/8 \Rightarrow a = b$.

(e) $a \neq b \Rightarrow a^2 \neq b^2$. (f) $a = b \Rightarrow a^2 = b^2$.

(g) $a^2 \neq b^2 \Rightarrow a \neq b$. (h) $a^2 = b^2 \Rightarrow a = b$.

(i) $a = b \Leftrightarrow a^2 = b^2$.

(j) $x = 3y$ for some integer $y \Leftrightarrow 5x = 3z$ for some integer z.

1.3   REAL NUMBERS AND THE NUMBER LINE

Ordinarily, the first numbers that a person uses are the counting numbers

$$1, 2, 3, 4, \ldots$$

This collection of numbers is sometimes called the set of **natural numbers** or **positive integers**. In the study of algebra one quickly realizes that this collection of numbers is inadequate to solve some problems. For instance, there is no positive integer m which satisfies the equation

(I.3.1) $m + 3 = 2.$

To solve this problem, a larger system of numbers is devised—namely, the system of **integers**:

$$\ldots, -4, -3, -2, -1, 0, 1, 2, 3, 4, \ldots$$

Equation I.3.1 does have a solution $m = -1$ in the system of integers.

For the purpose of mathematical analysis it is helpful to have a geometric model of a given system of numbers, and so let us construct one for the integers. Let **L** be a straight line, and choose any point, called the **origin**, to represent the integer 0. Designate the length of a certain line segment as the **unit length** and let the point which is a unit length to the right of 0 represent the integer 1 (see Figure I.3). Multiples of the unit length determine the locations of the remaining integers (see Figure I.4).

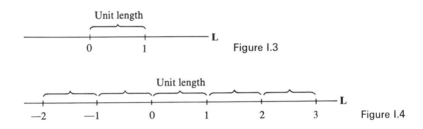

Figure I.3

Figure I.4

Notice that there is a natural ordering of the integers: for any two integers m, n we say that **m is less than n**, written $m < n$, if $m + p = n$ for some positive integer p. Geometrically, $m < n$ means that m is to the left of n on the number line **L**. In summary, for integers m and n

(I.3.2) $m < n \Leftrightarrow m + p = n$ for some positive integer p.

EXAMPLE 1 $-7 < -4$ since $-7 + 3 = -4$, where 3 is a positive integer.

However, the system of integers is also inadequate to solve certain problems. For instance, the equation

(I.3.3) $2x = 5$

is not satisfied by any integer x. This leads us to a consideration of the **rational numbers**—those numbers that can be expressed as the quotient of two integers. Then Equation I.3.3 does have a solution $x = \frac{5}{2}$ in the system of rational numbers.

Geometrically, the rational numbers can be located on the line **L** by using fractional parts of the unit length. For example,

$$\tfrac{2}{3}, \tfrac{3}{2}, -\tfrac{5}{4}, -3$$

are shown in Figure I.5. Also, the above-mentioned ordering of the integers can be extended to an ordering of the rational numbers as follows. For positive rational numbers m/n and p/q,

(I.3.4)
$$\frac{m}{n} < \frac{p}{q} \Leftrightarrow mq < np.$$

Figure I.5

EXAMPLE 2 $\tfrac{2}{3} < \tfrac{5}{4}$ since $(2)(4) = 8 < (3)(5) = 15$. Observe that $\tfrac{2}{3}$ lies to the left of $\tfrac{5}{4}$ on the line **L**.

Recall that addition of rational numbers is given by

$$\frac{m}{n} + \frac{p}{q} = \frac{mq + np}{nq}.$$

It is interesting to note that

$$\frac{1}{2}\left[\frac{mq + np}{nq}\right]$$

is a rational number between m/n and p/q. More precisely, if $m/n < p/q$, then it can be shown that (see Exercise 5)

$$\frac{m}{n} < \frac{1}{2}\left[\frac{mq + np}{nq}\right] < \frac{p}{q};$$

that is, between any two rational numbers there is another rational number.

EXAMPLE 3   We know that $\tfrac{2}{3} < \tfrac{5}{4}$; hence

$$\tfrac{2}{3} < \tfrac{1}{2}[\tfrac{2}{3} + \tfrac{5}{4}] < \tfrac{5}{4};$$

that is,

$$\frac{2}{3} < \frac{1}{2}\left[\frac{8 + 15}{12}\right] < \frac{5}{4};$$

or

$$\tfrac{2}{3} < \tfrac{23}{24} < \tfrac{5}{4}.$$

One could easily continue this example and find another rational number between $\frac{2}{3}$ and $\frac{23}{24}$, say m/n; then find another rational number between $\frac{2}{3}$ and m/n; etc. This suggests a very important property of the rational numbers:

> *Between any two rational numbers there are infinitely many rational numbers.*

Therefore, for a long time it was thought that every point on the line **L** could be associated with a rational number.

The Greek mathematician Pythagoras (born in 572 B.C.) proved that for any right triangle the square of the length of the hypotenuse equals the sum of the squares of the lengths of the legs. Thus, if one considers a right triangle whose legs are of length 1 (see Figure I.6), then the length x of the hypotenuse AB satisfies the equation

$$x^2 = 1 + 1,$$

or

$$x^2 = 2.$$

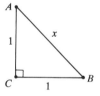

Figure I.6

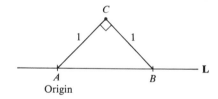

Figure I.7

But, as Pythagorean mathematicians exhibited, there is no rational number whose square is 2 (see Exercise 18). Therefore, if we draw the above triangle so that hypotenuse AB lies on the line **L** with point A at the origin (see Figure I.7), then it follows that point B cannot be associated with a rational number. The number x which satisfies

$$x^2 = 2$$

is called the *square root of 2* and is written as

$$\sqrt{2}.$$

This is the number which is associated with point B on line **L**. It is an example of a nonrational number, or an **irrational number**. We shall see other examples later in this section.

The collection of all rational numbers and irrational numbers together (that is, the numbers that give *all* lengths of line segments, along with their negatives) is called the collection of **real numbers**, denoted by **R**. Operations of addition, subtraction, multiplication, division, raising to powers, and taking roots can be performed on the real numbers subject to certain familiar rules of computation. An ordering of **R** is defined as follows: r is less than s if $s - r$ is a positive number. That is, for real numbers r, s,

(I.3.5) $r < s^* \Leftrightarrow s - r$ is positive.

The collection of all positive real numbers, denoted by $\mathbf{R}^+$, has several important properties worthy of statement:

1. If r is any real number, then r is positive, or $r = 0$, or $-r$ positive.
2. If r and s are positive, then the sum $r + s$ is positive.
3. If r and s are positive, then the product rs is positive.

These properties imply that the square of any number is nonnegative. In the next section certain facts about inequalities will be verified using these properties of $\mathbf{R}^+$.

It is the set of real numbers, and not the set of rational numbers, which completely covers the line **L**. More precisely, there exists a one-to-one correspondence between the collection of real numbers and the points on line **L**.

I.3.6
One-to-One
Correspondence
Between
Numbers
and Points

For each real number x there is one and only one point P on line **L** to which x corresponds. Conversely, for each point P on line **L** there is one and only one real number x corresponding to P.

If the real number x corresponds to the point P on **L**, we say that x is the **coordinate of P**. Line **L**, a geometric model of **R**, is sometimes called the **real line** or a **coordinatized line**. Occasionally we refer to real numbers and points on **L** interchangeably.

One interesting and useful fact about the real numbers is that each has a representation as a decimal. If P is a point on the real line, the decimal representation of the coordinate of P can be obtained by repeated division of intervals into ten pieces having the same length. Figure I.8 illustrates this

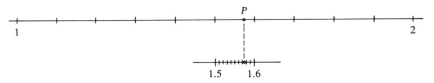

Figure I.8

* "$r < s$" may also be written "$s > r$," which is read as "s is greater than r."

procedure; there the coordinate of P, correct to two decimal places, is 1.57.

The decimal expansion for a rational number can be obtained simply by carrying out the process of division. For instance,*

$$\tfrac{3}{4} = 0.750\overline{0}\ldots$$

$$5 = \tfrac{5}{1} = 5.00\overline{0}\ldots$$

$$\tfrac{2}{3} = 0.6666\overline{6}\ldots$$

$$\tfrac{2}{7} = 0.285714\overline{285714}\ldots$$

In each of these examples, beyond a certain point the expansion consists entirely of the repetition of one particular digit or block of digits (such a decimal expansion is called **repeating**). Conversely, suppose that we are given a repeating decimal, say

$$0.148148\overline{148}\ldots$$

Let

$$r = 0.148148\overline{148}\ldots;$$

then

$$1000r = 148.148\overline{148}\ldots$$

$$1000r - r = 148.00\overline{0}\ldots$$

$$999r = 148$$

$$r = \tfrac{148}{999}.$$

Therefore

$$0.148148\overline{148}\ldots$$

represents the rational number $\tfrac{148}{999}$.

In general each rational number can be represented by a repeating decimal, and, conversely, every repeating decimal represents a rational number.

Thus those numbers which are not rational have decimal representations which are nonrepeating, and, conversely, nonrepeating decimal expansions cannot represent rational numbers. Such numbers, of course, are the irrational numbers. As mentioned above, $\sqrt{2}$ is an irrational number. One can compute its decimal expansion to as many places as desired (here we shall do it to five places) as follows:

$$1^2 = 1 < 2 < 4 = 2^2$$
$$1.4^2 = 1.96 < 2 < 2.25 = 1.5^2$$
$$1.41^2 = 1.9881 < 2 < 2.0164 = 1.42^2$$
$$1.414^2 = 1.999396 < 2 < 2.002225 = 1.415^2$$
$$1.4142^2 = 1.99996164 < 2 < 2.00024449 = 1.4143^2$$
$$1.41421^2 = 1.9999899241 < 2 < 2.0000182084 = 1.41422^2$$
$$\vdots \qquad\qquad \vdots \qquad\qquad \vdots$$

---

* A horizontal line over a digit or block of digits indicates that that digit or block of digits repeats indefinitely.

The numbers on the left form a collection of rational numbers

$$1$$
$$1.4 = 1 + 4/10 = 14/10$$
$$1.41 = 1 + 41/100 = 141/100$$
$$1.414 = 1 + 414/1000 = 1414/1000$$
$$1.4142 = 1 + 4142/10000 = 14142/10000$$
$$1.41421 = 1 + 41421/100000 = 141421/100000$$
$$\vdots$$

which closes in on $\sqrt{2}$; in fact, we can get as close to $\sqrt{2}$ as we wish by continuing this collection far enough. Hence we write the decimal expansion to five places of $\sqrt{2}$ as

$$\sqrt{2} = 1.41421\ldots \quad .$$

A similar procedure will yield decimal expansions for certain other irrational numbers, such as

$$\sqrt{3} = 1.73205\ldots$$

$$\sqrt{5} = 2.23606\ldots \quad .$$

However, another technique must be used to get

$$\pi = 3.14159\ldots \quad ,$$

which is a different kind of irrational number because it cannot be expressed as the root of some integer.

EXAMPLE 4   The decimal expansions to five places of π and $\sqrt{3}$ are given above. Using these we can form the following table:

r	3	3.1	3.14	3.141	3.1415	3.14159	$\cdots$
s	1	1.7	1.73	1.732	1.7320	1.73205	$\cdots$
$r + s$	4	4.8	4.87	4.873	4.8735	4.87364	$\cdots$

The collection of rational numbers

$$4, 4.8, 4.87, 4.873, 4.8735, 4.87364, \ldots$$

closes in on $\pi + \sqrt{3}$. From this we have

$$4.8736\ldots$$

as a decimal expansion to four places of

$$\pi + \sqrt{3}.$$

EXAMPLE 5   Knowing the decimal expansion of $\sqrt{2}$, we can form the table

r	1	1.4	1.41	1.414	1.4142	1.41421	$\cdots$
$4r$	4	5.6	5.64	5.656	5.6568	5.65684	$\cdots$

The collection of rational numbers

$$4, 5.6, 5.64, 5.656, 5.6568, 5.65684, \ldots$$

closes in on $4\sqrt{2}$, and

$$5.6568\ldots$$

is the decimal expansion to four places of

$$4\sqrt{2}.$$

The ideas presented in the second half of this section provide an introduction to one of the basic ideas of the calculus—the notion of *limit*. Loosely speaking, whenever we have a mathematical expression which can be made as close as we wish to a number b, then we say that the limit of the expression is b. These thoughts will be examined more carefully in Chapter 3.

EXERCISES

1. Draw a coordinatized line and locate the points $-5, 0, \frac{7}{3}, 6, -1, -\frac{1}{4}, \frac{11}{5}$.

2. Determine which of the following statements are true.

(a) $4 < 7$.   (b) $-2 < -3$.

(c) $-3 < -2$.   (d) $0 < -\frac{1}{2}$.

(e) $0 < \frac{1}{2}$.   (f) $-\frac{12}{7} < -\frac{12}{7}$.

(g) $0 < 10 < 12$.   (h) $-2 < -5 < 10$.

(i) $1 < \sqrt{2} < \pi$.   (j) $\frac{23}{6} < \frac{32}{5}$.

(k) $-\frac{7}{2} < -5$.   (l) $-\frac{2}{3} < \frac{11}{7} < \frac{13}{10}$.

3. (a) Find an integer m satisfying $\frac{7}{10} < m/5 < \frac{5}{6}$.

(b) Find two integers which satisfy $-5 < m/3 < -4$.

(c) Is there an integer m satisfying $\frac{2}{3} < m/5 < \frac{3}{4}$?

4. Find three rational numbers between $\frac{5}{3}$ and $\frac{9}{4}$.

5. Verify the following statement: If m/n, p/q are rational numbers where $m/n < p/q$, then $m/n < \frac{1}{2}[(mq + np)/nq] < p/q$. [*Hint:* Use the definition of $<$.]

6. Find the decimal expansions of the numbers $4, -\frac{1}{2}, \frac{13}{3}, -\frac{25}{99}, \frac{1}{7}, \frac{8}{7}$.

7. In each of the following cases find a rational number which is equal to the given repeating decimal.

(a) $0.66\overline{6}\ldots.$

(b) $-2.66\overline{6}\ldots.$

(c) $0.28\overline{28}\ldots.$

(d) $5.436\overline{436}\ldots.$

(e) $0.0001\overline{0001}\ldots.$

(f) $1.00\overline{0}\ldots.$

(g) $0.99\overline{9}\ldots.$

8. (a) Derive to four places the decimal expansion of $\sqrt{5}$.

(b) Using this computation, write a collection of rational numbers which closes in on $\sqrt{5}$.

9. Locate $\sqrt{5}$ exactly on the number line **L**.

[*Hint:* Use the right triangle in Figure I.9 and the method of locating $\sqrt{2}$ employed in the text.]

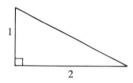

Figure I.9

10. Find to four places the decimal expansion of $\sqrt{2} + \sqrt{3}$.

11. (a) Given the table

r	3	3.1	3.14	3.141	3.1415	3.14159	3.141592	3.1415926	3.14159265	3.141592653
2^r	8	8.6	8.82	8.821	8.8244	8.82496	8.824974	8.8249775	8.82497781	8.824977823

use it to determine the decimal expansion to five places of 2^π.

(b) Find the decimal expansion to five places of $2^\pi - 1$.

12. Verify that the square of a real number is either zero or positive. [*Hint:* Use property 3 of $\mathbf{R}^+$.]

13. For any positive number b, the positive number s satisfying $s^2 = b$ is called the **square root of *b*** and is denoted by $\sqrt{b}$.

(a) Verify that for positive numbers r, s, $\sqrt{rs} = \sqrt{r}\sqrt{s}$ and $\sqrt{r/s} = \sqrt{r}/\sqrt{s}$.

(b) Is it true that for any positive numbers r, s, $\sqrt{r + s} = \sqrt{r} + \sqrt{s}$?

14. Simplify the following.

(a) $\sqrt{36}$.

(b) $\sqrt{20}$.

(c) $\sqrt{128}$.

(d) $\sqrt{\frac{4}{49}}$.

(e) $\sqrt{\frac{81}{25}}$.

(f) $\sqrt{3} + \sqrt{27}$.

15. Identify each of the following as rational or irrational.

(a) $\sqrt{12}$.

(b) $\sqrt{16}$.

(c) 4π.

(d) $\sqrt{72}$.

(e) $\sqrt{\pi/49}$.

(f) $\sqrt{2} + 1$.

16. Let r_1, r_2 be nonzero rational numbers and i_1, i_2 be irrational numbers. In each of the following, determine whether the given statement is true or false.

(a) $r_1 + r_2$ is a rational number.

(b) $i_1 + i_2$ is an irrational number.

(c) $r_1 + i_1$ is an irrational number.

(d) $r_1 r_2$ is a rational number.

(e) $i_1 i_2$ is an irrational number.

(f) $r_1 i_1$ is an irrational number.

(g) $1/i_1$ is an irrational number.

17. Review the definitions of even integer and odd integer given in Exercise 5 of Section I.2. We wish to prove the following statement: If k is an integer, k^2 even $\Rightarrow k$ even. This will be done by the method of contradiction.

Proof: Let k^2 be an even integer, and suppose that k is an odd integer. Then k can be written in the form $k = 2n + 1$ for some integer n. Why? Therefore, $k^2 = (2n + 1)^2 = 4n^2 + 4n + 1 = 2(2n^2 + 2n) + 1$.

Therefore, k^2 is odd (why?), which contradicts the fact that k^2 is even. Hence the supposition that k is odd is false. The reader should supply the requested reasons.

18. Below is a proof (using the method of contradiction) that $\sqrt{2}$ is an irrational number. The basic idea involved in this method of proof is to suppose that $\sqrt{2}$ is a rational number and deduce from this supposition a conclusion which is known to be false. This will allow us to conclude that the supposition (namely, that $\sqrt{2}$ is rational) is false. Recall that any fraction can be placed in reduced form, where the numerator and denominator have only 1 and -1 as common factors (e.g., $-\frac{2}{3}$ is the reduced form of $-\frac{6}{9}$).

Proof: Suppose that $\sqrt{2}$ is a rational number. Then $\sqrt{2}$ can be written in the form $p/q = \sqrt{2}$, where p, q are integers and p/q is in reduced form. Then the following are true.

(a) $p/q = \sqrt{2}$.

(b) $p^2/q^2 = 2$. Why?

(c) $p^2 = 2q^2$.

(d) p^2 is even.

(e) p is even. Why?

(f) $p = 2n$ for some integer n. Why?

(g) $(2n)^2 = 2q^2$. Why?

(h) $4n^2 = 2q^2$.

(i) $2n^2 = q^2$.

(j) q^2 is even.

(k) q is even. Why?

It is a contradiction to say that both p and q are even. Why? Therefore, $\sqrt{2}$ is not a rational number. The reader should supply the requested reasons.

I.4 SOME ALGEBRAIC FACTS AND TECHNIQUES

It is our purpose in this section to examine a few additional facts about the real number system and to develop some algebraic techniques which will be of particular help to us in the study of the calculus.

Recall from high school algebra that, given numbers* r and s, the quotient r/s is defined to be the number t such that

$$r = st.$$

That being the case, what meaning, if any, can be attached to the symbols

$$\tfrac{0}{1}, \tfrac{1}{0}, \tfrac{0}{0}?$$

First, since

$$0 = 1 \cdot t$$

is satisfied by $t = 0$, then we say that

$$\tfrac{0}{1} = 0.$$

But

$$1 = 0 \cdot t$$

is satisfied by no number t, and so we say that

$$\tfrac{1}{0}$$

makes no sense, or is not defined. Finally, since

$$0 = 0 \cdot t$$

is satisfied by any number t, then we say that

$$\tfrac{0}{0}$$

is not uniquely determined, or is **indeterminate**.

These observations suggest the following more general facts:

1. A fraction is not defined whenever its denominator is zero.
2. A fraction is zero whenever its numerator is zero and its denominator is not zero.

EXAMPLE 1   For what values of x is the fraction

$$\frac{x - 2}{2x + 3}$$

not defined? For what values of x is this same fraction equal to zero?

The fraction is not defined when

$$2x + 3 = 0,$$

that is,

$$x = -\tfrac{3}{2}.$$

---

* In the remainder of this text, whenever we refer to a *number* we mean a real number.

Whereas

$$\frac{x - 2}{2x + 3} = 0$$

when

$$x - 2 = 0.$$

that is,

$$x = 2.$$

Note that

$$2x + 3 = 7 \neq 0 \qquad \text{when } x = 2.$$

Another fact one usually encounters in high school algebra is

If a product of numbers is zero, then at least one of the factors is zero.

For, suppose that

$$rs = 0.$$

If $r \neq 0$, then $1/r$ is defined and

$$\frac{1}{r}(rs) = \frac{1}{r}(0);$$

that is,

$$s = 0.$$

EXAMPLE 2 For what values of x is

$$2x^2 - x - 10 = 0?$$

Factoring, we have

$$2x^2 - x - 10 = (2x - 5)(x + 2).$$

Therefore

$$2x^2 - x - 10 = 0$$

if

$$2x - 5 = 0 \qquad \text{or} \qquad x + 2 = 0,$$

that is, if

$$x = \tfrac{5}{2} \qquad \text{or} \qquad x = -2.$$

In the calculus we sometimes must determine those numbers at which a given expression is positive and those numbers at which it is negative. Thus it will be helpful to recall the **law of signs** for multiplying numbers:

1. The product of two positive numbers is positive.
2. The product of two negative numbers is positive.
3. The product of a positive number and a negative number is negative.

Each of these statements is also true if the word *product* is replaced by the word *quotient*.

In addition to the law of signs, we mention below some facts about inequalities which will allow us to manipulate inequalities algebraically. For instance, consider the inequality

$$2 < 3.$$

One can add -2 to each side and obtain

$$0 < 1,$$

which is true; also, one can multiply each side by 2, obtaining

$$4 < 6,$$

which is true. However, if one multiplies each side of the given inequality by -3, then we get

$$-6 < -9,$$

which is *not* true; in fact, the opposite,

$$-9 < -6,$$

is true. These are special cases of

**I.4.1
Facts About
Inequalities**

Let r, s be numbers.

1. If $r < s$ and t is any number, then

$$r + t < s + t.$$

2. If $r < s$ and t is any positive number, then

$$rt < st.$$

3. If $r < s$ and t is any negative number, then

$$st < rt.$$

Each of the three facts above is also true if the symbol $<$ is replaced everywhere by the symbol $\leq$, read "less than or equal to."*

To verify fact 2, recall that

$$r < s \Rightarrow s - r \text{ is positive.}$$

* "$r \leq s$" may also be written "$s \geq r$," read "s is greater than or equal to r."

Since t is positive, then

$$(s - r)t = st - rt$$

is also positive; that is,

$$rt < st.$$

EXAMPLE 3 For what values of x is

$$3x - 3 > x - 1?$$

Our aim here is to isolate x on one side of the inequality symbol $>$.

$$3x - 3 > x - 1$$

Add $-x$ to each side [fact 1]: $\quad 2x - 3 > -1$

Add 3 to each side [fact 1]: $\quad 2x > 2$

Multiply each side by $\frac{1}{2}$ [fact 2]: $\quad x > 1.$

The result is therefore

$$x > 1.$$

Given numbers r, s, the expression

$$r < x \quad \text{and} \quad x < s$$

is abbreviated by

$$r < x < s;$$

the expression

$$r \leq x \quad \text{and} \quad x \leq s$$

is abbreviated by

$$r \leq x \leq s;$$

and the expression

$$r < x \quad \text{and} \quad x \leq s$$

is abbreviated by

$$r < x \leq s.$$

I.4.2
Definitions Let r, s be numbers and $r < s$.

1. The **open interval** denoted by (r, s) is the collection of all points x satisfying

$$r < x < s.$$

2. The **closed interval** denoted by $[r, s]$ is the collection of all points x satisfying

$$r \leq x \leq s.$$

3. The **half-open intervals** $[r, s)$ and $(r, s]$ represent the collections of points x satisfying

$$r \leq x < s \quad \text{and} \quad r < x \leq s,$$

respectively.

EXAMPLE 4   See Figure I.10. The open interval $(-6, -4)$ is the collection of all points x satisfying

$$-6 < x < -4;$$

Figure I.10

the closed interval $[-3, -2]$ is the collection of all points x satisfying

$$-3 \leq x \leq -2;$$

the half-open interval $[0, 4)$ is the collection of all points x satisfying

$$0 \leq x < 4;$$

and the half-open interval $(5, 8]$ is the collection of all points x satisfying

$$5 < x \leq 8.$$

EXAMPLE 5   For what values of x is

$$x^2 - 4x + 3 < 0?$$

Factoring the expression on the left-hand side of the inequality gives

$$(x - 3)(x - 1) < 0.$$

Now a product is negative if one factor is positive and the other factor is negative (law of signs), and therefore we have two cases.

Case 1: $x - 3 < 0$ and $x - 1 > 0$. Then

$$x < 3 \quad \text{and} \quad x > 1;$$

that is,

$$1 < x < 3.$$

Case 2: x − 3 > 0 and *x − 1 < 0.* Then

$$x > 3 \quad \text{and} \quad x < 1;$$

however, it is impossible for a number *x* to satisfy both these inequalities.
The desired answer, then, is the collection of those numbers *x* satisfying

$$1 < x < 3,$$

that is, the closed interval [1, 3] (see Figure I.11).

Figure I.11

EXAMPLE 6   For what values of *x* is $(1 − x)/(2x^2 − x)$ positive, and for what values of *x*
is it negative? Let us begin by examining the numerator and denominator
separately.

Numerator:

$$1 − x > 0 \quad \text{when } x < 1$$
$$1 − x < 0 \quad \text{when } x > 1.$$

Schematically, we indicate this in Figure I.12.

Figure I.12

Denominator:

$$2x^2 − x = x(2x − 1) > 0.$$

Case 1:

$$x > 0 \quad \text{and} \quad 2x − 1 > 0;$$
$$x > 0 \quad \text{and} \quad x > \tfrac{1}{2};$$
$$x > \tfrac{1}{2}.$$

Case 2:

$$x < 0 \quad \text{and} \quad 2x − 1 < 0;$$
$$x < 0 \quad \text{and} \quad x < \tfrac{1}{2};$$
$$x < 0.$$

Hence, $2x^2 − x > 0$ for those *x* satisfying either

$$x > \tfrac{1}{2} \quad \text{or} \quad x < 0.$$

Now observe that $2x^2 - x = x(2x - 1) = 0$ only when

$$x = 0 \quad \text{or} \quad x = \tfrac{1}{2}.$$

Therefore, $2x^2 - x < 0$ for those x *other than*

$$x \geq \tfrac{1}{2} \quad \text{or} \quad x \leq 0.$$

That is,

$$2x^2 - x < 0 \quad \text{when } 0 < x < \tfrac{1}{2}.$$

Figure I.13 summarizes these results.

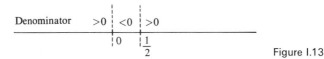

Figure I.13

Recall that a fraction is positive whenever its numerator and denominator have the same sign and negative when they have opposite signs. Combining the above two diagrams, we have Figure I.14. Therefore, the given fraction is positive when

$$x < 0 \quad \text{or} \quad \tfrac{1}{2} < x < 1$$

and negative when

$$0 < x < \tfrac{1}{2} \quad \text{or} \quad x > 1.$$

Figure I.14

The results of this example can be summarized by the following table:

x	$x < 0$	$x = 0$	$0 < x < \tfrac{1}{2}$	$x = \tfrac{1}{2}$	$\tfrac{1}{2} < x < 1$	$x = 1$	$x > 1$
$\dfrac{1 - x}{2x^2 - x}$	$+$	n.d.*	$-$	n.d.	$+$	0	$-$

* The abbreviation n.d. means not defined.

EXERCISES

1. In each case, determine those values of x for which the given fraction is not defined.

(a) $\dfrac{3x + 1}{x - 1}$.

(b) $\dfrac{6x + 2}{3x + 5}$.

(c) $\dfrac{2x - 7}{3x + 2}$.

(d) $\dfrac{x - 2}{x^2 + 1}$.

(e) $\dfrac{x - 2}{x^2 - 1}$.

(f) $\dfrac{3}{x^2 + 5x + 6}$.

2. In each case of Exercise 1, find those values of x for which the given fraction equals zero.

3. Solve the following equations.

(a) $4x^2 - 3x = 0$.

(b) $x^2 - 4 = 0$.

(c) $x^2 - x - 20 = 0$.

(d) $2x^2 + 2x - 12 = 0$.

4. Solve the following inequalities.

(a) $2x + 5 > 0$.

(b) $4x - 2 \geq 7x + 3$.

(c) $x \leq 5x - 6$.

(d) $2(x + 5) > -3(x - 2)$.

(e) $x^2 + 5x + 6 > 0$.

(f) $x^2 + 5x + 6 \leq 0$.

(g) $x^2 - 6x + 5 \geq 0$.

(h) $x^2 - x - 2 < 0$.

(i) $(x + 2)(x - 1)(x + 5) < 0$.

(j) $x^4 - 16 > 0$.

(k) $x^3 - x \leq 0$.

5. Draw the following intervals and identify each as open, closed, or neither: $[0, 3]$, $(0, 3)$, $(-5, 2)$, $[-5, 2)$, $[\frac{1}{2}, \frac{7}{3}]$, and $(-\frac{21}{5}, -1]$.

6. For each of the following, determine where the given expression is positive, negative, zero, and not defined. Summarize your results by using tables similar to the table in Example 6 in the text.

(a) $\dfrac{x}{x - 1}$.

(b) $\dfrac{2x + 1}{x^2 + 3x + 2}$.

(c) $\dfrac{1}{x - 2} + \dfrac{1}{x}$.

(d) $\dfrac{1}{3x + 2} + x$.

(e) $\dfrac{4}{x^2 - x - 6}$.

(f) $\dfrac{x + 1}{x^2 - 1}$.

(g) $\dfrac{x - 1}{x^2 + 2}$.

7. Give an example showing that the following statement is not true: For any number t, $r < s \Rightarrow tr < ts$.

8. Verify that $r < s$ and $s < t \Rightarrow r < t$.

9. Verify that $r < s$ and $t < w \Rightarrow r + t < s + w$.

10. Give an example showing that the following statement is not true: $r < s$ and $t < w \Rightarrow r - t < s - w$.

11. Verify the following statement: For positive numbers r, s, t, and w, $r < s$ and $t < w \Rightarrow rt < sw$.

12. Verify the following statement: $0 < r < s \Rightarrow r^2 < s^2$. More generally, for any positive integer m, $0 < r < s \Rightarrow r^m < s^m$.

13. Verify the following statement: $0 < r < s \Rightarrow \sqrt{r} < \sqrt{s}$. More generally, for any positive integer m, $0 < r < s \Rightarrow \sqrt[m]{r} < \sqrt[m]{s}$.

14. Verify the following statement: $rs > 0$ and $r < s \Rightarrow 1/r > 1/s$. [*Hint:* Rewrite $1/r - 1/s$.]

I.5 ABSOLUTE VALUE AND DISTANCE

Before considering the notion of distance between two points on a co-ordinatized line, we shall introduce the following piece of new notation.

I.5.1

Notation   For any number r, the nonnegative number

$$|r| = \begin{cases} r, & \text{if } r > 0, \\ 0, & \text{if } r = 0, \\ -r, & \text{if } r < 0, \end{cases}$$

is called the **absolute value of r**.

EXAMPLE 1 $|4| = 4, |-3| = 3, |0| = 0, |-\tfrac{5}{2}| = \tfrac{5}{2}, |\sqrt{2}| = \sqrt{2}.$

The following is a list of three important facts concerning absolute value. Others appear in the Exercises at the end of this section.

I.5.2

Facts About Absolute Value   Let r, s, x be numbers.

1. $|r| = |-r|$.

2. If $r > 0$, then
$$|x| < r \Leftrightarrow -r < x < r.$$

3. $|r + s| \leq |r| + |s|$ (triangle inequality).

Fact 2 remains true if the symbol $<$ is replaced everywhere by $\leq$.

One way of verifying the triangle inequality is as follows: Observe that for any number x,

$$x \leq |x| \qquad \text{and} \qquad x^2 = |x|^2.$$

Hence

$$|r + s|^2 = (r + s)^2$$

$$= r^2 + 2rs + s^2$$

$$= |r|^2 + 2rs + |s|^2$$

$$\leq |r|^2 + 2|r||s| + |s|^2$$

$$= (|r| + |s|)^2.$$

Since

$$|r + s| \geq 0 \quad \text{and} \quad |r| + |s| \geq 0,$$

then

$$|r + s| \leq |r| + |s|.$$

EXAMPLE 2   For what values of x is

$$5 - |x + 3| > 0?$$

First, rearrange terms so that just the absolute value expression is on one side of the inequality:

$$|x + 3| < 5.$$

Then, using fact 2 of I.5.2,

$$-5 < x + 3 < 5;$$

and so

$$-8 < x < 2.$$

In interval form, the answer is $(-8, 2)$.

EXAMPLE 3   For what values of x is

$$|x - 2| \geq 3?$$

We can solve the problem by first finding those values of x for which $|x - 2| \geq 3$ is false; that is, find those values of x for which $|x - 2| < 3$. Therefore

$$|x - 2| < 3$$

$$-3 < x - 2 < 3 \quad [\text{I.5.2(2)}]$$

$$-1 < x < 5.$$

Thus our answer is all those x satisfying either $x \geq 5$ or $x \leq -1$.

As mentioned above, the absolute value notation provides a convenient device for dealing with distance.

I.5.3

Definition

Let r and s be numbers. Then the **distance between the points r and s** on a coordinatized line is the nonnegative number

$$|r - s|.$$

Geometrically, $|r - s|$ is the length of that portion of the coordinatized line between the points r and s; in particular, $|r| = |r - 0|$ is the length of that portion of the line between r and zero. This geometric interpretation makes reasonable the statement of fact 2 in I.5.2, for to say that "the distance from zero to x is less than the positive number r" is exactly the same as saying that "x lies between $-r$ and r" (see Figure I.15).

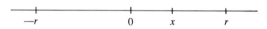

$-r$ \qquad\qquad 0 \qquad x \qquad r \qquad\qquad **Figure I.15**

EXAMPLE 4

$$|\tfrac{3}{2}| = \tfrac{3}{2}, \qquad |-2| = 2$$
$$|\tfrac{3}{2} - (-2)| = |\tfrac{3}{2} + 2| = \tfrac{7}{2}.$$

Therefore the distance between zero and $\tfrac{3}{2}$ is $\tfrac{3}{2}$, the distance between zero and -2 is 2, and the distance between -2 and $\tfrac{3}{2}$ is $\tfrac{7}{2}$ (see Figure I.16).

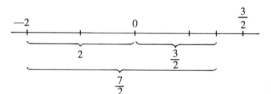

Figure I.16

We conclude this section by applying the concept of distance to find the coordinate of the midpoint of an interval.

r \qquad M \qquad\qquad s \qquad **Figure I.17**

To this end, consider numbers r, s, where $r < s$, as in Figure I.17. We seek the number M which is the midpoint of the interval $[r, s]$. In order that M be midway between r and s, it must be that the distance between r and M equals the distance between M and s. That is,

$$M - r = s - M$$
$$2M = s + r$$
$$M = \frac{s + r}{2}.$$

I.5.4
Midpoint of an Interval

The midpoint of the interval $[r, s]$ is

$$\frac{s + r}{2}.$$

EXAMPLE 5

Given the points -1 and $\frac{5}{3}$, the distance between them is

$$\left|-1 - \tfrac{5}{3}\right| = \left|-\tfrac{8}{3}\right| = \tfrac{8}{3},$$

and the midpoint of $\left[-1, \tfrac{5}{3}\right]$ is

$$\frac{\tfrac{5}{3} + (-1)}{2} = \frac{1}{3}.$$

See Figure I.18.

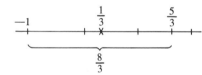

Figure I.18

EXERCISES

1. Determine the following.

    (a) $|5|$.   (b) $|29|$.   (c) $|-1|$.   (d) $\left|\tfrac{5}{9}\right|$.

    (e) $\left|-\tfrac{5}{9}\right|$.   (f) $|\pi|$.   (g) $\left|-1/\sqrt{3}\right|$.   (h) $|-492|$.

2. Evaluate $|2 - 3x + x^2|$ for $x = -1, 0, 1, 2, 3$.

3. Solve each of the following equations.

    (a) $|x + 3| = 0$.   (b) $|x| + 3 = 0$.

    (c) $|x| - 3 = 0$.   (d) $|x + 3| + |x + 5| = 0$.

4. Solve each of the following inequalities.

    (a) $|x| < 2$.   (b) $|2x + 1| < 5$.

    (c) $|x - 3| \le 1$.   (d) $0 < |2x + 1| < 5$.

5. Solve each of the following inequalities.

    (a) $|x| > 3$.   (b) $|x| \ge 3$.

    (c) $|x - 4| > 1$.   (d) $|2x + 5| \ge 2$.

    (e) $|7 - x| \ge a \; (a > 0)$.

6. Plot each of the following pairs of points on the coordinatized line and find the distance between the two.

    (a) $0, 3$.   (b) $-5, 2$.   (c) $\tfrac{1}{2}, \tfrac{7}{6}$.   (d) $-\tfrac{21}{5}, -1$.

7. Find the midpoint of each of the following intervals.

(a) $[0, 3]$.   (b) $[-5, 2]$.   (c) $[\frac{1}{2}, \frac{7}{6}]$.   (d) $[-\frac{21}{5}, -1]$.

8. Verify that for any numbers r, s,

(a) $|r| = |-r|$.   (b) $|r - s| = |s - r|$.

9. Verify that for any number r,

(a) $r^2 = |r|^2$.   (b) $\sqrt{r^2} = |r|$.

10. Show that for any numbers r, s,

(a) $|rs| = |r||s|$.   (b) $|r/s| = |r|/|s| \ (s \neq 0)$.

11. Show that for any number r, $-|r| \le r \le |r|$.

12. Verify that for any numbers r, s, $|r - s| \le |r| + |s|$.

13. Verify that for any numbers r, s, and t, $|r + s + t| \le |r| + |s| + |t|$. [*Hint:* $r + s + t = (r + s) + t$.]

14. Show that for $r > 0$, $|x| < r \Leftrightarrow -r < x < r$. [*Hint:* Consider the cases $x = 0$, $x < 0$, $x > 0$ separately.]

15. Show that for any numbers r, s, $|r - s| \ge ||r| - |s||$.

16. Simplify each of the following.

(a) $\sqrt{9x^2}$.   (b) $\sqrt{x^2 + 6x + 9}$.   (c) $\sqrt{\dfrac{x^2}{x^2 + 2x + 1}}$.

Chapter 1

ELEMENTS OF ANALYTIC GEOMETRY

One of the important precalculus developments in mathematics was the introduction of coordinates into geometry. Plane coordinate geometry, or plane analytic geometry, establishes a one-to-one correspondence between ordered pairs of real numbers and points in the plane. The importance of this correspondence lies in the fact that a curve in the plane can be represented by the equation.

The Egyptians, Romans, and Greeks all used the idea of coordinates, but they did not know coordinate geometry as we know it today because they did not possess the necessary algebraic notation and technique. By the seventeenth century, however, there was enough algebra available for two French mathematicians, René Descartes (1596–1650) and Pierre de Fermat (1601–1665), to develop analytic geometry.

Descartes' contribution to the subject first appeared in 1637 in an appendix of a book on the philosophy of science entitled (briefly) *Discours*. There is a story that says that Descartes got his idea for introducing coordinates while lying in bed and observing the motion of a fly on the ceiling near a corner. He noticed that the position of the fly could be described by indicating its distance from the adjacent walls.

Fermat's contribution to analytic geometry came at approximately the same time that Descartes' work appeared. While Descartes considered the problem of writing an equation to describe a given locus, Fermat studied the converse problem of describing the locus of a given equation. In general,

Fermat did more work on curves. The equations of a general straight line and of a circle can be found in his writings. He also studied parabolas, ellipses, and hyperbolas.

1.1 COORDINATES IN THE PLANE

To place coordinates on points in a plane, we shall use two lines which are perpendicular to each other at a point which we shall call the origin and designate by O (see Figure 1.1). We set up coordinates as before on both lines, as indicated in Figure 1.2. The horizontal line is referred to as the **x-axis** and the vertical line as the **y-axis**.

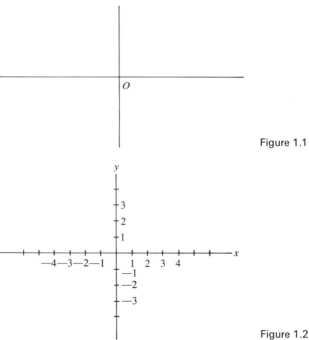

Figure 1.1

Figure 1.2

To describe a point in the plane, we draw a horizontal and vertical line through the given point (see Figure 1.3). The point will be designated by (a, b) if the vertical line crosses the x-axis at the point a and the horizontal line crosses the y-axis at the point b. When each point in the plane is designated in this fashion, we say that we have a **rectangular coordinate system**. The number a is called the **x-coordinate of** (a, b) and b is called the **y-coordinate of** (a, b). We shall sometimes refer to the point (a, b) by the letter P, and we shall write both together: $P(a, b)$. We should note that the point a on the x-axis is now denoted by $(a, 0)$ and the point b on the y-axis by $(0, b)$.

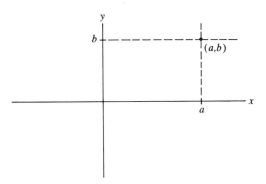

Figure 1.3

EXAMPLE 1   Figure 1.4 shows the points $(2, 3)$, $(-\frac{1}{2}, 4)$, $(-3, -3)$, and $(4, -\frac{5}{3})$ plotted in a rectangular coordinate system.

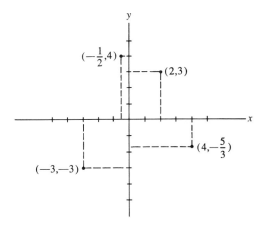

Figure 1.4

A rectangular coordinate system divides the plane into four parts called quadrants, which are numbered as in Figure 1.5.

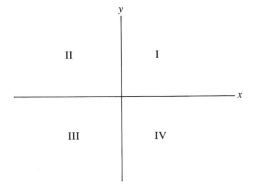

Figure 1.5

EXAMPLE 2   The point $(2, 3)$ is in quadrant I, $(-\frac{1}{2}, 4)$ is in II, $(-3, -3)$ is in III, and $(4, -\frac{5}{3})$ is in IV. (See Example 1.)

Suppose that we have two points (x_1, y_1) and (x_2, y_1) on a horizontal line, as in Figure 1.6. Then the distance between (x_1, y_1) and (x_2, y_1) is

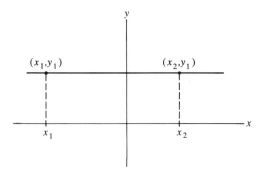

Figure 1.6

$|x_1 - x_2|$. Let (x_1, y_1) and (x_1, y_2) be two points on a vertical line, as in Figure 1.7. Again we see that the distance between (x_1, y_2) and (x_1, y_1) is

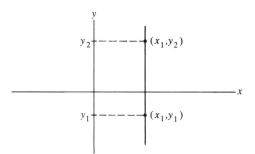

Figure 1.7

$|y_2 - y_1|$. Now we wish to derive a formula for the distance between $P_1(x_1, y_1)$ and $P_2(x_2, y_2)$, where $x_1 \neq x_2$ and $y_1 \neq y_2$. We shall denote the distance by $d(P_1, P_2)$ (see Figure 1.8).

Consider the right triangle above determined by P_1, P_2, and Q. The length of the segment between P_1 and Q is $|x_1 - x_2|$ and the length of the segment between Q and P_2 is $|y_1 - y_2|$. The theorem of Pythagoras states that the square of the hypotenuse of a right triangle is the sum of the squares of the legs. It then follows that

$$[d(P_1, P_2)]^2 = |x_1 - x_2|^2 + |y_1 - y_2|^2$$
$$= (x_1 - x_2)^2 + (y_1 - y_2)^2.$$

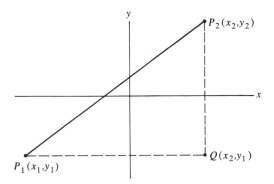

Figure 1.8

Taking the square root of both sides and noting that distance should be a nonnegative number, we have

$$d(P_1, P_2) = \sqrt{(x_1 - x_2)^2 + (y_1 - y_2)^2}.$$

Observe that $d(P_1, P_2) = d(P_2, P_1)$.

1.1.1
Distance
Formula

The **distance between two points** $P_1(x_1, y_1)$ and $P_2(x_2, y_2)$ is

$$d(P_1, P_2) = \sqrt{(x_1 - x_2)^2 + (y_1 - y_2)^2}.$$

EXAMPLE 3   Find the distance between $P(2, -4)$ and $Q(-1, 2)$.

$$d(P, Q) = \sqrt{(2 - (-1))^2 + (-4 - 2)^2} = \sqrt{9 + 36} = \sqrt{45}.$$

Now suppose that we consider the line determined by two points $P_1(x_1, y_1)$ and $P_2(x_2, y_2)$ and that we wish to know whether the point $P_0(x_0, y_0)$ is on the line. Since introducing coordinates and distance in the coordinatized plane, it is possible to state a condition involving the coordinates which gives the answer. Let us assume that the points P_1, P_0, P_2 are situated in the plane in such a way that

$$x_1 \le x_0 \le x_2.$$

Consider the two numbers

$$d(P_1, P_2)$$

and

$$d(P_1, P_0) + d(P_0, P_2).$$

An important inequality in mathematics gives the relationship between these two numbers, namely,

$$d(P_1, P_2) \leq d(P_1, P_0) + d(P_0, P_2).$$

This inequality is known as the **triangle inequality** in the plane.

The strict inequality

$$d(P_1, P_2) < d(P_1, P_0) + d(P_0, P_2)$$

expresses the principle from high school geometry which states that the length of the third side of a triangle is less than the sum of the lengths of the other two sides (see Figure 1.9). On the other hand,

$$d(P_1, P_2) = d(P_1, P_0) + d(P_0, P_2)$$

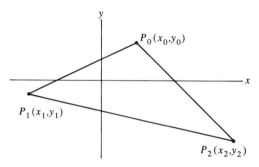

Figure 1.9

expresses the fact that P_0 lies on the line determined by two points P_1, P_2 (see Figure 1.10).

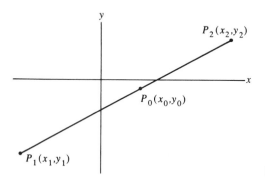

Figure 1.10

1.1.2
Condition for
Collinearity

Let $P_1(x_1, y_1)$, $P_0(x_0, y_0)$, and $P_2(x_2, y_2)$ be points in the plane where $x_1 \leq x_0 \leq x_2$. Then

P_0 is on the line determined by P_1 and P_2

$$\Leftrightarrow d(P_1, P_2) = d(P_1, P_0) + d(P_0, P_2).$$

It should be noted that if $x_0 \le x_1 \le x_2$, then 1.1.2 should read

$$d(P_0, P_2) = d(P_0, P_1) + d(P_1, P_2);$$

whereas if $x_1 \le x_2 \le x_0$, then 1.1.2 should read

$$d(P_1, P_0) = d(P_1, P_2) + d(P_2, P_0).$$

EXAMPLE 4   Show that $P_0(0, 4)$ is collinear with $P_1(-2, 8)$ and $P_2(3, -2)$:

$$d(P_1, P_0) = \sqrt{(-2 - 0)^2 + (8 - 4)^2} = \sqrt{4 + 16} = \sqrt{20} = 2\sqrt{5}$$

$$d(P_0, P_2) = \sqrt{(0 - 3)^2 + (4 - (-2))^2} = \sqrt{9 + 36} = \sqrt{45} = 3\sqrt{5}$$

$$d(P_1, P_2) = \sqrt{(-2 - 3)^2 + (8 - (-2))^2} = \sqrt{25 + 100} = \sqrt{125} = 5\sqrt{5}.$$

We see that

$$d(P_1, P_2) = d(P_1, P_0) + d(P_0, P_2).$$

For a line which is neither horizontal nor vertical, there is an interesting geometric way of rewriting the collinearity condition 1.1.2. With P_0 on the line determined by P_1 and P_2, we can draw two similar triangles, as in Figure 1.11. Since corresponding sides of similar triangles are proportional, we have

(1.1.3) $$\frac{y_0 - y_1}{x_0 - x_1} = \frac{y_2 - y_0}{x_2 - x_0} \qquad (x_0 \ne x_1, \ x_0 \ne x_2).$$

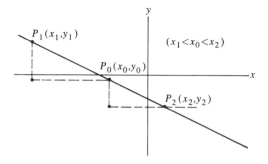

Figure 1.11

It should be pointed out that (1.1.3) is still valid if $x_0 < x_1$ or $x_2 < x_0$. Also, the condition (1.1.3) holds if $y_1 = y_0 = y_2$; that is, it holds for a horizontal line, too.

EXAMPLE 5   Show again that $P_0(0, 4)$ is collinear with $P_1(-2, 8)$ and $P_2(3, -2)$.

We did the problem in Example 4 by using the condition for collinearity and we now do it using (1.1.3):

$$\frac{4 - 8}{0 - (-2)} = \frac{-4}{2} = -2$$

$$\frac{-2 - 4}{3 - 0} = \frac{-6}{3} = -2.$$

Therefore (1.1.3) is satisfied.

EXAMPLE 6   Show that the points $P_1(-1, 5)$, $P_0(0, 0)$ and $P_2(2, 2)$ do not lie on the same straight line.

$$\frac{y_0 - y_1}{x_0 - x_1} = \frac{0 - 5}{0 - (-1)} = -5$$

$$\frac{y_2 - y_0}{x_2 - x_0} = \frac{2 - 0}{2 - 0} = 1.$$

Therefore (1.1.3) is not satisfied.

Next we would like to determine the midpoint of the line segment joining $P_1(x_1, y_1)$ and $P_2(x_2, y_2)$ (see Figure 1.12).

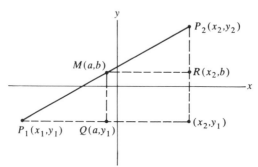

Figure 1.12

Let $M(a, b)$ be the midpoint of the segment joining P_1 and P_2. Since the triangle with vertices P_1, Q, M is similar to the triangle with vertices M, R, P_2, we have

$$\frac{d(P_1, Q)}{d(P_1, M)} = \frac{d(M, R)}{d(M, P_2)}.$$

Since

$$d(P_1, M) = d(M, P_2),$$

then

$$d(P_1, Q) = d(M, R)$$
$$a - x_1 = x_2 - a$$
$$2a = x_1 + x_2$$
$$a = \frac{x_1 + x_2}{2}.$$

Likewise,

$$\frac{d(M, Q)}{d(P_1, M)} = \frac{d(R, P_2)}{d(M, P_2)}$$

implies that

$$b = \frac{y_1 + y_2}{2}.$$

1.1.4
Midpoint of a
Line Segment

The midpoint of the line segment between (x_1, y_1) and (x_2, y_2) is

$$\left(\frac{x_1 + x_2}{2}, \frac{y_1 + y_2}{2}\right).$$

EXAMPLE 7 Find the midpoint of line segment between $(-1, 4)$ and $(2, -3)$.
  The midpoint is

$$\left(\frac{-1 + 2}{2}, \frac{4 - 3}{2}\right) = (\tfrac{1}{2}, \tfrac{1}{2}).$$

Suppose that we have an equation involving the variables x and y. The collection of points (x, y) satisfying the given equation is called the **graph of the equation**.

EXAMPLE 8 Sketch the graph of the equation $y = 5$. Since x is not explicitly mentioned, we plot those points (x, y), where x is arbitrary and $y = 5$. The graph is the horizontal line sketched in Figure 1.13.

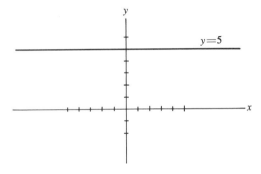

Figure 1.13

In the next few sections we shall consider graphs of several types of equations. In particular, we shall show that the graph of $Ax + By + C = 0$ is a straight line.

EXAMPLE 9

(*Optional*) If two medians of a triangle are equal in length, then the triangle is isosceles, and, conversely, an isosceles triangle has two medians of equal length.

We first recall that a median is a line drawn from the vertex of a triangle to the midpoint of the opposite side. We set the triangle in the coordinatized plane by placing one vertex at the origin and another vertex along the *x*-axis (see Figure 1.14). The dotted lines are the medians of the triangle.

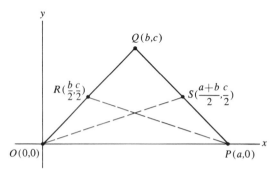

Figure 1.14

We wish to show that the two equations

$$d(O, Q) = d(P, Q) \qquad \text{(that is, triangle is isosceles)}$$

and

$$d(P, R) = d(O, S) \qquad \text{(that is, medians have equal lengths)}$$

are the same.

First we observe that $d(O, Q) = d(P, Q)$ can be written as

$$\sqrt{(b - 0)^2 + (c - 0)^2} = \sqrt{(b - a)^2 + (c - 0)^2}$$
$$b^2 + c^2 = (b - a)^2 + c^2$$
$$b^2 = b^2 - 2ab + a^2$$
$$a = 2b.$$

Second, we see that $d(P, R) = d(O, S)$ is the same as

$$\sqrt{(b/2 - a)^2 + (c/2 - 0)^2} = \sqrt{((a + b)/2 - 0)^2 + (c/2 - 0)^2}$$
$$(b/2 - a)^2 + (c/2)^2 = ((a + b)/2)^2 + (c/2)^2$$
$$b^2/4 - ab + a^2 + c^2/4 = a^2/4 + ab/2 + b^2/2 + c^2/4$$
$$-ab + a^2 = a^2/4 + ab/2$$
$$3a^2/4 - 3ab/2 = 0$$
$$a^2 - 2ab = 0$$
$$a = 2b.$$

Thus our two equations are the same.

EXERCISES

1. Draw a rectangular coordinate system and plot the points $(4, 1)$, $(\frac{7}{3}, 5)$, $(-\frac{1}{2}, 1)$, $(-5, -10)$, $(0, -3)$, $(-\frac{10}{7}, 0)$, $(5, -1)$, $(\frac{9}{2}, 0)$, $(\frac{1}{2}, -\frac{2}{3})$, $(-2, 3)$, $(-\sqrt{2}, -1)$, and $(0, \pi)$.

2. For each point in Exercise 1, determine either the axis on which it lies or the quadrant in which it lies.

3. In each of the following cases find the distance between the pair of given points.

    (a) $(2, 3)$ and $(-5, 4)$.   (b) $(-2, 4)$ and $(-2, \pi)$.

    (c) $(4, 7)$ and $(-2, -1)$.   (d) $(\sqrt{2}, -3)$ and $(0, -3)$.

    (e) $(\frac{3}{2}, 1)$ and $(4, -\frac{1}{2})$.   (f) $(7, 5)$ and $(12, 5)$.

    (g) $(6, 0)$ and $(0, 5)$.

4. In each of the case in Exercise 3, determine the midpoint of the line segment formed by the two given points.

5. Before plotting points, determine in each of the following cases if the third point given is collinear with the first two given.

    (a) $(0, -3)$, $(1, 1)$, $(-1, -7)$.   (b) $(2, 1)$, $(-5, 1)$, $(\frac{3}{2}, 1)$.

    (c) $(0, -1)$, $(\frac{1}{2}, 0)$, $(1, 1)$.   (d) $(2, 3)$, $(1, 0)$, $(0, 1)$.

    (e) $(-2, 1)$, $(-2, -\frac{3}{2})$, $(-2, 7)$.   (f) $(1, 0)$, $(2, 1)$, $(-5, -6)$.

6. In each of the following, determine whether or not the given points lie on the same straight line. If not, are these points vertices of a right triangle? Isosceles triangle?

    (a) $(0, 0)$, $(5, 0)$, $(4, 3)$.   (b) $(0, 0)$, $(1, -2)$, $(-5, -5)$.

    (c) $(-7, 8)$, $(1, 2)$, $(5, -1)$.   (d) $(-6, 3)$, $(-2, 0)$, $(-5, -4)$.

7. The three points $(2, 3)$, $(3, 1)$, $(5, 2)$ are vertices of a rectangle.

    (a) Find its fourth vertex and the lengths of its sides.

    (b) Find the point of intersection of the diagonals of this rectangle. (Recall that the diagonals of a rectangle bisect each other.)

8. Find the point three fourths of the way from $(1, 1)$ to $(4, 5)$.

9. Show that the diagonals of a rectangle are equal in length.

10. Show that the length of the line joining the midpoints of two sides of a triangle is equal to one half of the length of the third side.

11. Show that the midpoints of the sides of a right triangle are the vertices of a right triangle.

1.2 SLOPE OF A LINE

Associated with a nonvertical line, there is a number, called the slope of the line. The slope will indicate something geometrically about the line and how it relates to other lines. In particular, the slopes of lines would tell whether the lines are parallel or intersect. Moreover, if they intersect, the slopes would tell whether they are perpendicular.

1.2.1

Definition   Let (x_1, y_1) and (x_2, y_2) be any two points on a nonvertical line *l*.
The **slope *m* of the line *l*** is the number

$$m = \frac{y_2 - y_1}{x_2 - x_1} \qquad (x_1 \neq x_2).$$

We do not define the slope of a vertical line.

It should be pointed out that the number $(y_2 - y_1)/(x_2 - x_1)$ is the same number regardless of the two points chosen. Let (x_3, y_3) and (x_4, y_4) be two other points on the line. Then, since (x_3, y_3) is collinear with (x_1, y_1) and (x_2, y_2), we have

$$\frac{y_3 - y_2}{x_3 - x_2} = \frac{y_2 - y_1}{x_2 - x_1}.$$

Moreover, (x_4, y_4) is collinear with (x_2, y_2) and (x_3, y_3), and so

$$\frac{y_4 - y_3}{x_4 - x_3} = \frac{y_3 - y_2}{x_3 - x_2}.$$

It follows that

$$\frac{y_4 - y_3}{x_4 - x_3} = \frac{y_2 - y_1}{x_2 - x_1}.$$

The last equation can be seen geometrically from the similar triangles in Figure 1.15.

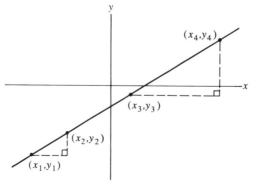

Figure 1.15

EXAMPLE 1   Let l be the line determined by the points $(1, -2)$ and $(-3, 4)$. Then the slope m of l is

$$m = \frac{4 - (-2)}{(-3) - 1} = \frac{6}{-4} = \frac{-3}{2}.$$

The slope of a line is a number which is positive, negative, or zero. Let us see what each case means geometrically.

1. If $m = (y_2 - y_1)/(x_2 - x_1) = 0$, then $y_1 = y_2$; that is, the y-coordinates of each point on the line are the same. Thus the line is a horizontal line (see Figure 1.16). We observe that there is a difference

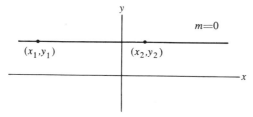

Figure 1.16

between saying that a line has zero slope and that a line has no slope. *A horizontal line has zero slope, while a vertical line has no slope.*

2. If $m = (y_2 - y_1)/(x_2 - x_1) > 0$, then $y_2 - y_1$ and $x_2 - x_1$ have the same sign. If $y_2 - y_1 > 0$ and $x_2 - x_1 > 0$, then $y_2 > y_1$ and $x_2 > x_1$. Let us interpret these two inequalities geometrically. The point (x_2, y_2) is to the right of the point (x_1, y_1) since $x_2 > x_1$, and it is above the point (x_1, y_1) since $y_2 > y_1$. In this case we say that the

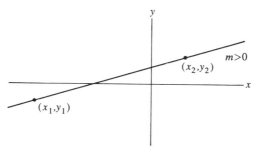

Figure 1.17

graph of the line **rises** from left to right (see Figure 1.17). If $y_2 - y_1 < 0$ and $x_2 - x_1 < 0$, then the labeling of these points is reversed, but the interpretation is the same.

3. If $m = (y_2 - y_1)/(x_2 - x_1) < 0$, then $y_2 - y_1$ and $x_2 - x_1$ have opposite signs. Let us say that $x_2 - x_1 > 0$ and $y_2 - y_1 < 0$; that is, $x_2 > x_1$ and $y_2 < y_1$. This time the point (x_2, y_2) is to the right of (x_1, y_1) but below it. We say that the graph **falls** from left to right (see Figure 1.18).

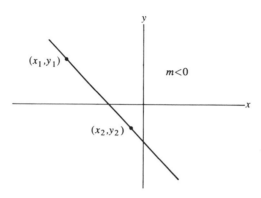

Figure 1.18

EXAMPLE 2   Draw through the point $(1, 2)$ lines of slope 2 and -3.

For the line of slope 2, begin at $(1, 2)$ and move to the right 1 unit and *up* 2 units, arriving at the point $(2, 4)$. The points $(1, 2)$ and $(2, 4)$ determine the line of slope 2 we wish to draw (see Figure 1.19).

For the line of slope -3, begin at $(1, 2)$ and move to the right 1 unit and *down* 3 units, arriving at $(2, -1)$. The points $(1, 2)$ and $(2, -1)$ determine the line of slope -3 we wish to draw (see Figure 1.20).

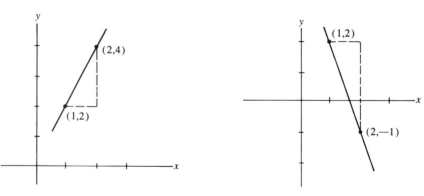

Figure 1.19   Figure 1.20

Two important ideas connected with a straight line are parallelism and perpendicularity.

1.2.2

Condition for Parallelism

Two vertical lines are parallel. If l_1 and l_2 are two nonvertical lines with slopes m_1 and m_2, respectively, then

$$l_1 \text{ is parallel to } l_2 \Leftrightarrow m_1 = m_2.$$

EXAMPLE 3

Let l_1 be a line determined by $(-3, 4)$ and $(1, -2)$ and let l_2 be a line determined by $(0, 1)$ and $(2, -2)$.

The slope m_1 of l_1 is

$$m_1 = \frac{4 - (-2)}{(-3) - (1)} = \frac{-3}{2}$$

and the slope m_2 of l_2 is

$$m_2 = \frac{(-2) - (1)}{(2) - (0)} = \frac{-3}{2}.$$

Thus l_1 is parallel to l_2 (see Figure 1.21).

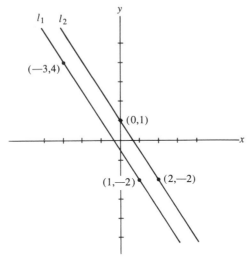

Figure 1.21

EXAMPLE 4

Let l_1 be the line as given in Example 3 and let l_3 be the line determined by $(2, 3)$ and $(-1, 6)$.

Then the slope m_1 of l_1 is $m_1 = -\frac{3}{2}$ and the slope m_3 of l_3 is

$$m_3 = \frac{(6) - (3)}{(-1) - (2)} = -1.$$

Thus l_1 is not parallel to l_3 (see Figure 1.22).

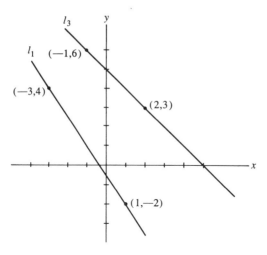

Figure 1.22

To determine the condition of perpendicularity, we shall assume without loss of generality that our lines pass through the origin and that neither is horizontal or vertical, as in Figure 1.23. Note that a vertical line is obviously perpendicular to a horizontal line.

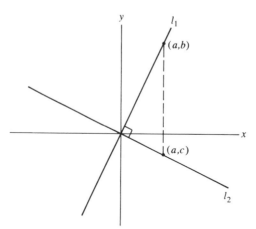

Figure 1.23

We first observe that the slope m_1 of l_1 is

$$m_1 = \frac{b-0}{a-0} = \frac{b}{a}$$

and that the slope m_2 of l_2 is

$$m_2 = \frac{c-0}{a-0} = \frac{c}{a}.$$

The triangle shown in Figure 1.23 is a right triangle whenever the theorem of Pythagoras is satisfied. Thus

$$(a^2 + b^2) + (a^2 + c^2) = (b - c)^2.$$

Dividing the equation by a^2, we obtain

$$1 + (b/a)^2 + 1 + (c/a)^2 = (b/a - c/a)^2$$
$$1 + m_1^2 + 1 + m_2^2 = (m_1 - m_2)^2$$
$$2 + m_1^2 + m_2^2 = m_1^2 - 2m_1 m_2 + m_2^2$$
$$2 = -2m_1 m_2$$
$$-1 = m_1 m_2.$$

1.2.3
Condition for
Perpendicularity

A horizontal line is perpendicular to a vertical line. If lines l_1 and l_2 are neither horizontal nor vertical with slopes m_1 and m_2, respectively, then

$$l_1 \text{ is perpendicular to } l_2 \Leftrightarrow m_1 m_2 = -1.$$

Thus two lines are perpendicular whenever their slopes are negative reciprocals of each other.

EXAMPLE 5   Let l_1 be a line determined by $(2, -2)$ and $(-2, 4)$ and let l_2 be a line determined by $(0, 0)$ and $(3, 2)$.

The slope m_1 of l_1 is

$$m_1 = \frac{(4) - (-2)}{(-2) - (2)} = -\frac{3}{2}$$

and the slope m_2 of l_2 is

$$m_2 = \frac{2 - 0}{3 - 0} = \frac{2}{3}.$$

Thus $m_1 m_2 = (-\frac{3}{2})(\frac{2}{3}) = -1$ and so the lines l_1 and l_2, drawn in Figure 1.24, are perpendicular.

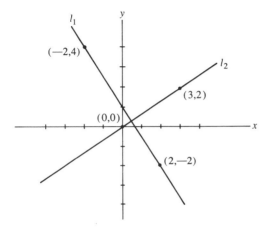

Figure 1.24

EXAMPLE 6   Let l_1 and l_3 be the lines as given in Example 4. Then

$$m_1 m_3 = (-\tfrac{3}{2})(-1) = \tfrac{3}{2} \neq -1.$$

Thus l_1 and l_3 are not perpendicular.

EXAMPLE 7   (*Optional*) The midpoints of the sides of any quadrilateral are vertices of a parallelogram.

We begin by placing a quadrilateral with one corner at the origin and one side along the positive x-axis. Then the other corners are labeled appropriately. The midpoints of sides are indicated in Figure 1.25.

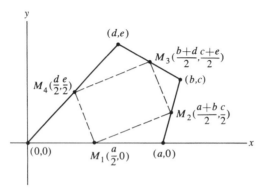

Figure 1.25

The slope of the line determined by M_1 and M_4 is

$$\frac{e/2 - 0}{d/2 - a/2} = \frac{e}{d - a},$$

while the slope of the line determined by M_2 and M_3 is

$$\frac{(c + e)/2 - c/2}{((b - d)/2) - ((a + b)/2)} = \frac{e}{d - a}.$$

Moreover, the slope of the line determined by M_4 and M_3 is

$$\frac{(c + e)/2 - e/2}{(b + d)/2 - d/2} = \frac{c}{b},$$

while the slope of the line determined by M_1 and M_2 is

$$\frac{c/2 - 0}{(a + b)/2 - a/2} = \frac{c}{b}.$$

Therefore M_1, M_2, M_3, M_4 are vertices of a parallelogram.

EXERCISES

1. Find the slope of the line determined by the given points and sketch the line.

(a) (4, 3) and (−1, 2).   (b) (5, 6) and (7, 9).

(c) ($\frac{1}{2}$, −$\frac{3}{4}$) and ($\frac{2}{3}$, 1).   (d) (4, 0) and (0, 3).

(e) (−2, −5) and (3, −5).   (f) (4, −1) and (4, 1).

2. In each of the following, determine whether the given points lie on the same straight line. If the given points are vertices of a triangle, find the slopes of the sides of the triangle. Indicate which triangles are right triangles.

(a) (0, −3), (1, 1), (−1, −1).   (b) (−6, 3), (−2, 0), (−5, −4).

(c) (−3, 4), (0, 4), (2, 4).   (d) (−7, 8), (1, 2), (5, −1).

(e) (0, 0), (1, −2), (−5, −5).

3. Draw through the point (−2, 7) lines of slope

(a) 1.   (b) −2.   (c) $\frac{3}{5}$.   (d) 0.   (e) −$\frac{4}{3}$.

4. Draw lines of slope $\frac{1}{2}$ through the following points.

(a) (0, 0).   (b) (0, 1).   (c) (−5, −3).   (d) (2, 6).

5. Two lines are determined by pairs of points below. Determine whether the lines are parallel or intersect. If they intersect, determine if they are perpendicular.

(a) L_1: (0, 2) and (3, 0).   (b) L_1: (4, −$\frac{2}{3}$) and (1, $\frac{3}{2}$).
$\quad$ L_2: (0, 4) and (6, 0). $\quad$ L_2: (1, $\frac{10}{3}$) and ($\frac{9}{2}$, −1).

(c) L_1: (2, $\frac{2}{3}$) and (−1, $\frac{9}{2}$).   (d) L_1: (0, 3) and (2, 0).
$\quad$ L_2: (2, 0) and (0, 3). $\quad$ L_2: ($\frac{1}{2}$, $\frac{1}{3}$) and (0, 0).

6. The three points (0, 0), (0, 3), (1, 2) are vertices of a parallelogram.

(a) Find its fourth vertex and the lengths of its sides.

(b) Find the point of intersection of the diagonals of this parallelogram.

Using the method introduced in Section 1.1, prove each of the following geometric statements (Exercises 7–11).

7. The diagonals of a rhombus are perpendicular. (A **rhombus** is a parallelogram whose sides are all of the same length.)

8. A parallelogram having diagonals of equal length is a rectangle.

9. The line joining the midpoints of nonparallel sides of a trapezoid is parallel to the bases, and its length equals one half of the sum of the lengths of the bases.

10. One median of an isosceles triangle is an altitude.

11. A quadrilateral whose diagonals bisect each other is a parallelogram.

1.3   EQUATIONS OF LINES

In dealing with lines in coordinate geometry, it is important to be able to represent them by means of an equation. Given certain information concerning a particular line, we shall show that the line is the graph of an equation, usually called **the equation of the line**.

First, let us consider the nonvertical line determined by the points (x_0, y_0) and (x_1, y_1), where $x_0 \neq x_1$. To say that (x, y) is on the line is the same as saying that (x, y) is collinear with (x_0, y_0) and (x_1, y_1); that is,

$$\frac{y - y_0}{x - x_0} = \frac{y_1 - y_0}{x_1 - x_0}$$

or

$$y - y_0 = \frac{y_1 - y_0}{x_1 - x_0}(x - x_0) \qquad (x_0 \neq x_1).$$

1.3.1
Two-Point
Equation of a
Line

The equation of the line passing through (x_0, y_0) and (x_1, y_1) is

$$y - y_0 = \frac{y_1 - y_0}{x_1 - x_0}(x - x_0) \qquad (x_1 \neq x_0).$$

It should be noted that the equation in 1.3.1 can be written as

$$y - y_1 = \frac{y_1 - y_0}{x_1 - x_0}(x - x_1).$$

Note also that the quotient $(y_1 - y_0)/(x_1 - x_0)$ in 1.3.1 is the slope of the line passing through (x_0, y_0) and (x_1, y_1). Therefore we have

1.3.2
Point Slope
Equation of a
Line

The equation of the line with slope m and passing through the point (x_0, y_0) is

$$y - y_0 = m(x - x_0).$$

If the line is a horizontal line through the point (x_0, y_0), then $m = 0$, and so by 1.3.2

$$y - y_0 = 0$$

or

$$y = y_0.$$

**1.3.3
Equation of a
Horizontal Line**

The equation of the horizontal line through (x_0, y_0) is

$$y = y_0.$$

To say that (x, y) is on a vertical line through the point (x_0, y_0) is the same as saying that

$$x = x_0.$$

**1.3.4
Equation of a
Vertical Line**

The equation of the vertical line through (x_0, y_0) is

$$x = x_0.$$

EXAMPLE 1   Find the equation of the line passing through $(-1, 2)$ and having slope 5.
By 1.3.2, we have

$$y - 2 = 5(x - (-1))$$

or

$$y - 2 = 5x + 5$$

or

$$5x - y + 7 = 0 \qquad \text{(see Figure 1.26)}.$$

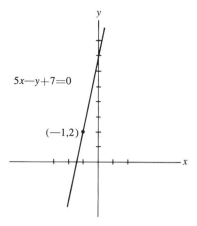

Figure 1.26

EXAMPLE 2   Find the equation of the line passing through the points $(2, -3)$ and $(-4, 1)$.
The slope m of the line is

$$m = \frac{1 - (-3)}{(-4) - 2} = -\frac{2}{3}.$$

Let $(x_0, y_0) = (2, -3)$. Then by 1.3.1,

$$y - (-3) = -\tfrac{2}{3}(x - 2)$$

$$3y + 9 = -2x + 4$$

$$2x + 3y + 5 = 0 \quad \text{(see Figure 1.27)}.$$

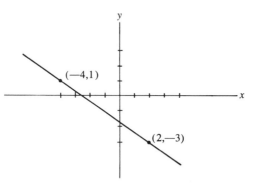

Figure 1.27

EXAMPLE 3   Find the equations of the horizontal and vertical lines passing through the point (4, 3).

From 1.3.4 and 1.3.3, we see that

$$x = 4 \quad \text{and} \quad y = 3$$

are the equations of the vertical and horizontal lines, respectively (see Figure 1.28).

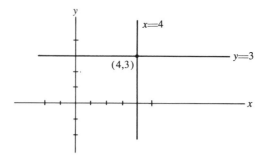

Figure 1.28

EXAMPLE 4   Find the equation of the line which is the perpendicular bisector of the line segment joining the points $(-3, 2)$ and $(1, 4)$.

The midpoint of the segment is $(-1, 3)$, and the slope of the line joining the given points is

$$\frac{4 - 2}{1 - (-3)} = \frac{2}{4} = \frac{1}{2}.$$

Therefore the line we seek has slope -2 (the negative reciprocal of $\frac{1}{2}$) and passes through $(-1, 3)$.

By 1.3.2, the equation of the line is

$$y - 3 = -2(x - (-1))$$

$$y - 3 = -2x - 2$$

$$2x + y - 1 = 0 \quad \text{(see Figure 1.29)}.$$

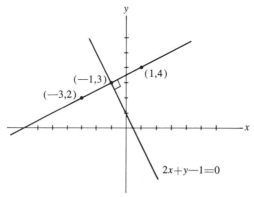

Figure 1.29

It is interesting to note that each equation of a line considered above has the form

$$Ax + By + C = 0 \quad (A \text{ and } B \text{ not both zero}).$$

1. For the equation $x = x_0$, we have $A = 1$, $B = 0$, and $C = -x_0$.
2. For the equation $y = y_0$, we have $A = 0$, $B = 1$, and $C = -y_0$.
3. For the equation $y - y_0 = m(x - x_0)$ or $y - y_0 = mx - mx_0$, we have $A = m$, $B = -1$, and $C = y_0 - mx_0$.

We would now like to ask the converse question: Is the graph of $Ax + By + C = 0$ a straight line? The answer is yes, for suppose that $B \neq 0$; then

$$By + C = -Ax$$

$$B(y + C/B) = -Ax$$

$$B(y - (-C/B)) = -Ax$$

$$y - (-C/B) = (-A/B)x$$

$$y - (-C/B) = (-A/B)(x - 0),$$

which we recognize from 1.3.2 as the equation of a line with slope $-A/B$ and passing through the point $(0, -C/B)$.

If $B = 0$, then

$$Ax + C = 0$$

$$x = -C/A,$$

which is the equation of a vertical line.

1.3.5

The graph of the equation $Ax + By + C = 0$ is a straight line (A, B not both zero).

It is useful to remember that by solving $Ax + By + C = 0$ for y and obtaining

$$y = (-A/B)x - C/B \qquad (B \neq 0),$$

we can obtain the slope of the line by reading the coefficient of x—namely, $-A/B$.

EXAMPLE 5   Sketch the graph of $2x + 3y - 6 = 0$.

Noting that $(0, 2)$ and $(3, 0)$ are points on the line, we have Figure 1.30.

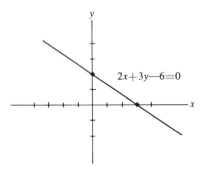

Figure 1.30

EXAMPLE 6   Find the equation of the line passing through $(-2, 3)$ and parallel to the line

$$2x + 3y - 6 = 0.$$

We rewrite the given equation

$$3y = -2x + 6$$

$$y = -\tfrac{2}{3}x + 2,$$

and so the slope of the given line is $-\frac{2}{3}$. The slope of the line we seek is $-\frac{2}{3}$, and hence its equation is

$$y - 3 = -\tfrac{2}{3}(x - (-2))$$

$$3y - 9 = -2x - 4$$

$$2x + 3y - 5 = 0 \qquad \text{(see Figure 1.31)}.$$

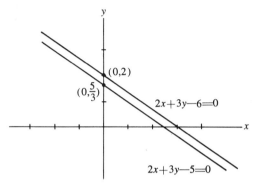

Figure 1.31

EXAMPLE 7 Find the point of intersection of the lines

$$2x - 3y = 7$$

$$5x + y - 1 = 0.$$

The point we seek satisfies both equations. Solve for y in the second equation. We obtain

$$y = 1 - 5x$$

and substitute into the first equation:

$$2x - 3(1 - 5x) = 7$$

$$2x - 3 + 15x = 7$$

$$17x = 10$$

$$x = \tfrac{10}{17}.$$

Since $y = 1 - 5x$, the y-coordinate of the point of intersection is

$$y = 1 - 5(\tfrac{10}{17}) = 1 - \tfrac{50}{17} = \tfrac{17}{17} - \tfrac{50}{17} = -\tfrac{33}{17}.$$

The point is

$$(\tfrac{10}{17}, -\tfrac{33}{17}).$$

EXAMPLE 8 Find the shortest distance from the point $(4, 1)$ to the line

$$4x - 3y + 2 = 0.$$

We seek the perpendicular distance from (4, 1) to the line $4x - 3y + 2 = 0$ (see Figure 1.32). Initially, we shall find the equation of the line perpendicular to $4x - 3y + 2 = 0$ and passing through the point (4, 1).

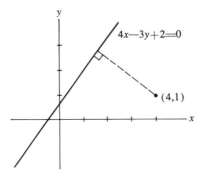

Figure 1.32

Since $\frac{4}{3}$ is the slope of $4x - 3y + 2 = 0$, the line we seek has slope $-\frac{3}{4}$ and its equation is

$$y - 1 = -\tfrac{3}{4}(x - 4)$$

or

$$3x + 4y = 16.$$

Using the technique of the preceding example, it can be shown that the point of intersection of the two lines

$$4x - 3y = 2$$
$$3x + 4y = 16$$

is

$$(\tfrac{8}{5}, \tfrac{14}{5}).$$

Hence the distance d is given by

$$d = \sqrt{(\tfrac{8}{5} - 4)^2 + (\tfrac{14}{5} - 1)^2}$$
$$= \frac{\sqrt{225}}{5}$$
$$= \tfrac{15}{5}$$
$$= 3.$$

EXERCISES

1. Find the equation of the line passing through the given points.

(a) $(4, -2)$ and $(-3, 2)$.   (b) $(6, 4)$ and $(1, -1)$.

(c) $(\tfrac{2}{3}, 1)$ and $(-2, \tfrac{4}{7})$.   (d) $(-2, -1)$ and $(-2, 4)$.

(e) $(-3, 4)$ and $(-2, 4)$.   (f) $(0, 0)$ and $(\sqrt{22}, 1)$.

2. Find the equation of the line passing through the given point and having the given slope m.

   (a) $(0, 0)$ and $m = 3$.   (b) $(2, 3)$ and $m = 5$.

   (c) $(\frac{1}{2}, -3)$ and $m = \frac{2}{3}$.   (d) $(0, 0)$ and $m = -4$.

   (e) $(5, 5)$ and $m = 0$.   (f) $(0, 1)$ and $m = -\frac{1}{2}$.

3. Find the equations of the horizontal and vertical lines passing through the given point.

   (a) $(1, 2)$.   (b) $(-3, 2)$.   (c) $(0, 1)$.   (d) $(1, 0)$.

4. Show that the equation of line with slope m and passing through point $(0, b)$ is $y = mx + b$. (This equation is called the **slope-intercept form** of the equation of a line. The number b is called the **y-intercept**.)

5. Find the equation of the line with slope m and y-intercept b if

   (a) $m = 2, b = 1$.   (b) $m = 0, b = -2$.

   (c) $m = -\frac{1}{2}, b = 0$.   (d) $m = -\sqrt{2}, b = \frac{1}{3}$.

6. Rewrite each of the following equations in slope-intercept form and determine the slope and y-intercept of each.

   (a) $6x - 3y + 2 = 0$.   (b) $3y - 4 = 0$.

   (c) $7x - \sqrt{3y} = 0$.   (d) $-x - 4y + 3 = 0$.

7. Show that the equation of the line through the points $(a, 0)$ and $(0, b)$ is $(x/a) + (y/b) = 1$ $(a \neq 0, b \neq 0)$. (The equation is called the **intercept form** of the equation of a line. The numbers a and b are called the **x- and y-intercepts**, respectively.)

8. Find the equation of the line whose x- and y-intercepts are given.

   (a) $a = 4, b = 5$.   (b) $a = -\frac{1}{2}, b = 4$.

   (c) $a = \frac{3}{4}, b = -\frac{1}{5}$.   (d) $a = -12, b = -\pi$.

9. Find the x- and y-intercepts of each of the following lines and sketch the graph.

   (a) $3x + 2y - 3 = 0$.   (b) $5x - 1/2y + 7 = 0$.

   (c) $-x - 4y + 3 = 0$.   (d) $4x - 7 = 0$.

   (e) $-2x + 3 = 2$.   (f) $x + y = 0$.

   (g) $7x - \sqrt{3y} = 0$.

10. Find the equation of the line which is the perpendicular bisector of the line segment joining the points given in Exercise 1.

11. Given the line $6x - 3y - 4 = 0$ and the point $P(5, 2)$, find

   (a) The equation of the line l_1 through the point P and parallel to the given line.

   (b) The equation of the line l_2 through P and perpendicular to the given line.

12. Find the equation of the line l through the point $(-1, 5)$ if l

(a) Has slope 4.   (b) Has slope $-\frac{7}{2}$.

(c) Has slope 0.   (d) Is parallel to the x-axis.

(e) Is parallel to the y-axis.   (f) Is parallel to $2x + y - 7 = 0$.

(g) Is perpendicular to $2x + y - 7 = 0$.

(h) Passes through the origin.

(i) Passes through the point $(4, -3)$.

Draw a diagram which includes all these lines.

13. Consider the parallelogram with vertices $(-1, -2), (5, -1), (0, 3), (6, 4)$. Find

(a) The equations of the sides.   (b) The equations of the diagonals.

14. Consider the triangle with vertices $(2, 3), (-1, -1), (0, -4)$. Find

(a) The equations of the three sides.

(b) The equations of the three lines through the vertices and parallel to the opposite sides.

(c) The equations of the three lines through the vertices and perpendicular to the opposite sides.

(d) The equations of the three lines through the vertices and through the midpoints of the opposite sides.

15. Prove the following.

(a) Any line parallel to the line $4x + 3y - 7 = 0$ has an equation of the form $4x + 3y + D = 0$ for some constant D.

(b) Any line parallel to the line $Ax + By + C = 0$ has an equation of the form $Ax + By + D = 0$ for some constant D.

16. Prove the following.

(a) Any line perpendicular to the line $4x + 3y - 7 = 0$ has an equation of the form $3x - 4y + D = 0$ for some constant D.

(b) Any line perpendicular to the line $Ax + By + C = 0$ has an equation of the form $Bx - Ay + D = 0$ for some constant D.

17. Find the shortest distance from the point $P(1, 1)$ to the line $2x + 3y + 7 = 0$. [*Hint:* Shortest distance is perpendicular distance.]

18. Find the distance between the parallel lines $4x - 3y + 2 = 0$ and $4x - 3y - 13 = 0$.

19. Find the point of intersection of each of the following pairs of lines.

(a) $2x + y = 7$ (b) $6x - 5y - 2 = 0$
    $x = 3.$ $y = 1.$

(c) $x = -1$ (d) $2x + 3 = 0$
    $y = 5.$ $-5y + 7 = 0.$

(e) $x + y = 1$ (f) $5x - 3y + 6 = 0$
    $x - y = 1.$ $-2x + 7y - 1 = 0.$

20. Find the equation of the line passing through $(1, 2)$ and the point of intersection of the lines $2x - 3y = 1$ and $x + 2y = -2$.

21. Show that the altitudes of a triangle are concurrent.

22. Show that the line joining one vertex of a parallelogram to the midpoint of an opposite side trisects a diagonal. (That is, in Figure 1.33, *AM* divides *BD* in the ratio 1 to 2.)

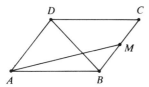

Figure 1.33

23. The medians of a triangle intersect at a point, and this point is called the **centroid** of the triangle. Find the centroid of the triangle with vertices $(0, 0)$, $(5, 0)$, $(4, 3)$.

1.4 CIRCLES AND PARABOLAS

At this point we want to discuss two useful types of curves in analytic geometry, namely circles and parabolas.

1.4.1

Definition

A **circle** is a collection of all points which are at a given distance from a fixed point.

Let $C(a, b)$ be the fixed point, called the **center of the circle**, and let r be the given distance, called the **radius of the circle**. Let $P(x, y)$ be any point on the circle (see Figure 1.34).

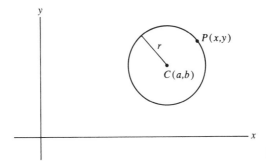

Figure 1.34

Then, to say that $P(x, y)$ is on the circle is the same as saying that

$$d(P, C) = r$$
$$\sqrt{(x - a)^2 + (y - b)^2} = r$$
$$(x - a)^2 + (y - b)^2 = r^2.$$

1.4.2

The circle with center (a, b) and radius r is the graph of the equation

$$(x - a)^2 + (y - b)^2 = r^2.$$

EXAMPLE 1   Find the equation of a circle with center $(2, -1)$ and radius 2.

By 1.4.2 we have

$$(x - 2)^2 + (y - (-1))^2 = 2^2$$
$$x^2 - 4x + 4 + y^2 + 2y + 1 = 4$$
$$x^2 + y^2 - 4x + 2y + 1 = 0 \quad \text{(see Figure 1.35)}.$$

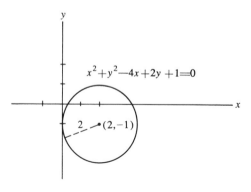

Figure 1.35

EXAMPLE 2   Suppose that we are given the equation

$$x^2 + y^2 - 4x + 2y + 1 = 0.$$

We can reverse the steps in Example 1 by using the **method of completion of squares** to show that the graph of this equation is a circle.

We first group with x and y terms separately and place the constant on the right-hand side of the equation:

$$(x^2 - 4x) + (y^2 + 2y) = -1.$$

The coefficient of x (namely, -4) is divided by 2 (we get -2) and then -2 is squared (we get 4). We then add 4 to both sides of the equation and place the number 4 with the x terms:

$$(x^2 - 4x + 4) + (y^2 + 2y) = (-1) + 4.$$

The same procedure is performed with the coefficient of y (namely, 2). In this case, we add 1 to both sides of the equation and place 1 with the y terms:

$$(x^2 - 4x + 4) + (y^2 + 2y + 1) = (-1) + (4) + (1)$$
$$(x - 2)^2 + (y + 1)^2 = 4$$
$$(x - 2)^2 + (y - (-1))^2 = 4$$
$$(x - 2)^2 + (y - (-1))^2 = 2^2.$$

Thus we see that the graph of the equation is a circle with center $(2, -1)$ and radius 2.

EXAMPLE 3   What is the graph of the equation

$$x^2 + y^2 - 4x - 6y + 13 = 0?$$

We shall use the method of completion of squares:

$$(x^2 - 4x) + (y^2 - 6y) = -13$$
$$(x^2 - 4x + 4) + (y^2 - 6y + 9) = -13 + 4 + 9$$
$$(x - 2)^2 + (y - 3)^2 = 0.$$

The last equation can be satisfied only if $x = 2$ and $y = 3$. Therefore the graph of the equation is exactly the point $(2, 3)$.

EXAMPLE 4   What is the graph of the equation

$$x^2 + y^2 + 2x - 2y + 3 = 0?$$

Again, by completing the square, we obtain

$$(x^2 + 2x + 1) + (y^2 - 2y + 1) = -3 + 1 + 1$$
$$(x + 1)^2 + (y - 1)^2 = -1.$$

Since it is impossible for the sum of squares to be negative, there are no points in the plane which satisfy the given equation.

Another interesting curve in the plane is the parabola.

1.4.3
Definition

A **parabola** is a collection of all points equidistant from a fixed point and a fixed line.

The fixed point is called the **focus of the parabola** and the fixed line is called the **directrix of the parabola**.

As in the case of the circle, we wish to find an equation whose graph is the parabola. To simplify the algebra, we shall choose the focus to be the point $F(a, b + p)$ and the directrix to be the horizontal line $y = b - p$. The two possible cases are shown in Figure 1.36.

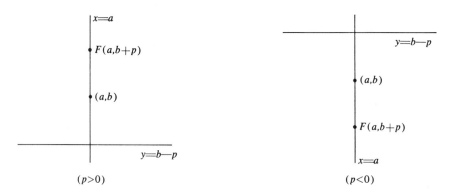

Figure 1.36

Then, according to 1.4.3, (a, b) is a point on the parabola. It is called the **vertex of the parabola**. Let $P(x, y)$ be any point on the parabola distinct from (a, b). If $P(x, y)$ is to satisfy the property stated in the definition, the graph of the parabola has one of the shapes given in Figure 1.37.

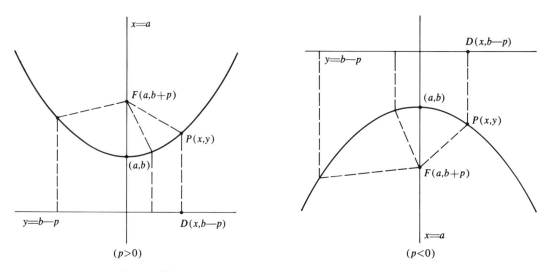

Figure 1.37

Then, to say that

$$P(x, y) \text{ is on the parabola}$$

is the same as saying that

$$d(P, F) = d(P, D)$$

$$\sqrt{(x - a)^2 + (y - b - p)^2} = \sqrt{(x - x)^2 + (y - b + p)^2}$$

$$(x - a)^2 + ([y - b] - p)^2 = ([y - b] + p)^2$$

$$(x - a)^2 + (y - b)^2 - 2p(y - b) + p^2 = (y - b)^2 + 2p(y - b) + p^2$$

$$(x - a)^2 = 4p(y - b).$$

1.4.4

The parabola with vertex (a, b) and focus $(a, b + p)$ is the graph of the equation

$$(x - a)^2 = 4p(y - b).$$

EXAMPLE 5   Find the equation of the parabola with vertex $(2, -3)$ and focus $(2, -1)$.
    Let $(a, b) = (2, -3)$ and $(a, b + p) = (2, -1)$. Then $b = -3$ and $b + p = -1$, and so $p = 2$. From 1.4.4, we have

$$(x - 2)^2 = 4(2)(y - (-3))$$

$$x^2 - 2x + 4 = 8(y + 3)$$

$$x^2 - 2x + 4 = 8y + 24$$

$$y = \tfrac{1}{8}x^2 - \tfrac{1}{4}x - \tfrac{5}{2} \qquad \text{(see Figure 1.38)}.$$

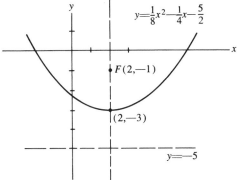

Figure 1.38

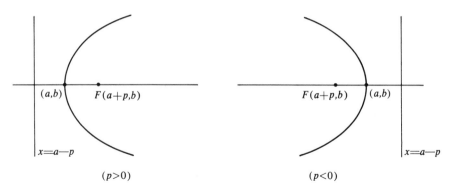

$(p>0)$ $(p<0)$

Figure 1.39

It is also possible to consider the situation where the directrix, vertex, and focus would appear as in Figure 1.39. If we carry out the computation as before, we are led to the following:

1.4.5

The parabola with vertex (a, b) and focus $(a + p, b)$ is the graph of the equation

$$(y - b)^2 = 4p(x - a).$$

EXAMPLE 6   Sketch the graph of

$$y^2 - 2y - 4x - 11 = 0.$$

First,

$$y^2 - 2y = 4x + 11$$

$$y^2 - 2y + 1 = 4x + 12$$

$$(y - 1)^2 = 4(x + 3).$$

Let $4p = 4$; then $p = 1$. We see that the graph of the equation is a parabola with vertex $(-3, 1)$ and focus $(-2, 1)$ (see Figure 1.40).

It should be pointed out that it is possible to have a directrix of a parabola be a line which is not parallel to the x- or y-axis. Let us look at an example.

EXAMPLE 7   Sketch the collection of points (x, y) equidistant from $(2, 2)$ and the line $y = -x$.

The graph is, of course, a parabola with directrix $y = -x$ and focus $(1, 1)$.

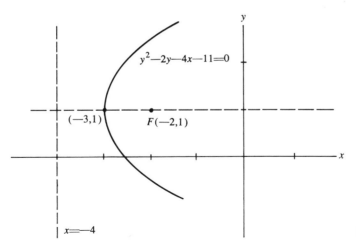

Figure 1.40

First, consider a point (a, b) which is equidistant from $(2, 2)$ and the line $y = -x$, as in Figure 1.41. Observe that the line l through (a, b) having slope 1 is perpendicular to the line $y = -x$, since the product of their slopes is -1. The equation of line l is

$$y - b = 1(x - a)$$

or

$$y = x - a + b.$$

One can verify that the intersection of lines

$$y = x - a + b$$

$$y = -x$$

is the point

$$\left(\frac{b - a}{-2}, \frac{b - a}{2} \right).$$

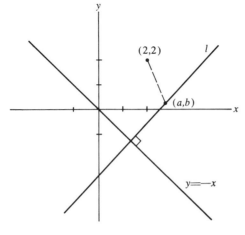

Figure 1.41

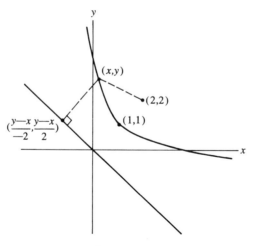

Figure 1.42

Therefore, if (x, y) is a point on the desired parabola (see Figure 1.42), then the equation of this parabola is

$$\sqrt{(x - 2)^2 + (y - 2)^2} = \sqrt{\left(x - \frac{y - x}{-2}\right)^2 + \left(y - \frac{y - x}{2}\right)^2}$$

$$(x - 2)^2 + (y - 2)^2 = \left(\frac{x + y}{2}\right)^2 + \left(\frac{x + y}{2}\right)^2$$

$$x^2 - 2xy + y^2 - 8x - 8y + 16 = 0.$$

Notice that the equation of this parabola does not have the form given in 1.4.4 or 1.4.5.

EXAMPLE 8   Find the points of intersection of the parabola $y = 16 - x^2$ and the line $y = x + 14$.

Solving the two equations simultaneously, we have

$$16 - x^2 = x + 14$$

or

$$x^2 + x - 2 = 0$$

$$(x + 2)(x - 1) = 0.$$

Therefore

$$x = -2 \quad \text{or} \quad x = 1.$$

The points of intersection are $(-2, 12)$ and $(1, 15)$ (see Figure 1.43).

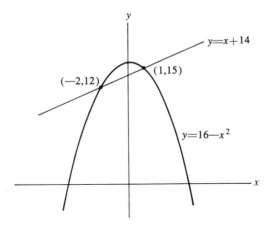

Figure 1.43

EXAMPLE 9   (*Optional*) Sketch the graph of the equation $y = Ax^2 + Bx + C$ ($A \neq 0$). Using the method of completing the square, we have

$$y = A\left[x + \frac{B}{2A}\right]^2 - \frac{B^2 - 4AC}{4A}$$

or

$$\left[x - \frac{-B}{2A}\right]^2 = 4\left(\frac{1}{4A}\right)\left(y - \frac{4AC - B^2}{4A}\right).$$

Thus the graph of our equation is a parabola with vertex

$$\left(-\frac{B}{2A}, \frac{4AC - B^2}{4A}\right)$$

and axis $x = -(B/2A)$. Our parabola opens up or down according as $A > 0$ or $A < 0$. If $A > 0$, the situation appears as one of the three cases in Figure 1.44.

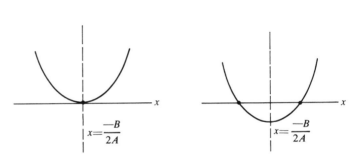

Figure 1.44

To distinguish the above cases, we must determine where the parabolas cross the x-axis. Setting $y = 0$, we have

$$0 = A\left[x + \frac{B}{2A}\right]^2 - \frac{B^2 - 4AC}{4A}$$

or

$$\left[x + \frac{B}{2A}\right]^2 = \frac{B^2 - 4AC}{4A^2}.$$

First, we observe that this equation has no solution if $B^2 - 4AC < 0$. It has one solution, namely $x = -B/2A$, if $B^2 - 4AC = 0$. Finally, if $B^2 - 4AC > 0$, we have that

$$x + \frac{B}{2A} = \frac{\pm\sqrt{B^2 - 4AC}}{2A}$$

or

$$x = \frac{-B \pm \sqrt{B^2 - 4AC}}{2A}.$$

Thus the roots of the equation

$$Ax^2 + Bx + C = 0$$

are

$$x = \frac{-B \pm \sqrt{B^2 - 4AC}}{2A},$$

which is called the **quadratic formula**.

We summarize our observations:

1. If $B^2 - 4AC = 0$, then $x = -B/2A$ is a double root and hence the graph crosses the x-axis exactly once at the point $(-b/2a, 0)$.
2. If $B^2 - 4AC > 0$, then both roots are real and unequal. In this case, the graph crosses the x-axis in two distinct points.
3. If $B^2 - 4AC < 0$, then the roots are not real (in fact, they are complex conjugates), and in this case, the graph does not cross the x-axis.

EXERCISES

1. Determine the equation of the circle with
   (a) Radius 3 and center $(0, 0)$.　　　　(b) Radius 2 and center $(1, 2)$.
   (c) Radius $\frac{1}{2}$ and center $(-\frac{7}{3}, 4)$.　　(d) Radius $\sqrt{6}$ and center $(0, -5)$.

2. The graphs of the following equations are circles. In each case determine the center and radius and graph it.
   (a) $x^2 + y^2 = 49$.　　　　　　(b) $(x - 3)^2 + (y + 5)^2 = 4$.
   (c) $x^2 + y^2 - 4x - 6y + 11 = 0$.　(d) $x^2 + y^2 + 6x + 5 = 0$.
   (e) $x^2 + y^2 - 3y = 0$.　　　　(f) $2x^2 + 2y^2 - 4x + 12y + 18 = 0$.

3. (a) Graph the circle $(x - 4)^2 + (y - 3)^2 = 25$ and verify that the points $(0, 0)$, $(8, 6)$, $(1, -1)$, and $(7, -1)$ lie on this circle.

   (b) For each of the four points given in part (a), determine the equation of the line tangent to the circle at that point.

4. Determine the graph of each of the following equations.

   (a) $(x + 3)^2 + (y - 7)^2 = 0.$
   (b) $(x - 5)^2 + (y + 2)^2 = -2.$
   (c) $x^2 + y^2 + 6x + 14y + 58 = 0.$
   (d) $x^2 + y^2 + 2x - 2y + 3 = 0.$

5. Find the equation of the parabola with vertex V and focus F given below. What is the directrix in each case?

   (a) $V(3, 2)$, $F(3, 4)$. (b) $V(3, 2)$, $F(4, 2)$.
   (c) $V(-2, -1)$, $F(-2, -2)$. (d) $V(-5, 1)$, $F(-6, 1)$.

6. Find the equation of the parabola with focus F and directrix given below.

   (a) $F(2, -3)$, $y = -1.$ (b) $F(-3, -4)$, $x = 1.$
   (c) $F(-1, 2)$, $y = -4.$ (d) $F(-3, 3)$, $y = x.$
   (e) $F(0, 0)$, $x + y = 1.$

7. Graph each of the following parabolas.

   (a) $y = -4x^2.$ (b) $x = 6y^2.$
   (c) $y + 3 = 2(x - 1)^2.$ (d) $x + 2 = 5(y + 2)^2.$
   (e) $y = x^2 + 2x + 4.$ (f) $x = -8y^2 - 32y - 29.$

8. Determine the focus, directrix, axis, and vertex of each parabola in Exercise 7.

9. Find the points of intersection of the circle $(x - 1)^2 + y^2 = 2$ and the line $y = x.$

10. (a) Show that the circle $x^2 + y^2 = 1$ and the line $x + y = 2$ do not intersect

    (b) For what values of k does the line $x + y = k$ intersect the circle $x^2 + y^2 = 1$?

11. Find the points of intersection of the parabola $x^2 - y - 1 = 0$ and the line $x - y + 1 = 0.$

12. (a) Find the points of intersection of the parabolas $y = x^2$ and $x = y^2.$

    (b) Draw the region enclosed by these two curves.

13. Find the real roots, if any, of each of the following equations.

    (a) $x^2 - 2x - 1 = 0.$ (b) $x^2 + 2x + 5 = 0.$
    (c) $4x^2 + 12x + 9 = 0.$

14. Let P be any point on the circle $x^2 + y^2 = r^2$, as in Figure 1.45. Show that the line AP is perpendicular to the line BP.

An **ellipse** is a collection of points the sum of whose distances from two fixed points is a constant. The fixed points are called the **foci** of the ellipse.

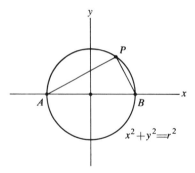

Figure 1.45

15. (a) If the foci are taken to be $F_1(-c, 0)$ and $F_2(c, 0)$ and the sum of the distances is denoted by $2a$, show that the coordinates of a point $P(x, y)$ on the ellipse satisfy the equation $\sqrt{(x + c)^2 + y^2} + \sqrt{(x - c)^2 + y^2} = 2a$.

(b) With $b^2 = a^2 - c^2$, show that the equation in part (a) can be written $(x^2/a^2) + (y^2/b^2) = 1$. The graph of this equation appears in Figure 1.46.

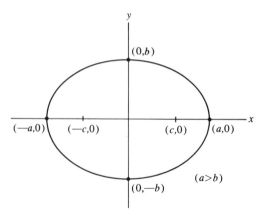

Figure 1.46

The point $(0, 0)$ is called the **center** of the ellipse. The part of the axis between $(-a, 0)$ and $(a, 0)$ is called the **major axis** of the ellipse and the part of the axis between $(0, -b)$ and $(0, b)$ is called the **minor axis**.

It should be noted that if $b > a$, the major and minor axes are reversed.

16. Sketch the graphs of the following ellipses.

(a) $x^2/4 + y^2/9 = 1$.   (b) $x^2/9 + y^2/4 = 1$.

(c) $16x^2 + 25y^2 = 400$.   (d) $3x^2 + 2y^2 = 6$.

17. (a) Show that the graph of $[(x - h)^2/a^2] + [(y - k)^2/b^2] = 1$ is an ellipse. [The point (h, k) is called the **center** of the ellipse.]

(b) Sketch the graphs of the following.

(1) $2x^2 + 3y^2 - 4x + 6y - 1 = 0$.

(2) $x^2 + 9y^2 - 54y + 72 = 0$.

(3) $2x^2 + 3y^2 - 4x + 6y + 5 = 0$.

(4) $x^2 + 2y^2 + 2x - 8y + 10 = 0$.

A **hyperbola** is a collection of points the difference of whose distances from two fixed points is a constant. The fixed points are called the **foci** of the hyperbola.

18. (a) If the foci are taken to be $F_1(-c, 0)$ and $F_2(c, 0)$ and the difference of the distances is denoted by $2a$, show that the coordinates of a point $P(x, y)$ on the hyperbola satisfy the equation $|\sqrt{(x + c)^2 + y^2} - \sqrt{(x - c)^2 + y^2}| = 2a$.

(b) Letting $b^2 = c^2 - a^2$, show that the equation in part (a) can be written $x^2/a^2 - y^2/b^2 = 1$. The graph of this equation appears in Figure 1.47.

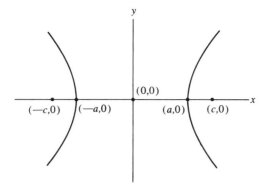

Figure 1.47

The point $(0, 0)$ is called the **center** of the hyperbola. [*Remark:* If instead one begins with foci at the points $(0, -c)$ and $(0, c)$, then the resulting equation is $y^2/a^2 - x^2/b^2 = 1$, where $b^2 = c^2 - a^2$; its graph is given in Figure 1.48.]

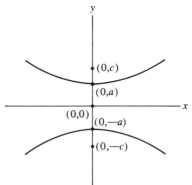

Figure 1.48

19. Sketch the graphs of each of the following hyperbolas.

(a) $x^2/4 - y^2/9 = 1$.

(b) $x^2/9 - y^2/4 = 1$.

(c) $y^2/4 - x^2/9 = 1$.

(d) $16x^2 - 25y^2 = 400$.

(e) $2x^2 - 3y^2 = -6$.

20. (a) Show that the graph of $[(x - h)^2/a^2] - [(y - k)^2/b^2] = 1$ is a hyperbola. [The point (h, k) is called the **center** of the hyperbola.]

(b) Sketch the graphs of the following.

(1) $x^2 - 2x - 4y^2 + 16y - 31 = 0$.

(2) $9x^2 + 36x - 25y^2 + 150y - 414 = 0$.

(3) $x^2 - y^2 - 2y - 2 = 0$.

(4) $16y^2 - 128y - 9x^2 - 36x + 76 = 0$.

1.5   THE GRAPHS OF SOME NONLINEAR EQUATIONS

In addition to the graphs of linear equations which we studied in Section 1.3, we have also studied in Section 1.4 the graphs of certain nonlinear equations such as the circle and the parabola. Here we want to study the graphs of some additional nonlinear equations. We begin by looking at some piecewise linear curves.

EXAMPLE 1   Sketch the graph of $y = |x|$.

Since $|x| = x$ if $x > 0$ and $|x| = -x$ if $x < 0$, we observe that the graph consists of that part of the line $y = x$ where $x > 0$ and that part of the line $y = -x$ where $x < 0$ (see Figure 1.49).

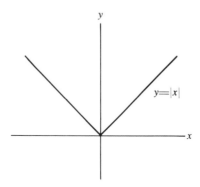

Figure 1.49

EXAMPLE 2   Sketch the graph of $|x| + |y| = 1$.

We first observe that the equation takes on different forms in each quadrant:

$$x + y = 1 \qquad \text{in quadrant I}$$

$$-x + y = 1 \qquad \text{in quadrant II}$$

$$-x - y = 1 \qquad \text{in quadrant III}$$

$$x - y = 1 \qquad \text{in quadrant IV}.$$

Each is the equation of a line in its respective quadrant (see Figure 1.50).

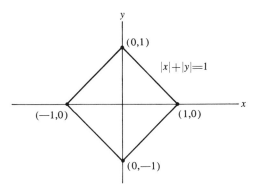

Figure 1.50

EXAMPLE 3   Sketch the graph of $y = [x]$, where $[x]$ is the largest integer less than or equal to x.

We first list some values:

x	$y = [x]$
0	0
$\frac{1}{2}$	0
$\frac{2}{3}$	0
1	1
$\frac{3}{2}$	1
$\frac{9}{2}$	4
$-\frac{5}{2}$	-3
$\vdots$	$\vdots$

It should be observed that $[x]$ is a fixed constant between successive integers. More precisely, if $n \leq x < n + 1$, then $[x] = n$ (see Figure 1.51).

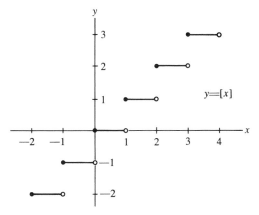

Figure 1.51

EXAMPLE 4   Sketch the graphs of $y = x^3$ and $x = y^3$.
We first tabulate some values (x, y) which satisfy $y = x^3$:

x	$y = x^3$		x	$y = x^3$
0	0		-1	-1
1	1		-2	-8
2	8		-3	-27
3	27		-4	-64
4	64		$\vdots$	$\vdots$
$\vdots$	$\vdots$			

From the table on the left, we see that as x increases positively, y also increases positively, and, from the table on the right, we observe that as x increases negatively, y also increases negatively. The graph of $y = x^3$ is given in Figure 1.52. Observe that some of the pairs which satisfy $y = x^3$

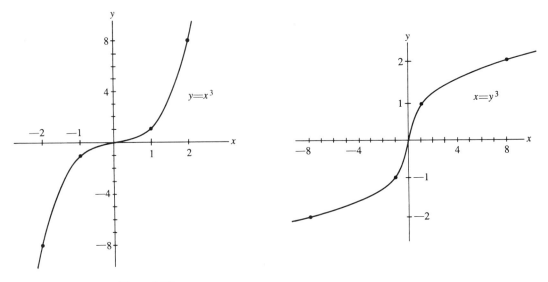

Figure 1.52

include $(0, 0)$, $(1, 1)$, $(-1, -1)$, $(2, 8)$, $(-2, -8)$, $(3, 27)$, $(-3, -27)$, $(4, 64)$, and $(-4, -64)$. By interchanging the x- and y-coordinates, we see that the pairs $(0, 0)$, $(1, 1)$, $(-1, -1)$, $(8, 2)$, $(-8, -2)$, $(27, 3)$, $(-27, -3)$, $(64, 4)$, and $(-64, -4)$ satisfy the equation $x = y^3$ (see Figure 1.52).

EXAMPLE 5   Sketch the graph of $y = 1/x$.
We begin by tabulating some values (x, y) which satisfy $y = 1/x$:

x	$1/x$		x	$1/x$
1	1		-2	$-\frac{1}{2}$
2	$\frac{1}{2}$		-7	$-\frac{1}{7}$
10	$\frac{1}{10}$		-94	$-\frac{1}{94}$
88	$\frac{1}{88}$		-997	$-\frac{1}{997}$
1000	$\frac{1}{1000}$		-5000	$-\frac{1}{5000}$
2500	$\frac{1}{2500}$		$\vdots$	$\vdots$
$\vdots$	$\vdots$			

Therefore we see that as x gets large positively, $1/x$ gets close to zero. Moreover, as x gets large negatively, $1/x$ gets close to zero. Notice also that

x	$1/x$		x	$1/x$
1	1		-1	-1
$\frac{1}{2}$	2		$-\frac{1}{2}$	-2
$\frac{1}{3}$	3		$-\frac{1}{3}$	-3
$\frac{1}{100}$	100		$-\frac{1}{225}$	-225
$\frac{1}{3000}$	3000		$-\frac{1}{2400}$	-2400
$\vdots$	$\vdots$		$\vdots$	$\vdots$

Hence as x gets close to zero from the right, $1/x$ gets large positively, and as x gets close to zero from the left, $1/x$ gets large negatively. The graph of $y = 1/x$ is given in Figure 1.53.

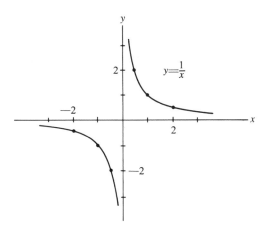

Figure 1.53

EXAMPLE 6   Sketch the graph of $y = \sqrt{x}$ and $y = \sqrt[3]{x}$.

The graph of $y = \sqrt{x}$ is that part of the parabola $y^2 = x$ where $y > 0$, and the graph of $y = \sqrt[3]{x}$ is the same as the graph of $y^3 = x$ (see Figure 1.54).

For reference, we list in Figure 1.55 some graphs of equations which occur frequently.

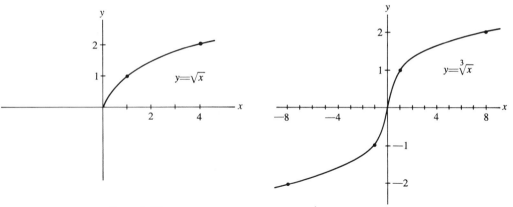

Figure 1.54

EXERCISES

1. Sketch the graph of each of the following.

(a) $y = -|x|$.   (b) $y = |-x|$.

(c) $x = |y|$.   (d) $y = |x + 5|$.

(e) $y = |x| + 5$.   (f) $2y = |3x + 4|$.

(g) $y = |x|/x$.   (h) $|2x| + |3y| = 1$.

2. Sketch the graph of each of the following.

(a) $y = [2x]$.   (b) $y = 2[x]$.

(c) $y = |[x]|$.   (d) $y = [|x|]$.

(e) $[x] + [y] = 1$.

3. Sketch the graph of the following.

(a) $y = -4x^3$.   (b) $y = x^3 + 1$.

(c) $y = x^{2/3}$.   (d) $y = \sqrt[3]{x} - 2$.

(e) $y = -\sqrt{x}$.   (f) $y = x^4$.

(g) $y = -1/x$.   (h) $y = 1/x^2$.

4. Sketch the graph of the equation $(x + y - 2)(2x - 3y + 1) = 0$.

5. Sketch the graphs of the following.

(a) $x^4 - y^4 = 0$.   (b) $x^2 + 2xy + y^2 = 0$.

(c) $4x^2 - 9y^2 = 0$.   (d) $4x^2 - y^2 - 4x - 4y - 3 = 0$.

(e) $x^3 + 3x^2y + 3xy^2 + y^3 = 0$.

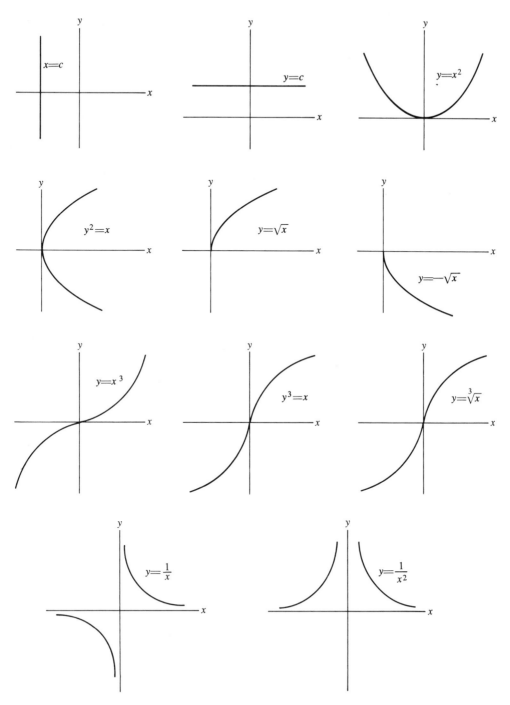

Figure 1.55

Chapter 2

THE CONCEPT
OF A FUNCTION

2.1 DEFINITION AND EXAMPLES

One of the fundamental concepts in mathematics is that of function. Basically, a function is a relationship or correspondence between two collections of numbers. For instance, knowing the radius r of a sphere, one can compute the volume V by using the formula

$$V = \tfrac{4}{3}\pi r^3.$$

This formula establishes a relationship between one collection of numbers (representing radii of spheres) and another collection of numbers (representing volumes of spheres). In Figure 2.1 we indicate a few of the correspondences formed between numbers, where the numbers on the left represent radii and those on the right represent volumes.

Mathematics used this notion for hundreds of years, but it was not until 1694 that Leibniz introduced the term *function*. He first used the term to denote certain qualities connected with a curve, such as slope and radius of curvature. The explanation of the meaning of *function* underwent refinement for over 100 years. In the nineteenth century, Lejeune Dirichlet (1805–1859) gave a formulation of the definition of *function* which is used regularly in modern-day mathematics. It is essentially Dirichlet's definition which we shall study and use in this chapter

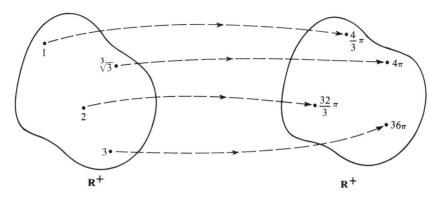

Figure 2.1

2.1.1
Definition

Let A and B be collections of numbers. A **function** (denoted by the letter f) **from A to B** is a correspondence which associates with each number x in the collection A one and only one number y in the collection B.

Figure 2.2 illustrates the definition of function.

To express the fact that f is a function from A to B, we use the notation

$$f: A \rightarrow B.$$

We say that f is defined on A, and frequently we call A the **domain of f**. If one sets

$$y = f(x),$$

read

"y equals f at x,"

then one has associated x and y. Alternatively, this may be expressed as

"y is the value of f at x"

or

"y is the image of x under f."

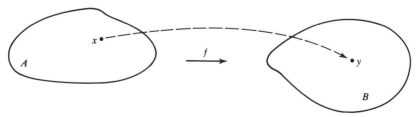

Figure 2.2

2.1.2
Definition

If A is a collection of numbers and f is a function from A to $\mathbf{R}$, then the **graph of f** is the graph of the equation

$$y = f(x),$$

where x is in A.

EXAMPLE 1

Let $f: \mathbf{R} \to \mathbf{R}$ be defined by the equation $f(x) = 5$; that is, $y = 5$.
Each value of f is the same, namely 5. Thus we have

$$f(0) = 5, \qquad f(1) = 5, \qquad f(-1) = 5, \ldots$$

The graph of f is, of course, the horizontal line $y = 5$ (see Figure 2.3).

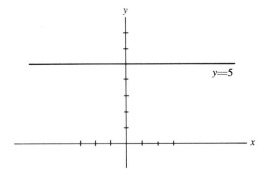

Figure 2.3

EXAMPLE 2

Let $f: \mathbf{R} \to \mathbf{R}$ be defined by the equation $f(x) = ax + b$; that is, $y = ax + b$, where a and b are fixed numbers, $a \neq 0$.
Note that

$$f(0) = b \qquad \text{and} \qquad f\left(-\frac{b}{a}\right) = 0;$$

that is, $(0, b)$ and $(-b/a, 0)$ are points on the graph of f. Its graph is, of course, a straight line with y-intercept $(0, b)$ and x-intercept $(-b/a, 0)$ (see Figure 2.4).

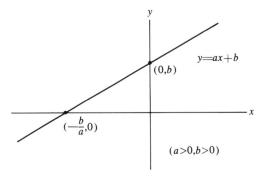

Figure 2.4

EXAMPLE 3   Let $f: \mathbf{R} \to \mathbf{R}$ be defined by the equation $f(x) = x^2$; that is, $y = x^2$. The following table indicates some of the values of f.

x	0	2	-2	3	-3	$\frac{1}{5}$	$\sqrt{7}$	4
x^2	0	4	4	9	9	$\frac{1}{25}$	7	16

Therefore we can write

$$f(0) = 0, \quad f(2) = 4, \quad f(-2) = 4, \quad f(\tfrac{1}{5}) = \tfrac{1}{25}, \ldots$$

Here we observe the fact that two numbers in $\mathbf{R}$ may have the same image. The graph of f is a parabola (recall Section 1.4), as drawn in Figure 2.5.

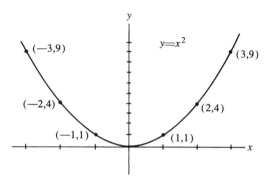

Figure 2.5

EXAMPLE 4   Is the graph of the equation $x = y^2$ the graph of a function f where $y = f(x)$? Let us set up a table of values (x, y) where $x = y^2$.

x	0	1	4	9	16	25
y	1	1 or -1	2 or -2	3 or -3	4 or -4	5 or -5

For each $x > 0$, there are two possible values of y such that $x = y^2$. Thus a function f where $y = f(x)$ is not defined by the given equation.

The graph of $x = y^2$ is the parabola given in Figure 2.6. Note that each vertical line $x = c$, where $c > 0$ crosses the graph of $x = y^2$ in two distinct points.

The remark made at the end of Example 4 can be generalized. The definition of a function f with domain A ensures that each vertical line

$$x = x_1 \qquad (x_1 \text{ in } A)$$

crosses the graph of f precisely once. Therefore the curve in Figure 2.7 is the graph of a function with domain $[a, b]$, whereas the curve in Figure 2.8 is *not* the graph of a function with domain $[a, b]$.

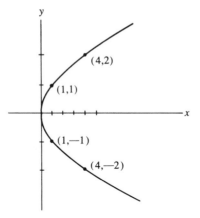

Figure 2.6

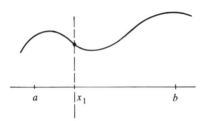

Figure 2.7   Figure 2.8

EXAMPLE 5   The graph of the circle $x^2 + y^2 = 9$ is not the graph of a function f, where $y = f(x)$. In fact, solving for y, we have $y = \pm\sqrt{9 - x^2}$.

However, parts of the graph of $x^2 + y^2 = 9$ are graphs of functions. The top semicircle is the graph of the function f defined by

$$f(x) = \sqrt{9 - x^2} \qquad \text{(see Figure 2.9)}.$$

The bottom semicircle is the graph of the function g, defined by

$$g(x) = -\sqrt{9 - x^2} \qquad \text{(see Figure 2.10)}.$$

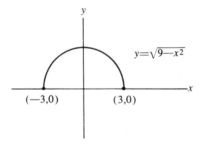

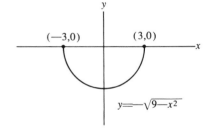

Figure 2.9   Figure 2.10

The function h defined by the equations

$$h(x) = \begin{cases} \sqrt{9 - x^2} & \text{if } 0 \leq x \leq 3, \\ -\sqrt{9 - x^2} & \text{if } -3 \leq x < 0, \end{cases}$$

has the graph given in Figure 2.11.

EXAMPLE 6   The postage rate on first-class mail is 8 cents per ounce or fraction thereof. If x is the weight in ounces, then let $f(x)$ be the postage rate. Thus, $f(x)$ is an integer which is some integral multiple of 8. Examples of values of f are given in the following table.

x	1	$\sqrt{3}$	3	$\frac{1}{2}$	$\frac{4}{3}$	4	$\frac{9}{2}$	$\frac{13}{3}$	5	$\frac{21}{4}$	6
$f(x)$	8	16	24	8	16	32	40	40	40	48	48

Figure 2.12 gives the graph of f.

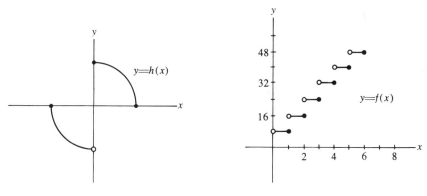

Figure 2.11   Figure 2.12

EXAMPLE 7   A certain sociological test attempts to measure job dissatisfaction within a particular group of workers. Based on a total of ten points, the higher the score, the higher the degree of job dissatisfaction. After administering the test to this group, it is observed that for each individual his job dissatisfaction score d and his number x of years of formal education are inversely proportional. More specifically, for each individual

$$d = \frac{10}{x}.$$

Also, formal education among the group members ranged from 1 to 16 years.

These results can be used to define a job dissatisfaction function

$$D: [1, 16] \rightarrow [0, 10]$$

given by

$$D(x) = \frac{10}{x}$$

where x is in $[1, 16]$. Tabulating some values of D, we have

x	1	2	3	5	8	10	13	16
$D(x)$	10	5	$\frac{10}{3}$	2	$\frac{5}{4}$	1	$\frac{10}{13}$	$\frac{5}{8}$

Therefore some points on the graph of D are

$$(1, 10), \quad (2, 5), \quad (3, \tfrac{10}{3}), \quad (5, 2), \quad (8, \tfrac{5}{4}), \quad (10, 1), \quad (13, \tfrac{10}{13}), \quad (16, \tfrac{5}{8});$$

others can be found in a similar fashion. The graph of D is drawn in Figure 2.13.

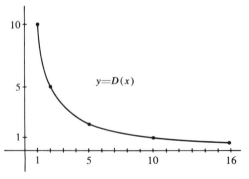

Figure 2.13

EXAMPLE 8   A projectile is shot upward from the surface of the earth and its distance s from the earth at any time t can be given by the quadratic function

$$s(t) = 320t - 16t^2.$$

The graph of s gives some interesting information:

$$s = 320t - 16t^2$$

$$s = -16(t^2 - 20t + 100) + 1600$$

$$s - 1600 = -16(t - 10)^2.$$

Therefore the graph of s is a parabola (see Figure 2.14). The graph indicates that the projectile reaches its maximum height of 1600 feet after 10 seconds.

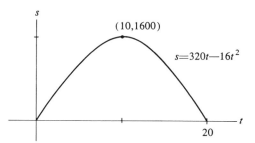

Figure 2.14

EXAMPLE 9 A function can also arise when a table of values is given. For example, in the table below, the column on the left indicates the number x of items of a certain commodity to be sold, while the one on the right indicates the cost c of producing those items.

x	c
5	95.12
10	96.46
15	97.39
20	98.14
25	98.77
30	99.32
35	99.81
40	100.26
45	100.67
50	101.05
55	101.40
60	101.74
65	102.06
70	102.36
75	102.65
80	102.92
85	103.19
90	103.44
95	103.69
100	103.92

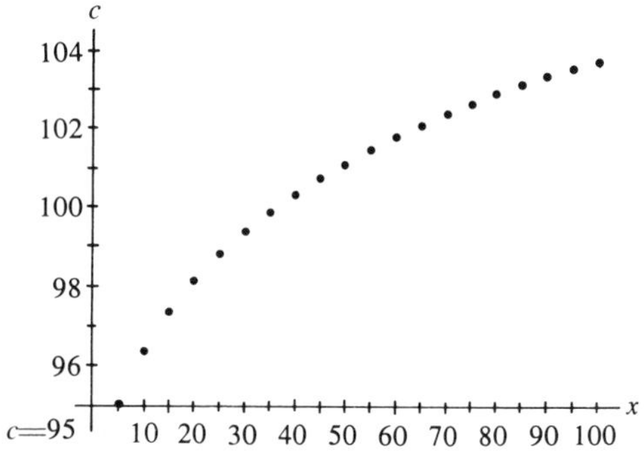

Figure 2.15

In Figure 2.15 we plot these values in a coordinatized plane.

For the purpose of mathematical analysis of a problem, it is often useful to draw a continuous curve joining the points, as done in Figure 2.16.

EXAMPLE 10 The cost of producing a commodity gives rise to a function usually referred to as a cost function. For example, the cost function might be $C(x) = 90 + 3\sqrt[3]{x}$, where x is the number of articles produced.

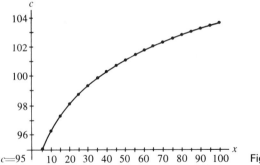

Figure 2.16

If the article sells for \$2, then we have a revenue function R, where

$$R(x) = 2x.$$

A profit function P is defined by the equation

$$P(x) = R(x) - C(x).$$

In our example, if 64 items are sold, then the profit is $P(64) = R(64) - C(64) = 128 - 102 = \$26$.

The concluding two examples of this section introduce functions which will appear in our later development of the calculus.

EXAMPLE 11   Consider the graph of the equation $y = x^2$ and let (x, x^2) be any point on this graph distinct from point $(2, 4)$ (see Figure 2.17). The line through these two points has slope

$$\frac{x^2 - 4}{x - 2}.$$

Thus the function S defined by

$$S(x) = \frac{x^2 - 4}{x - 2}$$

gives the slope of the line through (x, x^2) and $(2, 4)$. Observe that the domain of S is the set of all numbers with 2 deleted.

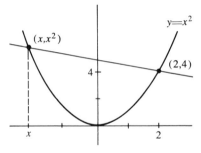

Figure 2.17

EXAMPLE 12   Consider the graph of the equation

$$y = \tfrac{1}{2}x + 1$$

and let x be any positive number (see Figure 2.18). Recalling that the area of such a trapezoid is half the altitude times the sum of the lengths of the bases, then the area of the shaded region is

$$\tfrac{1}{2}x(1 + \tfrac{1}{2}x + 1) = \tfrac{1}{4}x^2 + x.$$

Therefore the function A defined by

$$A(x) = \tfrac{1}{4}x^2 + x$$

gives the area of the shaded trapezoidal region for any positive number x. The domain of A is $\mathbf{R}^+$.

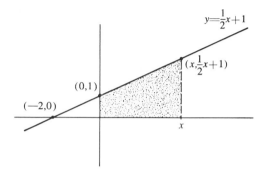

Figure 2.18

EXERCISES

1. Let f, g, h, k be functions defined by the following equations:

$$f(x) = 7x + 3 \qquad h(x) = 20\sqrt{25 - x^2}$$

$$g(x) = \frac{2x}{x^2 + 9} \qquad k(x) = x^2.$$

Determine

(a) $f(6), g(-1), h(4), k(\tfrac{1}{3})$.   (b) $f(x^2), g(1/x), h(x + 2), k(5x)$.

(c) $f(g(x)), g(f(x))$.   (d) $[f(x) - f(2)]/(x - 2)$.

2. Each of the following equations defines a function f. In each case write at least eight ordered pairs (x, y) where $y = f(x)$, plot these points, and obtain the graph of f.

(a) $f(x) = -7$.   (b) $f(x) = 3x - 5$.

(c) $f(x) = 3x^2$.   (d) $f(x) = -4x^3$.

(e) $f(x) = \sqrt{x} + 5$.   (f) $f(x) = 2|x|$.

(g) $f(x) = 5/x$.   (h) $f(x) = -1/x^2$.

(i) $f(x) = \sqrt{-x}$.   (j) $f(x) = 2 + \sqrt[3]{x}$.

(k) $f(x) = x/x$.

[*Note:* In some cases it may be helpful to refer to the graphs given in Section 1.5.]

3. In each case, graph the given function.

(a) $f(x) = \begin{cases} -1, & x \leq 0, \\ 1, & x > 0. \end{cases}$ (b) $g(x) = \begin{cases} 0, & x < -2, \\ 1, & -2 \leq x \leq 3, \\ -5, & x > 3. \end{cases}$

(c) $h(x) = \begin{cases} x, & x < 0, \\ \frac{1}{2}, & x = 0, \\ 1, & x > 0. \end{cases}$ (d) $k(x) = \begin{cases} 2x - 5, & x < 3, \\ 4 - x, & x \geq 3. \end{cases}$

(e) $l(x) = \begin{cases} -x, & x \leq -1, \\ x^2, & -1 < x < 1, \\ 1, & x \geq 1. \end{cases}$

4. Which of the curves in Figure 2.19 are graphs of functions with domain $[1, 3]$?

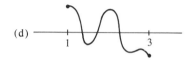

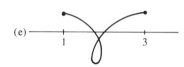

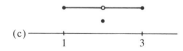

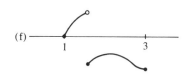

Figure 2.19

5. In each of the following, determine the domain of the function defined by the given equation.

(a) $f(x) = 7x + 3$.   (b) $g(x) = 2x/(x^2 + 9)$.

(c) $h(x) = 2x/(x^2 - 9)$.   (d) $k(x) = 20\sqrt{25 - x^2}$.

(e) $l(x) = 1/\sqrt[3]{x}$.   (f) $F(x) = x^2$.

(g) $G(x) = 2^x$.   (h) $H(x) = \sqrt{-x}$.

(i) $K(x) = \sqrt{|x|}$.

6. Define five functions f with domain $[-6, 6]$ such that $y = f(x)$ and $x^2 + y^2 = 36$.

7. Graph each of the functions f defined as follows.

(a) $f(x) = x + |x|$.   (b) $f(x) = x - |x|$.

(c) $f(x) = x|x|$.   (d) $f(x) = |x|/x$.

8. (a) Let r be the radius of a circle. Express the circumference and area as functions of r.

(b) Let s be the side of a square. Express the perimeter and area as functions of s.

9. A shoe salesman's gross weekly salary is \$150 plus \$2 commission for each pair of shoes he sells over 30 pairs. Express his gross weekly salary S as a function of the number of pairs sold.

10. The number of people on welfare in a certain city is 15,000 and growing by 600 persons per month. Express the number N of welfare recipients as a function of the month.

11. A projectile is shot upward from the surface of the earth and its distance s from the earth at any given time t is given by the quadratic function $s(t) = 400t - 40t^2$. Graph function s, and from the graph determine how long it takes the projectile to reach its maximum height.

12. A department store has tabulated the number of refrigerators it has sold in each of the preceding 12 months as follows.

m	1	2	3	4	5	6	7	8	9	10	11	12
n	51	70	93	100	88	64	50	41	32	25	29	40

Here $m = $ month and $n = $ number sold.

(a) Plot the 12 points (m, n) in a coordinatized plane, taking the m-axis to be horizontal and the n-axis vertical.

(b) Connect these points by an unbroken curve in such a fashion that the result is the graph of a function with domain [1, 12].

13. Consider the graph of the equation $y = x^3$. Find a function which gives the slope of the line through (1, 1) and any other distinct point (x, x^3) on this graph.

14. Consider the graph of the equation $y = |2x + 4|$ (see Figure 2.20). Find a function which gives the area of the shaded region for any $x \geq -2$.

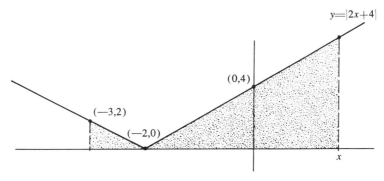

Figure 2.20

15. Express the postage stamp function of Example 6 in terms of the greatest integer function. [*Note:* The **greatest integer function** f is defined by the equation $f(x) = [x]$. See Example 3 of Section 1.5.]

16. A function $f: \mathbf{R} \to \mathbf{R}$ is called **one to one** if distinct numbers have distinct images; that is, $x_1 \neq x_2 \Rightarrow f(x_1) \neq f(x_2)$. Graph each the following functions f, and using that graph, determine whether or not f is one to one.

(a) $f(x) = 5$.   (b) $f(x) = x$.

(c) $f(x) = x^2$.   (d) $f(x) = x^3$.

(e) $f(x) = \sqrt[3]{x}$.

17. (a) Let $f(x) = x^2$. Graph $y = f(x)$ and $y = f(x - 2)$. Observe that the graph of $y = f(x - 2)$ is the graph of $y = f(x)$ translated two units to the right.

(b) Let $g(x) = |x|$. Graph $y = g(x)$ and $y = g(x + 5)$. Observe that the graph of $y = g(x + 5)$ is the graph of $y = g(x)$ translated 5 units to the left.

18. A function $f: \mathbf{R} \to \mathbf{R}$ is said to have a number p as a **period** if for each x in $\mathbf{R}$, $f(x + p) = f(x)$. For each of the following functions f, determine a period.

(a) $f(x) = -2$.

(b) $f(x) = \begin{cases} 1, & x \text{ an even integer,} \\ 0, & \text{otherwise.} \end{cases}$

(c) $f(x) = x - [x]$.

19. A function f defined on $\mathbf{R}$ is said to be an **even function** if $f(-x) = f(x)$ for each x in $\mathbf{R}$, and it is said to be an **odd function** if $f(-x) = -f(x)$ for each x in $\mathbf{R}$.

(a) Determine whether each of the following is even, odd, or neither.

(1) $f(x) = x^2$.   (2) $f(x) = |x|$.

(3) $f(x) = x$.   (4) $f(x) = x^3$.

(5) $f(x) = |x|^3$.   (6) $f(x) = x^2 + x$.

(b) Given function f, let $\phi_1(x) = [f(x) + f(-x)]/2$. Show that ϕ_1 is even.

(c) Given function f, let $\phi_2(x) = [f(x) - f(-x)]/2$. Show that ϕ_2 is odd.

(d) Show that any function is the sum of an even function and an odd function.

2.2 COMPOSITION OF FUNCTIONS

Often it is both natural and useful to piece together two given functions to form another, composite function. For instance, suppose that a spherical balloon is being inflated. Then the radius r is changing with time, say

$$r(t) = 2 + \tfrac{1}{3}t.$$

Consequently, the surface area

$$A(r) = 4\pi r^2$$

is also changing with time, and the dependence of A on t can be determined explicitly:

$$A(r(t)) = A(2 + \tfrac{1}{3}t)$$
$$= 4\pi(2 + \tfrac{1}{3}t)^2.$$

The last function is called the composition of r and A.

More generally,

2.2.1

Definition

Let f and g be functions, and let A be a collection of numbers such that for each x in A, $g(x)$ is a number in the domain of f. Then the function ϕ defined by the equation

$$\phi(x) = f(g(x))$$

is called the **composition of g and f**. A is called the **domain of ϕ**.

One may picture the composition of g and f as in Figure 2.21.

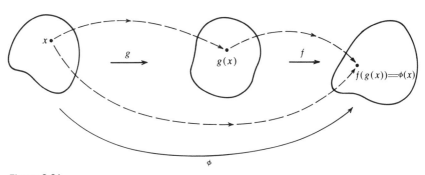

Figure 2.21

EXAMPLE 1

Compute $f(g(x))$ and $g(f(x))$ if

$$f(x) = 5x \qquad \text{and} \qquad g(x) = x^4.$$
$$f(g(x)) = f(x^4) = 5x^4, \text{ and } g(f(x)) = g(5x) = (5x)^4.$$

Note in Example 1 that

$$f(g(x)) \neq g(f(x));$$

that is, the composition of g and f is not the same as the composition of f and g.

EXAMPLE 2   Recall Example 10 of Section 2.1. There, revenue function R and cost function C where given by

$$R(x) = 2x$$

$$C(x) = 90 + 3\sqrt[3]{x},$$

respectively, and profit function P was defined by

$$P(x) = R(x) - C(x),$$

where x is the number of articles produced.

In addition, if x depends on time t, say

$$x(t) = 100 - \tfrac{1}{5}t,$$

then, R, C, and P also depend on time. In fact,

$$R(x(t)) = R(100 - \tfrac{1}{5}t)$$

$$= 2(100 - \tfrac{1}{5}t),$$

$$C(x(t)) = C(100 - \tfrac{1}{5}t)$$

$$= 90 + 3\sqrt[3]{100 - \tfrac{1}{5}t},$$

and

$$P(x(t)) = R(x(t)) - C(x(t))$$

$$= 2(100 - \tfrac{1}{5}t) - 90 - 3\sqrt[3]{100 - \tfrac{1}{5}t}$$

$$= 110 - \tfrac{2}{5}t - 3\sqrt[3]{100 - \tfrac{1}{5}t}.$$

As we shall see in our study of the derivative (Chapter 4), it is often helpful to recognize that a given function is the composition of two simpler ones, for then certain questions about the given complicated function may be answered by examining its simpler components.

EXAMPLE 3   Let $\phi(x) = (3x^2 + 1)^{20}$. If $g(x) = 3x^2 + 1$, then

$$\phi(x) = (g(x))^{20}.$$

Also, if $F(x) = x^{20}$, then

$$\phi(x) = F(g(x)).$$

Hence, ϕ is the composition of g and F.

EXAMPLE 4   Let $\phi(x) = \sqrt{x^2 + 5}$. With $g(x) = x^2 + 5$, we have

$$\phi(x) = \sqrt{g(x)}.$$

With $F(x) = \sqrt{x}$, we can write

$$\phi(x) = F(g(x)).$$

Again, ϕ is the composition of g and F.

EXERCISES

1. In each of the following, compute $f(g(x))$ and $g(f(x))$.

   (a) $f(x) = x^5$, $g(x) = 2x + 1$.
   (b) $f(x) = |x|$, $g(x) = 5x^2 + 2x + 1$.
   (c) $f(x) = x^4 + 9$, $g(x) = \sqrt{x}$.
   (d) $f(x) = 1/(x + 1)$, $g(x) = 1/x$.
   (e) $f(x) = 3x + 6$, $g(x) = \frac{1}{3}x - 2$.

2. Let $f(x) = 1/(2x - 4)$ and $g(x) = x - 3$.

   (a) Find $f(g(x))$ and $g(f(x))$.
   (b) Determine $f(g(0))$, $g(f(0))$, $f(g(1))$, $g(f(1))$, $f(g(-1))$, $g(f(-1))$, $f(g(2))$, $g(f(2))$, $f(g(5))$, $g(f(5))$, $f(g(\frac{13}{6}))$, and $g(f(\frac{13}{6}))$.

3. For each of the following functions ϕ, find simpler functions f and g such that $\phi(x) = f(g(x))$.

   (a) $\phi(x) = (4x^5 - 2)^{12}$.   (b) $\phi(x) = (x^3 - 2x^2 + 9)^5$.
   (c) $\phi(x) = (7x + 2)^{2/3}$.   (d) $\phi(x) = (x^2 - 3)^{4/3}$.
   (e) $\phi(x) = \sqrt{x^2 + 1}$.   (f) $\phi(x) = 3/\sqrt{2x}$.

4. If in Example 2 of this section $x(t) = 10 + 5/t^2$, $R(x) = 5x^{3/2}$, and $C(x) = 100 + \sqrt{2x}/3$, then determine R, C, and P as functions of t.

5. A rock is dropped into a pond at time $t = 0$ and it produces a circular wave of radius r. Given that for any time $t > 0$, $r = \sqrt{t}/2$, express the area A enclosed by the wave as a function of time t.

6. A student works his way through college by selling ice in the summer months. He knows that on a given day his profit P depends on the quantity Q sold and that the quantity sold in turn depends on the temperature T that particular day. In fact, he determines that $P(Q) = 0.02Q + 10$ and that $Q(T) = 3(T - 60)^2$. Express the profit P as a function of temperature T.

7. Functions f and g are said to be **inverses** if $f(g(x)) = x$ and $g(f(x)) = x$. Show that each of the following pairs are inverses.

   (a) $f(x) = 1/x$ and $g(x) = 1/x$.
   (b) $f(x) = x^2$ and $g(x) = \sqrt{x}$.
   (c) $f(x) = x^3$ and $g(x) = \sqrt[3]{x}$.
   (d) $f(x) = (2x - 1)/3$ and $g(x) = (3x + 1)/2$.

8. Let f and g be one-to-one functions, and let $\phi(x) = f(g(x))$. Show that ϕ is a one-to-one function.

9. Let g be a function which has p as a period, and let $\phi(x) = f(g(x))$, where f is a function. Show that ϕ has p as a period.

10. (a) Let f and g be odd functions, and let $\phi(x) = f(g(x))$. Show that ϕ is an odd function.

(b) Let g be an even function, and let $\phi(x) = f(g(x))$, where f is a function. Show that ϕ is an even function.

Chapter 3

LIMITS AND CONTINUITY

3.1 THE CONCEPT OF LIMIT

In the Introduction we made acquaintance with the limit concept. In discussing decimal representation of numbers, we saw that the irrational number $\sqrt{2}$, which can be represented as an infinite nonrepeating decimal, could be used to construct an unending sequence

$$1, 1.4, 1.41, 1.414, 1.4142, \ldots$$

of rational numbers closing in on $\sqrt{2}$. The farther out in the sequence we go, the nearer we get to $\sqrt{2}$. More precisely, we can find a rational number as close to $\sqrt{2}$ as we wish by going out far enough in the sequence.

In Chapter 2 we studied the concept of a function, and now we wish to study the behavior of functional values $f(x)$ subject to a certain behavior of the variable x. In particular, for a number c, it is often of interest to ask: What happens to $f(x)$ as x approaches c? To answer this question, we shall first deal with two related questions:

What happens to $f(x)$ as x approaches c through values less than c (from below)?

What happens to $f(x)$ as x approaches c through values greater than c (from above)?

If the symbols

$$\text{``}x \uparrow c\text{''} \qquad \text{and} \qquad \text{``}x \downarrow c\text{''}$$

replace the phrases

>   "as x approaches c through values less than c"

and

>   "as x approaches c through values greater than c,"

respectively, then we can rewrite the above questions as

$$\text{What is } \lim_{x \uparrow c} f(x)?$$

and

$$\text{What is } \lim_{x \downarrow c} f(x)?$$

The symbol

$$\lim_{x \uparrow c} f(x)$$

is read

>   "the limit of $f(x)$ as x approaches c from below,"

and the symbol

$$\lim_{x \downarrow c} f(x)$$

is read

>   "the limit of $f(x)$ as x approaches c from above."

Let us pause here to analyze a few examples.

EXAMPLE 1   Consider $F(x) = 2x + 8$ and the point $c = 5$ in the domain of F. It seems obvious that as we select values of x close to 5 the functional values $f(x)$ should get close to 18. This suspicion is confirmed in the following tables, which give values of F at several points close to 5.

x	$2x + 8$	x	$2x + 8$
4	16	5.7	19.4
4.5	17	5.01	18.02
4.92	17.84	5.006	18.012
4.99	17.98	5.0002	18.0004
4.994	17.988	5.00009	18.00018
4.998	17.996	5.00001	18.00002
4.9995	17.999	5.000004	18.000008
4.99999	17.99998	5.0000002	18.0000004

An inspection of the chart* on the left prompts us to say the following.

1. We can get $F(x)$ within 2 of 18 (that is, between 16 and 18) by choosing x within 1 of 5, $x < 5$ (that is, between 4 and 5).

---

* In the x columns of these charts, and those which follow in this chapter, we have not used a pattern in approaching c. Also, functional values have been rounded off.

2. We can get $F(x)$ within 1 of 18 (that is, between 17 and 18) by choosing x within 0.5 of 5, $x < 5$ (that is, between 4.5 and 5).

3. We can get $F(x)$ within 0.16 of 18 (that is, between 17.84 and 18) by choosing x within 0.08 of 5, $x < 5$ (that is, between 4.92 and 5).

It appears that, were we to continue this analysis of the chart on the left, we could get $F(x)$ as close to 18 as we like by choosing any x within a certain distance of 5, $x < 5$. We express this fact by the symbols

$$\lim_{x \uparrow 5}(2x + 8) = 18.$$

Similarly, inspecting the list on the right prompts us to say the following.

1. We can get $F(x)$ within 1.4 of 18 (that is, between 18 and 19.4) by choosing x within 0.7 of 5, $x > 5$ (that is, between 5 and 5.7).

2. We can get $F(x)$ within 0.02 of 18 (that is, between 18 and 18.02) by choosing x within 0.01 of 5, $x > 5$ (that is, between 5 and 5.01).

3. We can get $F(x)$ within 0.012 of 18 (that is, between 18 and 18.012) by choosing x within 0.006 of 5, $x > 5$ (that is, between 5 and 5.006).

Again, it appears that we can continue this procedure, thereby getting $F(x)$ as close to 18 as we like by choosing any x within a certain distance of 5, $x > 5$. We express this fact by the symbols

$$\lim_{x \downarrow 5}(2x + 8) = 18.$$

Note that we have drawn these conclusions without checking the value of $F(x) = 2x + 8$ at $x = 5$; it happens to be 18. Geometrically, the results are evident from the graph of F given in Figure 3.1.

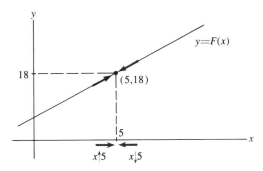

Figure 3.1

EXAMPLE 2 Let us proceed in a similar fashion with the function

$$g(x) = \frac{x^2 - 5x + 6}{x - 3}$$

and the point $c = 3$.

x	$\dfrac{x^2 - 5x + 6}{x - 3}$	x	$\dfrac{x^2 - 5x + 6}{x - 3}$
2.7	0.7	3.4	1.4
2.9	0.9	3.01	1.01
2.98	0.98	3.007	1.007
2.993	0.993	3.003	1.003
2.9962	0.9962	3.0008	1.0008
2.999995	0.999995	3.00001	1.00001
2.999999	0.999999	3.000005	1.000005

Again, we make some observations from the chart on the left.

1. We can get $g(x)$ within 0.3 of 1 (that is, between 0.7 and 1) by choosing x within 0.3 of 3, $x < 3$ (that is, between 2.7 and 3).

2. We can get $g(x)$ within 0.1 of 1 (that is, between 0.9 and 1) by choosing x within 0.1 of 3, $x < 3$ (that is, between 2.9 and 3).

3. We can get $g(x)$ within 0.02 of 1 (that is, between 0.98 and 1) by choosing x within 0.02 of 3, $x < 3$ (that is, between 2.98 and 3).

Were we to continue, we could get $g(x)$ as close to 1 as we like by choosing any x within a certain distance of 3, $x < 3$. This is expressed by the symbols

$$\lim_{x \uparrow 3} \frac{x^2 - 5x + 6}{x - 3} = 1.$$

Now turning our attention to the chart on the right:

1. We can get $g(x)$ within 0.4 of 1 (that is, between 1 and 1.4) by choosing x within 0.4 of 3, $x > 3$ (that is, between 3 and 3.4).

2. We can get $g(x)$ within 0.01 of 1 (that is, between 1 and 1.01) by choosing x within 0.01 of 3, $x > 3$ (that is, between 3 and 3.01).

3. We can get $g(x)$ within 0.007 of 1 (that is, between 1 and 1.007) by choosing x within 0.007 of 3, $x > 3$ (that is, between 3 and 3.007).

Again, the continuance of this procedure indicates that we can get $g(x)$ as close to 1 as we like by choosing x within a certain distance of 3, $x > 3$. We express this fact as

$$\lim_{x \downarrow 3} \frac{x^2 - 5x + 6}{x - 3} = 1.$$

It is important to observe here that we have made sense of

$$\lim_{x \uparrow 3} g(x) \qquad \text{and} \qquad \lim_{x \downarrow 3} g(x)$$

despite the fact that g is not defined at 3. $[g(x) = (x^2 - 5x + 6)/(x - 3) = x - 2$ for $x \neq 3$; remember that division by zero is not allowed.]

The above conclusions are suggested by the graph of g given in Figure 3.2.

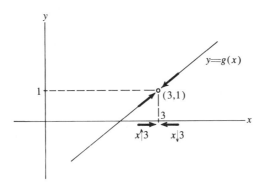

Figure 3.2

EXAMPLE 3 Consider the function

$$h(x) = \begin{cases} 2 - x^2, & x \le 0, \\ 3 + x, & x > 0, \end{cases}$$

and the point $c = 0$ in its domain.

x	$2 - x^2$	x	$3 + x$
−0.6	1.64	0.5	3.5
−0.3	1.91	0.4	3.4
−0.15	1.9775	0.25	3.25
−0.08	1.9936	0.13	3.13
−0.009	1.999919	0.06	3.06
−0.004	1.999984	0.01	3.01
−0.0005	1.99999975	0.002	3.002
−0.0001	1.99999999	0.003	3.003
−0.00002	1.9999999996	0.00001	3.0001

If we analyze these charts in a fashion similar to the two preceding examples, we are led to two conclusions.

First, the chart on the left indicates that we can get $h(x)$ as close to 2 as we like by choosing x within a certain distance of 0, $x < 0$. Therefore, symbolically, we write

$$\lim_{h\uparrow 0} h(x) = 2.$$

Second, the chart on the right suggests that we can get $h(x)$ as close to 3 as we like by choosing x within a certain distance of 0, $x > 0$. This is expressed by writing

$$\lim_{h\downarrow 0} h(x) = 3.$$

Figure 3.3 gives the graph of h in the vicinity of the point $(0, 2)$.

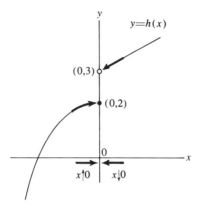

Figure 3.3

3.1.1
One-Sided Limits

We say that

$$\lim_{x \uparrow c} f(x) = L$$

if we can get $f(x)$ as close to L as we like by choosing any x within a certain distance of c, $x < c$. Also,

$$\lim_{x \downarrow c} f(x) = L$$

if we can get $f(x)$ as close to L as we like by choosing any x within a certain distance of c, $x > c$.

These limits are sometimes referred to as the **left-hand limit of f at c** and the **right-hand limit of f at c**, respectively.

If we do not wish to restrict the variable x to approaching point c from one side only, then we use the symbol

$$\text{``}x \to c\text{''}$$

to mean

$$\text{``as } x \text{ approaches } c.\text{''}$$

The symbol

$$\text{``}\lim_{x \to c} f(x)\text{''}$$

is read

$$\text{``the limit of } f(x) \text{ as } x \text{ approaches } c.\text{''}$$

3.1.2
Two-Sided Limit

We say that

$$\lim_{x \to c} f(x) = L$$

if

$$\lim_{x \uparrow c} f(x) = L \quad \text{and} \quad \lim_{x \downarrow c} f(x) = L.$$

If

$$\lim_{x \to c} f(x) = L \text{ for } no \text{ number } L,$$

then we say that

$$\lim_{x \to c} f(x) \textbf{ does not exist.}$$

Observe that 3.1.2 is equivalent to the following longer statement.

3.1.3
Two-Sided Limit
(alternative
statement)

We say that

$$\lim_{x \to c} f(x) = L$$

if we can get $f(x)$ as close to L as we like by choosing any x within a certain distance of c, $x \neq c$.

EXAMPLE 4

Recalling the functions considered in Examples 1–3 of this section, we have

$$\lim_{x \to 5} F(x) = 18$$

$$\lim_{x \to 3} g(x) = 1$$

and

$$\lim_{x \to 0} h(x) \text{ does not exist.}$$

Remarks on Two-Sided Limits

1. The restriction $x \neq c$ is included in 3.1.3 because it may be that f is not defined at c and yet there is a number L with the property that

$$\lim_{x \to c} f(x) = L.$$

We have already seen such a function in Example 2.

2. In 3.1.1 and 3.1.2 it is assumed that f is defined in some open interval about c, although not necessarily at c.

3. If f is defined at the point c, it is not necessarily true that

$$\lim_{x \to c} f(x) = f(c).$$

This particular topic will be investigated in Section 3.3.

4. It is sometimes convenient to write

$$\lim_{x \to c} f(x) = L$$

in the equivalent form

$$\lim_{h \to 0} f(c + h) = L.$$

The second form is obtained from the first by making the substitution

$$x = c + h$$

and noting that

$$x \to c \Leftrightarrow h \to 0.$$

5. If $\lim\limits_{x \to c} f(x) = L$, we sometimes say

$$\text{``}\lim\limits_{x \to c} f(x) \text{ exists''}$$

when we have no need of specifying the number L.

6. Even if $\lim\limits_{x \to c} f(x) = L$, it need not be true that

"as x gets closer to c, $f(x)$ gets closer to L."

An example is afforded by the function

$$f(x) = \begin{cases} 0, & x \text{ rational,} \\ x, & x \text{ irrational,} \end{cases}$$

where $c = 0$ and $L = 0$.

We shall forgo a more precise mathematical formulation of the notion of limit; as it stands it is sufficient for our purposes. Those who desire a more exact statement are referred to the optional section at the end of this chapter (Section 3.5).

EXERCISES

1. Consider $F(x) = 3x - 2$ and the point $c = 2$ in the domain of F. The following tables give values of F at points close to 2.

x	$3x - 2$	x	$3x - 2$
1	1	3	7
1.5	2.5	2.4	5.2
1.9	3.7	2.1	4.3
1.95	3.85	2.06	4.18
1.99	3.97	2.02	4.06
1.993	3.979	2.007	4.021
1.998	3.994	2.003	4.009
1.999	3.997	2.0008	4.0024
1.9997	3.9991	2.0003	4.0009
1.99992	3.99976	2.00001	4.00003

(a) Use the list on the left to fill in the following blanks.

   (1) We can get $F(x)$ within 3 of 4 (that is, between 1 and 4) by choosing x within _____ of 2, $x < 2$ (that is, between _____ and 2). We can

get $F(x)$ within 1.5 of 4 (that is, between 2.5 and 4) by choosing x within _____ of 2, $x < 2$ (that is, between _____ and 2). We can get $F(x)$ within 0.3 of 4 (that is, between 3.7 and 4) by choosing x within _____ of 2, $x < 2$ (that is, between _____ and 2).

(2) It appears as though we can get $F(x)$ as close to _____ as we like by choosing any x within a certain distance of _____, $x < 2$; therefore, $\lim_{x\uparrow-} F(x) =$ _____.

(b) Use the list on the right to fill in the following blanks.

    (1) We can get $F(x)$ within 3 of 4 (that is, between 4 and 7) by choosing x within _____ of 2, $x > 2$ (that is, between 2 and _____). We can get $F(x)$ within 1.2 of 4 (that is, between 4 and 5.2) by choosing x within _____ of 2, $x > 2$ (that is, between 2 and _____). We can get $F(x)$ within 0.3 of 4 (that is, between 4 and 4.3) by choosing x within _____ of 2, $x > 2$ (that is, between 2 and _____).

    (2) It appears as though we can get $F(x)$ as close to _____ as we like by choosing any x within a certain distance of _____, $x > 2$; therefore, $\lim_{x\downarrow-} F(x) =$ _____.

(c) Does $\lim_{x\to 2} F(x)$ exist? If so, what is it?

(d) Graph F.

2. Let $F(x) = (x^2 - 64)/(x - 8)$. The tables below give values of F at points close to 8.

x	$\dfrac{x^2 - 64}{x - 8}$	x	$\dfrac{x^2 - 64}{x - 8}$
7.3	15.3	8.5	16.5
7.8	15.8	8.1	16.1
7.94	15.94	8.08	16.08
7.98	15.98	8.02	16.02
7.993	15.993	8.007	16.007
7.9985	15.9985	8.001	16.001
7.99979	15.99979	8.0004	16.0004
7.999997	15.999997	8.00003	16.00003
		8.0000006	16.0000006

(a) Use the chart on the left to fill in the following blanks.

    (1) We can get $F(x)$ within 0.7 of 16 (that is, between 15.3 and 16) by choosing x within _____ of 8, $x < 8$ (that is, between _____ and 8). We can get $F(x)$ within 0.2 of 16 (that is, between 15.8 and 16) by choosing x within _____ of 8, $x < 8$ (that is, between _____ and 8). We can get $F(x)$ within 0.06 of 16 (that is, between 15.94 and 16) by choosing x within _____ of 8, $x < 8$ (that is, between _____ and 8).

(2) It appears that we can get $F(x)$ as close to _____ as we like by choosing any x within a certain distance of _____ , $x < 8$; therefore, $\lim\limits_{x \uparrow \_} F(x) =$ _____.

(b) Use the chart on the right to fill in the following blanks.

(1) We can get $F(x)$ within 0.5 of 16 (that is, between 16 and 16.5) by choosing x within _____ of 8, $x > 8$ (that is, between 8 and _____). We can get $F(x)$ within 0.1 of 16 (that is, between 16 and 16.1) by choosing x within _____ of 8, $x > 8$ (that is, between 8 and _____). We can get $F(x)$ within 0.08 of 16 (that is, between 16 and 16.08) by choosing x within _____ of 8, $x > 8$ (that is, between 8 and _____).

(2) It appears that we can get $F(x)$ as close to _____ as we like by choosing any x within a certain distance of _____, $x > 8$; therefore, $\lim\limits_{x \downarrow \_} F(x) =$ _____.

(c) Does $\lim\limits_{x \to 8} F(x)$ exist? If so, what is it?

(d) Graph F.

3. The following tables give values of function F defined by $F(x) = \sqrt{x + 2}$ at points close to 47.

x	$\sqrt{x + 2}$	x	$\sqrt{x + 2}$
45	6.86	50	7.21
46	6.93	49	7.14
46.5	6.96	48	7.07
46.9	6.993	47.7	7.05
46.92	6.994	47.1	7.007
46.97	6.998	47.05	7.0036
46.99	6.9993	47.025	7.0018
46.998	6.99986	47.002	7.00014
46.9995	6.999964	47.0001	7.00000714
46.99993	6.999995	47.00006	7.00000429

(a) Use the list on the left to fill in the following blanks.

(1) We can get $F(x)$ within 0.14 of 7 (that is, between 6.86 and 7) by choosing x within _____ of 47, $x < 47$ (that is, between _____ and 47). We can get $F(x)$ within 0.04 of 7 (that is, between 6.96 and 7) by choosing x within _____ of 47, $x < 47$ (that is, between _____ and 47). We can get $F(x)$ within 0.002 of 7 (that is, between 6.998 and 7) by choosing x within _____ of 47, $x < 47$ (that is, between _____ and 47).

(2) It seems that we can get $F(x)$ as close to _____ as we like by choosing any x within a certain distance of _____, $x < 47$; therefore, $\lim\limits_{x \uparrow \_} F(x) =$ _____.

(b) Use the chart on the right to fill in the following blanks.

(1) We can get $F(x)$ within 0.21 of 7 (that is, between 7 and 7.21) by choosing x within _____ of 47, $x > 47$ (that is, between 47 and _____). We can get $F(x)$ within 0.05 of 7 (that is, between 7 and 7.05) by choosing x within _____ of 47, $x > 47$ (that is, between 47 and _____). We can get $F(x)$ within 0.0018 of 7 (that is, between 7 and 7.0018) by choosing x within _____ of 47, $x > 47$ (that is, between 47 and _____).

(2) It seems that we can get $F(x)$ as close to _____ as we like by choosing any x within a certain distance of _____, $x > 47$; therefore, $\lim\limits_{x\downarrow} F(x) = $ _____.

(c) Does $\lim\limits_{x\to 47} F(x)$ exist? If so, what is it?

(d) Graph F.

4. Let function F be defined by

$$F(x) = \begin{cases} 2x - 4, & x < 1, \\ \sqrt{x}, & x \geq 1. \end{cases}$$

The following tables give values of F at points close to 1.

x	$2x - 4$	x	$\sqrt{x}$
0.5	-3	2	1.4
0.9	-2.2	1.6	1.26
0.97	-2.06	1.2	1.095
0.991	-2.018	1.05	1.025
0.998	-2.004	1.01	1.005
0.9996	-2.0008	1.007	1.003
0.9999	-2.0002	1.0004	1.0002
0.99995	-2.0001	1.0001	1.00005
0.999994	-2.000012	1.00006	1.00003

(a) Use the chart on the left to fill in the following blanks.

(1) We can get $F(x)$ within 1 of -2 (that is, between -3 and -2) by choosing x within _____ of 1, $x < 1$ (that is, between _____ and 1). We can get $F(x)$ within 0.06 of -2 (that is, between -2.06 and -2) by choosing x within _____ of 1, $x < 1$ (that is, between _____ and 1). We can get $F(x)$ within 0.0008 of -2 (that is, between -2.0008 and -2) by choosing x within _____ of 1, $x < 1$ (that is, between _____ and 1).

(2) It seems as though we can get $F(x)$ as close to _____ as we like by choosing any x within a certain distance of _____, $x < 1$; therefore, $\lim\limits_{x\uparrow} F(x) = $ _____.

(b) Use the chart on the right to fill in the following blanks.

(1) We can get $F(x)$ within 0.4 of 1 (that is, between 1 and 1.4) by choosing x within _____ of 1, $x > 1$ (that is, between 1 and _____). We can get $F(x)$ within 0.025 of 1 (that is, between 1 and 1.025) by choosing x

within _____ of 1, $x > 1$ (that is, between 1 and _____). We can get $F(x)$ within 0.00005 of 1 (that is, between 1 and 1.00005) by choosing x within _____ of 1, $x > 1$ (that is, between 1 and _____).

(2) It seems as though we can get $F(x)$ as close to _____ as we like by choosing any x with a certain distance of _____, $x > 1$; therefore, $\lim\limits_{x \downarrow -} F(x) =$ _____.

(c) Does $\lim\limits_{x \to 1} F(x)$ exist? If so, what is it?

(d) Graph F.

5. Consider the function $G(x) = (x - 3)/(x - 1)$ and the point $c = 5$.

x	$\dfrac{x-3}{x-1}$		x	$\dfrac{x-3}{x-1}$
3	0		6.5	0.63
4	0.33		5.7	0.57
4.5	0.429		5.1	0.51
4.9	0.487		5.01	0.501
4.92	0.4898		5.006	0.5007
4.99	0.4987		5.0002	0.50002
4.994	0.49925		5.00009	0.500011
4.998	0.49975		5.00001	0.500001
4.9993	0.499937		5.000004	0.5000005
4.99999	0.49999875		5.0000002	0.500000025

Because we can get $G(x) = (x - 3)/(x - 1)$ as close to _____ as we like by choosing any x within a certain distance of _____, we say that $\lim\limits_{x \to -} [(x - 3)/(x - 1)] =$ _____. (Fill in the blanks.)

6. Let $H(x) = x^3 - 5$ and $c = 0.5$.

x	$x^3 - 5$		x	$x^3 - 5$
-0.5	-5.125		2	3
-0.1	-5.001		1	-4
0.3	-4.973		0.75	-4.58
0.4	-4.936		0.6	-4.78
0.45	-4.9089		0.57	-4.815
0.48	-4.8894		0.53	-4.851
0.498	-4.87649		0.51	-4.867
0.499	-4.8757485		0.501	-4.874248499
0.4997	-4.87522487		0.5001	-4.874924985
0.499999	-4.87500075		0.5000001	-4.874999925

These charts indicate that we can get $H(x) = x^3 - 5$ as close to _____ as we like by choosing any x within a certain distance of _____. Hence $\lim\limits_{x \to -}(x^3 - 5) =$ _____. (Fill in the blanks.)

7. Let $G(x) = (x - 3)/(x - 1)$ and consider values of G near the point $c = -5$.

x	$\dfrac{x-3}{x-1}$	x	$\dfrac{x-3}{x-1}$
-7	1.25	-3.5	1.4
-5.9	1.29	-4.5	1.36
-5.2	1.32	-4.9	1.34
-5.1	1.328	-4.93	1.337
-5.01	1.3328	-4.99	1.334
-5.006	1.3330	-4.995	1.3336
-5.002	1.33322	-4.999	1.33339
-5.0005	1.333305	-4.9998	1.33334
-5.0001	1.333327	-4.9999	1.333339
-5.00004	1.333331111	-4.99997	1.333335

Since we can get $G(x) = (x - 3)/(x - 1)$ as close to _____ as we like by choosing any x within a certain distance of _____, then we say that $\lim\limits_{x \to -} [(x - 3)/(x - 1)] =$ _____. (Fill in the blanks.)

8. Consider function H defined by

$$H(x) = \begin{cases} x, & x \text{ rational,} \\ 0, & x \text{ irrational.} \end{cases}$$

(a) Does $\lim\limits_{x \to 1} H(x)$ exist?

(b) For $c \neq 0$, does $\lim\limits_{x \to c} H(x)$ exist?

(c) Does $\lim\limits_{x \to 0} H(x)$ exist?

3.2 LIMIT RULES

Complicated functions are often constructed by applying to simpler functions the fundamental algebraic operations of addition, subtraction, multiplication, division, and taking roots and powers. For example,

$$R(x) = \frac{x^3 + 9}{\sqrt[4]{x - 1}}$$

can be written in terms of the simpler functions $f(x) = x^3 + 9$ and $g(x) = x - 1$ as

$$R(x) = \frac{f(x)}{\sqrt[4]{g(x)}}.$$

This leads us to ask if one can determine

$$\lim_{x \to c} R(x)$$

by knowing

$$\lim_{x \to c}(x^3 + 9) \qquad \text{and} \qquad \lim_{x \to c}(x - 1).$$

It is the purpose of this section to deal with this type of problem, considering separately several of the operations just listed. In the process we shall develop a technique which allows us to compute many limits without resorting to the construction of lengthy charts, as in Section 3.1.

First, we observe the behavior of the function f defined by $f(x) = x^r$ (r is a rational number) near a point c in its domain. Consider the three special cases given in Figure 3.4. In each case, we can get x^r as close to c^r as we like by choosing any x within a certain distance of c. This suggests the result that

$$\lim_{x \to c} x^r = c^r.$$

In fact, a more general result is often quite useful.

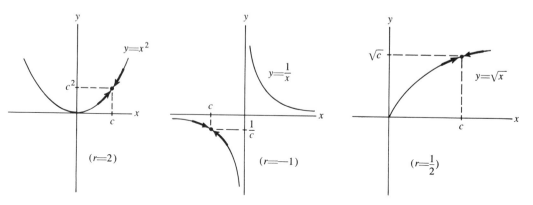

Figure 3.4

3.2.1
Limit Rule

1. If k is a constant and r is a rational number, then

$$\lim_{x \to c} kx^r = kc^r.$$

Note the following special cases of 3.2.1: When $r = 0$, then $x = c^r = 1$ and we get

$$\lim_{x \to c} k = k;$$

for $r = k = 1$, the rule becomes

$$\lim_{x \to c} x = c.$$

EXAMPLE 1

$$\lim_{x \to 1} 6 = 6$$

$$\lim_{x \to -3} x = -3$$

$$\lim_{x \to 2} 4x^5 = 4(2^5) = 128$$

$$\lim_{x \to 27} 3x^{-2/3} = 3(27)^{-2/3} = \tfrac{1}{3}.$$

To handle limits of more complicated functions—those built by sums, products, quotients, roots or powers—we need the following rules.

3.2.2

Limit Rules

Suppose that $\lim_{x \to c} f(x) = L$ and that $\lim_{x \to c} g(x) = M$.

2. If $L > 0$, $\lim_{x \to c} k[f(x)]^r = kL^r$, where r is rational.

3. $\lim_{x \to c} [f(x) + g(x)] = L + M$.

4. $\lim_{x \to c} f(x)g(x) = LM$.

5. $\lim_{x \to c} [f(x)/g(x)] = L/M, (M \neq 0)$.

Each of the five limit rules stated above is reasonable in light of the limit concept developed in the preceding section. For instance, regarding limit rule 3, if we can get $f(x)$ as close to L as we like by choosing any x within a certain distance of c ($x \neq c$) and if we can get $g(x)$ as close to M as we like by choosing any x within a certain distance of c ($x \neq c$), then it follows that we can get $f(x) + g(x)$ as close to $L + M$ as we like by choosing any x certain distance of c ($x \neq c$).

EXAMPLE 2   Compute $\lim_{x \to 4}(x^2 - 2x^{1/2} + 3)$.

Since by rule 1

$$\lim_{x \to 4} x^2 = 16, \qquad \lim_{x \to 4}(-2x^{1/2}) = -4, \qquad \lim_{x \to 4} 3 = 3,$$

then by rule 3,

$$\lim_{x \to 4}(x^2 - 2x^{1/2} + 3) = 16 - 4 + 3 = 15.$$

EXAMPLE 3   What is $\lim_{x \to 2} 5x\sqrt{9x - 2}$?

Since

$$\lim_{x \to 2} 5x = 10 \qquad \text{(rule 1)}$$

and

$$\lim_{x \to 2}(9x - 2) = \lim_{x \to 2} 9x + \lim_{x \to 2}(-2) \qquad \text{(rule 3)}$$

$$= 18 - 2 \qquad \text{(rule 1)}$$

$$= 16,$$

then

$$\lim_{x \to 2} \sqrt{9x - 2} = \sqrt{16} \qquad \text{(rule 2)}$$

$$= 4$$

and

$$\lim_{x \to 2} 5x\sqrt{9x - 2} = (10)(4) \qquad \text{(rule 4)}$$

$$= 40.$$

EXAMPLE 4   Compute $\lim_{x \to 10}(x^3 + 9)/\sqrt[4]{x - 1}$.

Since

$$\lim_{x \to 10}(x^3 + 9) = \lim_{x \to 10} x^3 + \lim_{x \to 10} 9 \qquad \text{(rule 3)}$$

$$= 1000 + 9 \qquad \text{(rule 1)}$$

$$= 1009$$

and

$$\lim_{x \to 10}(x - 1) = \lim_{x \to 10} x + \lim_{x \to 10}(-1) \qquad \text{(rule 3)}$$

$$= 10 - 1 \qquad \text{(rule 1)}$$

$$= 9,$$

then

$$\lim_{x \to 10} \sqrt[4]{x - 1} = \sqrt[4]{9} \qquad \text{(rule 2)}$$

$$= \sqrt{3}$$

and

$$\lim_{x \to 10} \frac{x^3 + 9}{\sqrt[4]{x - 1}} = \frac{1009}{\sqrt{3}} \qquad \text{(rule 5)}.$$

Remark

Although the five limit rules in 3.2.1 and 3.2.2 have been stated in terms of *two-sided* limits, these results are also correct when stated entirely in terms of *one-sided* limits. Therefore, if one replaces everywhere in rules 1–5 the symbol $x \to c$ by $x \downarrow c$, then one has another set of rules, which will be referred to as limit rules 1A, 2A, etc. Similarly, replacing $x \to c$ everywhere by $x \uparrow c$ gives another set of rules, designated as limit rules 1B, 2B, etc.

EXAMPLE 5   Determine

$$\lim_{x \uparrow -1} J(x), \quad \lim_{x \downarrow -1} J(x), \quad \lim_{x \uparrow 1} J(x), \quad \lim_{x \downarrow 1} J(x),$$

where J is defined by

$$J(x) = \begin{cases} 2x + 4, & x \leq -1, \\ x^2, & -1 < x < 1, \\ x, & x \geq 1. \end{cases}$$

$$\lim_{x \uparrow -1} J(x) = \lim_{x \uparrow -1} (2x + 4) \qquad \text{(definition of } J)$$

$$= \lim_{x \uparrow -1} 2x + \lim_{x \uparrow -1} 4 \qquad \text{(rule 3B)}$$

$$= -2 + 4 \qquad \text{(rule 1B)}$$

$$= 2$$

$$\lim_{x \downarrow -1} J(x) = \lim_{x \downarrow -1} x^2 \qquad \text{(definition of } J)$$

$$= (-1)^2 \qquad \text{(rule 1A)}$$

$$= 1$$

$$\lim_{x \uparrow 1} J(x) = \lim_{x \uparrow 1} x^2 \qquad \text{(definition of } J)$$

$$= 1^2 \qquad \text{(rule 1B)}$$

$$= 1$$

and

$$\lim_{x \downarrow 1} J(x) = \lim_{x \downarrow 1} x \qquad \text{(definition of } J)$$

$$= 1 \qquad \text{(rule 1A).}$$

Remark

One might wonder why we have bothered with the tedious process of applying limit rules 1–5 in Examples 1–4 when correct answers could be obtained by substituting c into $f(x)$. Points at which the substitution process *does* work are special points; these are the object of study in Section 3.3. In fact, the substitution process *does not* work for many of the important limits one studies in calculus.

In investigating

$$\lim_{x \to c} \frac{f(x)}{g(x)}$$

it occurs quite frequently that

$$\lim_{x \to c} g(x) = 0,$$

so that rule 5 is not applicable. In such a situation it may be helpful to algebraically transform (when possible) $f(x)/g(x)$.

EXAMPLE 6   Evaluate $\lim_{t \to -3} [(t^2 + 2t - 3)/(t + 3)]$.

Since $\lim_{t \to -3} (t + 3) = 0$, rule 5 is not applicable. However, it is possible to algebraically transform the given fraction:

$$\frac{t^2 + 2t - 3}{t + 3} = \frac{(t + 3)(t - 1)}{t + 3} = t - 1, \qquad t \neq -3.$$

Hence

$$\lim_{t \to -3} \frac{t^2 + 2t - 3}{t + 3} = \lim_{t \to -3} (t - 1)$$

$$= \lim_{t \to -3} t + \lim_{t \to -3} (-1) \qquad \text{(rule 3)}$$

$$= -3 - 1 \qquad \text{(rule 1)}$$

$$= -4.$$

EXAMPLE 7   Evaluate $\lim_{x \to 1} [(1 - x^3)/(2 - \sqrt{x^2 + 3})]$. We should first observe that

$$\lim_{x \to 1} (2 - \sqrt{x^2 + 3}) = 0.$$

Again, rule 5 is not applicable here. Our first step would be to **rationalize the denominator**:

$$\frac{1 - x^3}{2 - \sqrt{x^2 + 3}} = \frac{(1 - x^3)(2 + \sqrt{x^2 + 3})}{(2 - \sqrt{x^2 + 3})(2 + \sqrt{x^2 + 3})}$$

$$= \frac{(1 - x^3)(2 + \sqrt{x^2 + 3})}{1 - x^2}$$

$$= \frac{(1 - x)(1 + x + x^2)(2 + \sqrt{x^2 + 3})}{(1 - x)(1 + x)}$$

$$= \frac{(1 + x + x^2)(2 + \sqrt{x^2 + 3})}{1 + x} \qquad \text{for } x \neq 1.$$

Thus, by the concept of limit,

$$\lim_{x \to 1} \frac{1 - x^3}{2 - \sqrt{x^2 + 3}} = \lim_{x \to 1} \frac{(1 + x + x^2)(2 + \sqrt{x^2 + 3})}{1 + x} \, .$$

Now

$$\lim_{x \to 1}[(1 + x + x^2)(2 + \sqrt{x^2 + 3})]$$

$$= \left[\lim_{x \to 1}(1 + x + x^2)\right]\left[\lim_{x \to 1}(2 + \sqrt{x^2 + 3})\right] \qquad \text{(rule 4)}$$

$$= \left[\lim_{x \to 1} 1 + \lim_{x \to 1} x + \lim_{x \to 1} x^2\right]\left[\lim_{x \to 1} 2 + \sqrt{\lim_{x \to 1} x^2 + \lim_{x \to 1} 3}\right] \qquad \text{(rules 2, 3)}$$

$$= [1 + 1 + 1][2 + \sqrt{1 + 3}] \qquad \text{(rule 1)}$$

$$= 12.$$

On the other hand,

$$\lim_{x \to 1}(1 + x) = \lim_{x \to 1} 1 + \lim_{x \to 1} x \qquad \text{(rule 3)}$$

$$= 1 + 1 \qquad \text{(rule 1)}$$

$$= 2 \neq 0.$$

Hence

$$\lim_{x \to 1} \frac{1 - x^3}{2 - \sqrt{x^2 + 3}} = \lim_{x \to 1} \frac{(1 + x + x^2)(2 + \sqrt{x^2 + 3})}{(1 + x)}$$

$$= \frac{\lim\limits_{x \to 1}[(1 + x + x^2)(2 + \sqrt{x^2 + 3})]}{\lim\limits_{x \to 1}(1 + x)} \qquad \text{(rule 5)}$$

$$= \tfrac{12}{2}$$

$$= 6.$$

EXAMPLE 8   Let $f(x) = x^3$. If c is a fixed number, determine

$$\lim_{h \to 0} \frac{f(c + h) - f(c)}{h} \, .$$

$$\lim_{h \to 0} \frac{f(c + h) - f(c)}{h} = \lim_{h \to 0} \frac{(c + h)^3 - c^3}{h}$$

$$= \lim_{h \to 0} \frac{c^3 + 3c^2h + 3ch^2 + h^3 - c^3}{h}$$

$$= \lim_{h \to 0} \frac{(3c^2 + 3ch + h^2)h}{h}$$

$$= \lim_{h \to 0}(3c^2 + 3ch + h^2)$$

$$= \lim_{h \to 0} 3c^2 + \lim_{h \to 0} 3ch + \lim_{h \to 0} h^2 \qquad \text{(rule 3)}$$

$$= 3c^2 + 3c(0) + 0 \qquad \text{(rule 1)}$$

$$= 3c^2.$$

This type of limit will become familiar in Chapter 4.

We conclude this section with two facts that will be used at important places in Chapter 4.

3.2.3

Fact

Suppose that $\lim_{x \to c} f(x) = L$ and $L > 0$. Then there exists an open interval I containing c such that

$$f(x) > 0$$

for each x in I, $x \neq c$.

This fact is readily justified from our previous work. Since

$$\lim_{x \to c} f(x) = L > 0,$$

then we can get $f(x)$ as close to L as we like by choosing any x within a certain distance of c, $x \neq c$. In particular, we have

$$0 < \frac{L}{2} < f(x) < \frac{3L}{2}$$

provided that x is within a particular distance of c, $x \neq c$. Hence there must be an open interval I containing c (see Figure 3.5) such that

$$0 < f(x)$$

whenever x is in I, $x \neq c$.

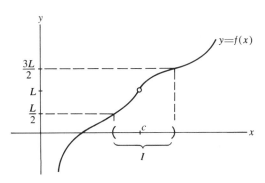

Figure 3.5

3.2.4

Fact   Suppose that $\lim\limits_{x \to c} f(x)$ exists, and there exists an open interval I containing c such that

$$f(x) \leq L$$

for each x in I, $x \neq c$, where L is a fixed number. Then

$$\lim_{x \to c} f(x) \leq L.$$

One can verify this by using Fact 3.2.3 as follows. Assume that the hypotheses of 3.2.4 are satisfied, but suppose that

$$\lim_{x \to c} f(x) > L.$$

Then

$$\lim_{x \to c} [f(x) - L] > 0,$$

and so by 3.2.3 there exists an open interval I containing c such that for each x in I, $x \neq c$,

$$f(x) - L > 0$$

or

$$f(x) > L.$$

However, this contradicts our hypotheses. Therefore it cannot be that

$$\lim_{x \to c} f(x) > L;$$

rather it must be that

$$\lim_{x \to c} f(x) \leq L.$$

It can be shown that Facts 3.2.3 and 3.2.4 are also valid if stated in terms of one-sided limits or if stated using the opposite inequalities.

EXERCISES

1. Compute the following limits.

(a) $\lim_{x \to 3} 4x$.

(b) $\lim_{x \downarrow 0} 5\sqrt{x}$.

(c) $\lim_{x \to 5}(2x + 8)$.

(d) $\lim_{x \to 0}(2 - x^2)$.

(e) $\lim_{x \downarrow 4/5} \sqrt{5x - 4}$.

(f) $\lim_{x \to 5}[(x - 3)/(x - 1)]$.

(g) $\lim_{x \to 0.5}(x^3 - 5)$.

(h) $\lim_{x \to -1}(2x^3 - 5x^2 + x - 12)$.

(i) $\lim_{x \to -2}(x^2 + x)(2x^3 - 7)$.

(j) $\lim_{x \downarrow 0}(3 - \sqrt{x})\sqrt[4]{16 + 6x^5}$.

(k) $\lim_{x \to 1/3}[\sqrt[3]{1 + 26x}/(x^2 + 1)]$.

(l) $\lim_{x \uparrow 3} \sqrt{9 - x^2}$.

(m) $\lim_{x \uparrow -4} |x + 4|$.

2. Determine the following limits.

(a) $\lim_{x \to 3}[(x^2 + 2x - 15/(x - 3)]$.

(b) $\lim_{x \to 0}[(x^2 + 3x)/(4x - 3x^2)]$.

(c) $\lim_{x \to 8}[(x - 8)/(x - 64)]$.

(d) $\lim_{x \to -1}(x^2 - 1)/(x + 1)]$.

(e) $\lim_{x \to 1}[(x^3 - 3x^2 + 3x - 1)/(x - 1)]$.

(f) $\lim_{x \to 2}\{(x^4 - 16)/[(x^2 + 4)(x - 2)]\}$.

(g) $\lim_{h \to 0}\{[(2 + h)^2 - 4]/h\}$.

(h) $\lim_{x \to 4}[(\sqrt{x} - 2)/(x - 4)]$.

3. Find $\lim_{h \to 0}\{[f(x_0 + h) - f(x_0)]/h\}$, where

(a) $f(x) = 2x$.

(b) $f(x) = 3x + 5$.

(c) $f(x) = x^2$.

(d) $f(x) = x^2 - 1$.

(e) $f(x) = 4x^3$.

(f) $f(x) = x^3 + 2$.

(g) $f(x) = 1/x$.

(h) $f(x) = 1/(x - 1)$.

(i) $f(x) = \sqrt{x}$.

4. Find $\lim_{x \to x_0}\{[f(x) - f(x_0)]/(x - x_0)\}$ for any three parts of Exercise 3.

5. Compute the following.

(a) $\lim_{x \to 9}[(x - 9)/(\sqrt{x} - 9)]$.

(b) $\lim_{x \to 2}[(x - 2)/(\sqrt{x^2 + 12} - 4)]$.

(c) $\lim_{\to 2}[(x^2 - 4)/(\sqrt{x^2 + 12} - 4)]$.

(d) $\lim_{x \to -1}[(1 + x^3)/(3 - \sqrt{x^2 + 8})]$.

(e) $\lim_{x \to 8}[(8 - x)/(\sqrt[3]{x} - 2)]$.

(f) $\lim_{x \to -1}[(1 + x)/(1 + \sqrt[3]{x})]$.

(g) $\lim_{x \downarrow 25}[(\sqrt{x} - 5)/(\sqrt{x} - 25)]$.

6. For each of the following, determine whether or not the limit exists. If so, what is the limit? If not, why not?

(a) $\lim_{x \to 0}(x/|x|)$.

(b) $\lim_{x \to 0}(x/x)$.

(c) $\lim_{x \to 1} f(x)$, where $f(x) = \begin{cases} 2, & x < 1, \\ x, & x > 1. \end{cases}$

(d) $\lim_{x \to 0} f(x)$, where $f(x) = \begin{cases} \sqrt{1 + x}, & x < 0, \\ 1 - x, & x > 0. \end{cases}$

(e) $\lim_{x \to -1} f(x)$, where $f(x) = \begin{cases} 3x, & x \le -1, \\ 3x^2 + 1, & x > -1. \end{cases}$

(f) $\lim_{x \to -1.01} f(x)$, where f is the function defined in part (e).

(g) $\lim_{x \to 3} f(x)$, where $f(x) = \begin{cases} 2x - 1, & x \ne 3, \\ \frac{9}{2}, & x = 3. \end{cases}$

(h) $\lim_{x \to -2} [x]$.

(i) $\lim_{x \to 0} [|x|]$.

(j) $\lim_{x \to 0} |[x]|$.

7. (a) State Fact 3.2.3 using the opposite inequality $<$.

(b) Verify the statement in part (a).

8. (a) State Fact 3.2.4 for the case $x \uparrow c$.

(b) State Fact 3.2.4 for the case $x \downarrow c$.

3.3 CONTINUITY

At certain places in our study thus far we have encountered functions whose graphs have gaps or jumps at particular points (see Figure 3.6). If one can find all the gaps and jumps of a function, then one can determine whether or not the graph of that function is unbroken on any given interval. These important considerations are related to another of the basic ideas of the calculus—the concept of continuity.

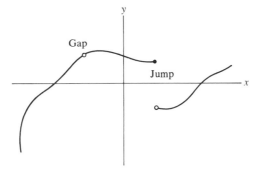

Figure 3.6

3.3.1
Continuity at a
Point

Function f is continuous at point c if

1. f is defined at c
2. $\lim\limits_{x \to c} f(x)$ exists
3. $\lim\limits_{x \to c} f(x) = f(c)$.

EXAMPLE 1   Let $F(x) = x^3 + 1$ and $c = 2$. Then we have

$$F(2) = 9$$

$$\lim_{x \to 2} F(x) = \lim_{x \to 2}(x^3 + 1) = 9$$

and

$$\lim_{x \to 2} F(x) = F(2);$$

that is, F is continuous at 2.

More generally, for arbitrary c,

$$F(c) = c^3 + 1$$

$$\lim_{x \to c} F(x) = \lim_{x \to c}(x^3 + 1) = c^3 + 1$$

and hence

$$\lim_{x \to c} F(x) = F(c);$$

that is, F is continuous at each point c. Observe in Figure 3.7 that the graph of F is an unbroken curve.

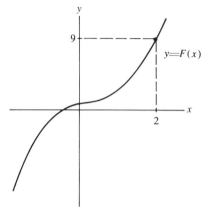

Figure 3.7

EXAMPLE 2   Consider the function G defined by

$$G(x) = \frac{x^3 - 3x^2}{x - 3}.$$

Since G is not defined at 3, property (1) of 3.3.1 is not fulfilled and hence G is not continuous at 3. However, if c is any number distinct from 3, observe that

$$G(c) = \frac{c^3 - 3c^2}{c - 3} = \frac{c^2(c - 3)}{(c - 3)} = c^2$$

$$\lim_{x \to c} G(x) = \lim_{x \to c} \frac{x^3 - 3x^2}{x - 3}$$

$$= \lim_{x \to c} x^2$$

$$= c^2,$$

and thus

$$\lim_{x \to c} G(x) = G(c);$$

that is, G is continuous at each c, $c \neq 3$.

Note that there is a gap in the graph of G above 3 but that the curve is unbroken elsewhere (see Figure 3.8).

EXAMPLE 3   Define function g by

$$g(x) = \begin{cases} x^2, & x \neq 3, \\ 5, & x = 3. \end{cases}$$

Then g is quite similar to function G of the preceding example (see Figure 3.9).

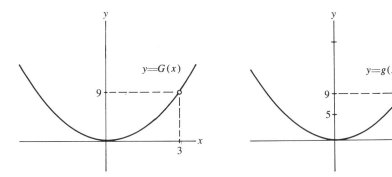

Figure 3.8   Figure 3.9

Now

$$g(3) = 5$$

$$\lim_{x \to 3} g(x) = \lim_{x \to 3} x^2 = 9$$

and

$$\lim_{x \to 3} g(x) = 9 \neq 5 = g(c);$$

therefore condition (3) of 3.3.1 is not fulfilled, and g is not continuous at 3.
   For $c \neq 3$, we have

$$\lim_{x \to c} g(x) = \lim_{x \to c} x^2 = c^2 = g(c),$$

and hence g is continuous at each point c, $c \neq 3$.
   Note the jump in the graph of g above 3, whereas the curve is unbroken elsewhere.

EXAMPLE 4   Let the function K be defined by

$$K(x) = \frac{5}{x^2}.$$

Since K is not defined at $x = 0$, we at once conclude that property (1) of 3.3.1 is not fulfilled and hence that K is not continuous at zero.
   For $c \neq 0$, we have

$$\lim_{x \to c} K(x) = \lim_{x \to c} \frac{5}{x^2} = \frac{5}{c^2} = K(c),$$

and so K is continuous at each point $c \neq 0$. In this case there is an obvious break in the graph at $x = 0$ (see Figure 3.10).

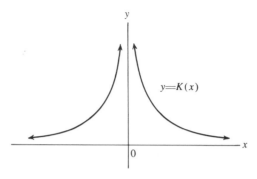

$y = K(x)$

Figure 3.10

EXAMPLE 5   Consider the function H defined by

$$H(x) = \begin{cases} 2 - x^2, & x \leq -1, \\ 3 + x, & x > -1 \end{cases} \quad \text{(see Figure 3.11).}$$

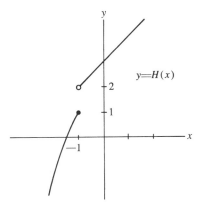

$y=H(x)$

Figure 3.11

The definition of H suggests that we examine its behavior at $x = -1$. Now

$$\lim_{x \uparrow -1} H(x) = \lim_{x \uparrow -1} (2 - x^2) = 1$$

and

$$\lim_{x \downarrow -1} H(x) = \lim_{x \downarrow -1} (3 + x) = 2,$$

and so $\lim_{x \to -1} H(x)$ does not exist; that is, property (2) of 3.3.1 is not satisfied, and hence H is not continuous at -1.

It may be verified that H is continuous at each point c, $c \neq -1$. Observe that there is a jump in the graph above -1.

Recall the definition of continuity at c given in 3.3.1. In 3.3.1 we assume that f is defined in an open interval containing c since the definition involves a two-sided limit.

However, if f is defined for $x \geq c$, then it makes sense to investigate the one-sided limit

$$\lim_{x \downarrow c} f(x).$$

If, in fact, this limit exists and

$$\lim_{x \downarrow c} f(x) = f(c),$$

then we say that ***f* is continuous from the right at** c. Similarly, if f is defined for $x \leq c$, $\lim_{x \uparrow c} f(x)$ exists, and

$$\lim_{x \uparrow c} f(x) = f(c);$$

then we say that ***f* is continuous from the left at** c.

Note that, in light of 3.1.2,

f is continuous at c

is equivalent to

f is continuous from the left at c and f is continuous from the right at c.

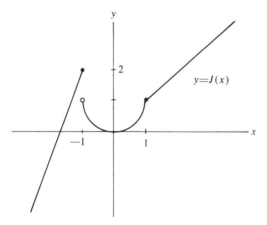

Figure 3.12

EXAMPLE 6   Consider the function J defined by

$$J(x) = \begin{cases} 2x + 4, & x \le -1, \\ x^2, & -1 < x < 1, \\ x, & x \ge 1 \end{cases} \quad \text{(see Figure 3.12)}.$$

This function was examined in Example 5 of Section 3.2, where we determined that

$$\lim_{x \uparrow -1} J(x) = 2 \ne 1 = \lim_{x \downarrow -1} J(x)$$

and that

$$\lim_{x \uparrow 1} J(x) = 1 = \lim_{x \downarrow 1} J(x).$$

Observe that $J(-1) = 2$ and $J(1) = 1$. Then J is continuous from the left at -1, but *not* continuous from the right at -1; hence J is *not* continuous at -1. But J is continuous from both the left and the right at 1, and so J is continuous at 1.

The concepts of continuity from the right and continuity from the left allow us to define continuity on an interval I. In particular, we say that f is **continuous on $[a, b]$** if

1. f is continuous at each c, where $a < c < b$.
2. f is continuous from the right at a.
3. f is continuous from the left at b.

Continuity on other kinds of intervals is defined analogously, keeping in mind the appropriate type of one-sided continuity at each end point.

If f is defined and continuous at each point in **R**, then we say that f is **continuous on R**.

EXAMPLE 7   Let B be defined by

$$B(x) = \sqrt{x}$$

for x in $[0, 4]$ (see Figure 3.13). If $0 < c < 4$, then

$$\lim_{x \to c} B(x) = \lim_{x \to c} \sqrt{x} = \sqrt{\lim_{x \to c} x} = \sqrt{c} = B(c).$$

Also,

$$\lim_{x \downarrow 0} B(x) = \lim_{x \downarrow 0} \sqrt{x} = \sqrt{\lim_{x \downarrow 0} x} = \sqrt{0} = B(0)$$

$$\lim_{x \uparrow 4} B(x) = \lim_{x \uparrow 4} \sqrt{x} = \sqrt{\lim_{x \uparrow 4} x} = \sqrt{4} = 2 = B(4).$$

Hence B is continuous on $[0, 4]$.

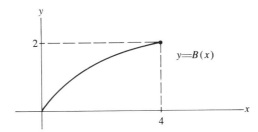

Figure 3.13

EXAMPLE 8   Consider a function C defined on $[-1, 1]$ by

$$C(x) = \begin{cases} 0, & \text{if } x = \pm 1, \\ x, & \text{if } -1 < x < 1 \end{cases} \quad \text{(see Figure 3.14).}$$

Then

$$\lim_{x \uparrow 1} C(x) = \lim_{x \uparrow 1} x = 1 \neq 0 = C(1)$$

and, likewise,

$$\lim_{x \downarrow -1} C(x) = \lim_{x \downarrow -1} x = -1 \neq 0 = C(-1).$$

Hence C is not continuous on $[-1, 1]$.

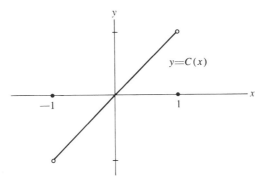

Figure 3.14

For any fixed positive integer n and constants $a_0, a_1, a_2, \ldots, a_n$, the function p defined by

$$p(x) = a_0 + a_1 x + a_2 x^2 + \cdots + a_n x^n \qquad (a_n \neq 0)$$

is called a **polynomial function of degree n**.

3.3.2 Fact

Each polynomial function is continuous on **R**.

One can verify this fact by using limit rules 1 and 3 of Section 3.2 to show that for every number c

$$\lim_{x \to c} P(x) = P(c).$$

A function R defined by

$$R(x) = \frac{P(x)}{Q(x)},$$

where P and Q are polynomials, is called a **rational function**.

3.3.3 Fact

Let R be a rational function defined by $R(x) = P(x)/Q(x)$.
Then R is continuous at each point c except where $Q(c) = 0$.

The validity of this fact can be seen by observing that

$$\lim_{x \to c} R(x) = \lim_{x \to c} \frac{P(x)}{Q(x)} = \frac{\lim_{x \to c} P(x)}{\lim_{x \to c} Q(x)} = \frac{P(c)}{Q(c)} = R(c)$$

provided that $Q(c) \neq 0$.

EXAMPLE 9   Determine where the function S defined by

$$S(x) = \frac{3x^6 + x^5 - 7x^2 + 2x}{x^2 - 8}$$

is continuous. S is continuous except where $x^2 - 8 = 0$; that is, $x = \sqrt{8}$ and $x = -\sqrt{8}$.

We conclude this section on continuity by stating and briefly discussing two classic theorems which are fundamental in the development of the calculus.

3.3.4
Important Facts
About Functions
Continuous on a
Closed Interval

Let function f be continuous on $[a, b]$.

1. **Maximum-Minimum Theorem.** There exist numbers p and q in $[a, b]$ such that for each x in $[a, b]$.

$$f(p) \le f(x) \le f(q).$$

2. **Intermediate Value Theorem.** For each number l between $f(a)$ and $f(b)$ there exists a number p in $[a, b]$ such that

$$f(p) = l.$$

The Maximum-Minimum Theorem says that if f is continuous on $[a, b]$, then of all the functional values on $[a, b]$ there is a minimum one and there is a maximum one. For instance, let g be the function continuous on $[a, b]$ as drawn in Figure 3.15. Then $g(x_1)$ is the maximum value of g on $[a, b]$ and $g(a) = g(x_2)$ is the minimum value of g on $[a, b]$.

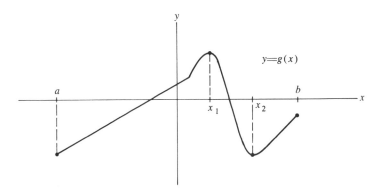

Figure 3.15

The Intermediate Value Theorem says that a continuous function on $[a, b]$ attains all values between $f(a)$ and $f(b)$. Geometrically, this means that each horizontal line $y = l$ [l between $f(a)$ and $f(b)$] crosses the graph of f in at least one point (p, l). It indicates that *there are no gaps in the graph of a function continuous on a closed interval and justifies drawing the graph of f on that interval as an unbroken curve* (see Figure 3.16).

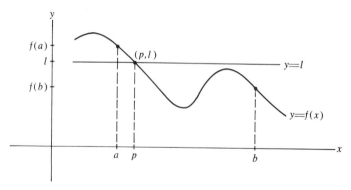

Figure 3.16

The two results in 3.3.4 seem apparent from the graphs of the elementary continuous functions which we have seen thus far. However, there exist continuous functions whose graphs are not easily drawn. For these functions the results in 3.3.4 are not apparent.

EXERCISES

1. Let K be defined by $K(x) = 5/x^2$. Show that K is continuous at each point c other than zero. [*Hint:* Since K is not continuous at zero (see Example 4), consider the two cases $c > 0$ and $c < 0$.]

2. Let J be given by

$$J(x) = \begin{cases} 2x + 4, & x \le -1, \\ x^2, & -1 < x < 1, \\ x, & x \ge 1. \end{cases}$$

Show that J is continuous at each point c other than -1. [*Hint:* Since J is not continuous at -1 and continuous at 1 (see Example 6), consider the following three cases: $c < -1$, $-1 < c < 1$, $c > 1$.]

3. Consider the absolute value function A defined by $A(x) = |x|$. Show that A is continuous at each point c. [*Hint:* Consider the three cases $c < 0$, $c = 0$, $c > 0$.]

4. Let S be defined by

$$S(x) = \begin{cases} -1, & x \le -2, \\ 4, & -2 < x < 3, \\ 2, & x \ge 3. \end{cases}$$

Show that S is continuous at each point c except -2 and 3. [*Hint:* Consider the five cases $c < -2$, $c = -2$, $-2 < c < 3$, $c = 3$, $c > 3$.]

5. Sketch the graph of each of the following functions, locate the points of discontinuity, and justify your observations.

(a) $f(x) = 2$ for all x.

(b) $g(x) = \begin{cases} 5 & \text{if } x < 1, \\ 4 & \text{if } x > 1. \end{cases}$

(c) $f(x) = \begin{cases} 5 & \text{if } x < 1, \\ 4 & \text{if } x \geq 1. \end{cases}$

(d) $f(x) = (x^2 - 49)/(x - 7).$

(e) $f(x) = \begin{cases} x + 7 & \text{if } x \neq 7, \\ 10, & \text{if } x = 7, \end{cases}$

(f) $f(x) = \begin{cases} -x - 2 & \text{if } x < -2 \\ 4 - x^2 & \text{if } -2 \leq x \leq 1, \\ 2.999 & \text{if } x > 1. \end{cases}$

(g) $f(x) = 5/(x - 1).$

(h) $f(x) = |x + 3|.$

(i) Postage stamp function (Section 2.1, Example 6).

(j) $f(x) = \begin{cases} x + 2 & \text{if } x > -1, \\ 3x + 3 & \text{if } x \leq -1. \end{cases}$

(k) $f(x) = x/|x| \ (x \neq 0).$

(l) $f(x) = \sqrt{x^2 + 1}.$

(m) $f(x) = (x^2 - 1)/|x - 1|.$

6. Give a rough sketch of the graph of each of the following functions, locate the points of discontinuity, and justify your observations.

(a) $f(x) = \begin{cases} 1 & \text{if } x \text{ is rational,} \\ 0 & \text{if } x \text{ is irrational.} \end{cases}$

(b) $f(x) = \begin{cases} x & \text{if } x \text{ is rational,} \\ 0 & \text{if } x \text{ is irrational.} \end{cases}$

(c) $f(x) = \begin{cases} x & \text{if } x \text{ is rational,} \\ 1 - x & \text{if } x \text{ is irrational.} \end{cases}$

7. Classify each of the discontinuities in Exercises 5 and 6 as one of the following types of discontinuities.

(a) $\lim_{x \to c} f(x)$ exists but either f is not defined at c or f is defined at c and $\lim_{x \to c} f(x) \neq f(c)$. [Such a discontinuity is called **removable** because defining, or redefining, f at c to be $\lim_{x \to c} f(x)$ makes f continuous at c.]

(b) $\lim_{x \to c} f(x)$ does not exist.

(1) Both $\lim_{x \uparrow c} f(x)$ and $\lim_{x \downarrow c} f(x)$ exist but $\lim_{x \uparrow c} f(x) \neq \lim_{x \downarrow c} f(x)$. (Such a discontinuity is called a **jump** discontinuity.)

(2) Either $\lim_{x \uparrow c} f(x)$ or $\lim_{x \downarrow c} f(x)$ does not exist.

8. Sketch the graph of each of the following functions, locate the points of discontinuity, and justify your observations.

(a) $f(x) = [x].$

(b) $f(x) = [2x].$

(c) $f(x) = 2[x].$

(d) $f(x) = [-2x].$

(e) $f(x) = [-x/3].$

(f) $f(x) = [x] + x.$

(g) $f(x) = \sqrt{[x]}.$

9. For the functions in Exercise 5, determine whether they are continuous from the right or left at the points of discontinuity.

10. Prove Fact 3.3.2. [*Hint:* Let $P(x) = a_0 + a_1 x + \cdots + a_n x^n$ and use limit rules.]

11. Let the function f be defined on $[a, c]$ by the graph in Figure 3.17. Why is it correct to say that f is continuous on $[a, b]$ and that f is continuous on $(b, c]$, yet incorrect to say that f is continuous on $[a, c]$?

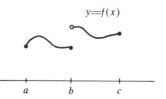

$y = f(x)$

a	b	c	

Figure 3.17

12. Let C be defined on $[-1, 1]$ by

$$C(x) = \begin{cases} 0 & \text{if } x = \pm 1 \\ x & \text{if } -1 < x < 1. \end{cases}$$

(This function was considered in Example 8.) Does C attain maximum and minimum values on $[-0.9, +0.98]$? If so, what are they?

13. If f and g are continuous at c, prove the following.

(a) $f(x) + g(x)$ is continuous at c.

(b) $f(x) - g(x)$ is continuous at c.

(c) $f(x)g(x)$ is continuous at c.

(d) $f(x)/g(x)$ is continuous at c provided that $g(c) \neq 0$.

14. Prove: Let f be continuous on $[a, b]$ and z_1, z_2 be in $[a, b]$. If z_1 and z_2 are consecutive zeros of f, then f is either always positive or always negative on $[z_1, z_2]$. [*Hint:* Use 3.3.4(2).]

15. Consider the following *fixed-point theorem*: If $f: [0, 1] \to [0, 1]$ is continuous, then there exists an x in $[0, 1]$ such that $f(x) = x$.

(a) Give a geometric justification.

(b) Prove it. [*Hint:* Let $g(x) = f(x) - x$ and use 3.3.4(2).]

3.4 LIMITS AND INFINITY

In our study of limits thus far we have not considered those limits which involve the notion of *infinity*. Yet, such limits are often of considerable interest.

Question 1

What is the behavior of $1/x$ as $x \to 0$?
  As before, let us construct lists of values for $1/x$.

x	$1/x$	x	$1/x$
-2	-0.5	1.5	0.67
-0.5	-2	0.7	1.43
-0.1	-10	0.4	2.5
-0.04	-25	0.1	10
-0.002	-500	0.02	50
-0.007	$-1,428$	0.009	111
-0.0008	$-12,500$	0.0001	$10,000$
-0.00001	$-100,000$	0.000025	$40,000$
-0.000003	$-333,333$	0.000003	$333,333$
-0.000000002	$-500,000,000$	0.00000001	$100,000,000$

3.4.1
Definitions

1. We say that

$$\lim_{x \downarrow c} f(x) = +\infty$$

if we can get $f(x)$ as large positively as we like by choosing any x within a certain distance of c, $x > c$.

2. We say that

$$\lim_{x \downarrow c} f(x) = -\infty$$

if we can get $f(x)$ as large negatively as we like by choosing any x within a certain distance of c, $x > c$.

The left-hand limits

$$\lim_{x \uparrow c} f(x) = +\infty \qquad \text{and} \qquad \lim_{x \uparrow c} f(x) = -\infty$$

are defined in a similar fashion.
  In Definition 3.4.1, to say that $f(x)$ is large positively means that

$$f(x) > 0 \qquad \text{and} \qquad f(x) \text{ is large.}$$

And to say that $f(x)$ is large negatively means that

$$f(x) < 0 \qquad \text{and} \qquad |f(x)| \text{ is large.}$$

An analysis of the chart on the left leads one to say

> "We can get $1/x$ as large negatively as we like by choosing any x within a certain distance of 0, $x < 0$."

We write this statement notationally as

$$\lim_{x\uparrow 0} \frac{1}{x} = -\infty.$$

Similarly, due to the chart on the right, we would say

> "We can get $1/x$ as large positively as we like by choosing any x within a certain distance of 0, $x > 0$."

Notationally we write this statement as

$$\lim_{x\downarrow 0} \frac{1}{x} = +\infty.$$

Remark

The symbols $+\infty$ (read "plus infinity") and $-\infty$ (read "minus infinity") are merely symbols used to express the ideas of "large positively" and "large negatively," respectively. They are *not* to be treated as numbers.

EXAMPLE 1   Examine the limits

$$\lim_{x\downarrow 3} \frac{x-4}{x-3} \quad \text{and} \quad \lim_{x\uparrow 3} \frac{x-4}{x-3}.$$

First, if $x < 3$, then

$$x - 3 < 0 \quad \text{and} \quad x - 3 \to 0 \quad \text{as } x \uparrow 3.$$

Hence $1/(x - 3)$ becomes large negatively as $x \uparrow 3$. Since $x - 4 \to -1$ as $x \uparrow 3$, we see that $(x - 4)/(x - 3) = (x - 4)[1/(x - 3)]$ gets large positively as $x \uparrow 3$; that is,

$$\lim_{x\uparrow 3} \frac{x-4}{x-3} = +\infty.$$

A similar argument shows that

$$\lim_{x\downarrow 3} \frac{x-4}{x-3} = -\infty.$$

EXAMPLE 2   Examine

$$\lim_{x\uparrow-2} \frac{-2x^2}{(x+2)^2} \quad \text{and} \quad \lim_{x\downarrow-2} \frac{-2x^2}{(x+2)^2}.$$

Note that $-2x^2/(x+2)^2 < 0$ for all $x \neq -2$. Again, we see that $1/(x+2)^2$ gets large positively as $x\uparrow-2$ or $x\downarrow-2$. Since $-2x^2 \to -4$ as $x\to-2$, we conclude that $-2x^2/(x+2)^2 = (-2x^2)[1/(x+2)^2]$ gets large negatively as $x\uparrow-2$ or $x\downarrow-2$. Hence

$$\lim_{x\uparrow-2} \frac{-2x^2}{(x+2)^2} = -\infty = \lim_{x\downarrow-2} \frac{-2x^2}{(x+2)^2}.$$

Now we wish to consider other kinds of limits which involve the symbols $+\infty$ and $-\infty$. Specifically, given a function f,

"What happens to $f(x)$ as x becomes large positively (written $x \to +\infty$)?"

Also,

"What happens to $f(x)$ as x becomes large negatively (written $x \to -\infty$)?"

Question 2

What is the behavior of $1/x$ as $x \to +\infty$ and as $x \to -\infty$?
We shall investigate the following tables.

x	$1/x$		x	$1/x$
-20	-0.05		10	0.1
-70	-0.014		20	0.05
-100	-0.010		50	0.02
-500	-0.002		100	0.01
$-1,500$	-0.0007		500	0.002
$-5,000$	-0.0002		1,000	0.001
$-12,000$	-0.00008		3,000	0.0003
$-100,000$	-0.00001		7,000	0.00014
$-4,000,000$	-0.00000025		800,000	0.00000125
$-9,000,000$	-0.00000011		12,000,000	0.0000000833

A look at the chart on the left prompts us to say

"We can get $1/x$ as close to zero as we like by choosing any x large enough negatively."

We write the last statement notationally as

$$\lim_{x \to -\infty} \frac{1}{x} = 0.$$

The chart on the right leads us to say

"We can get $1/x$ as close to zero as we like by choosing any x large enough positively."

Notationally, we write this statement as

$$\lim_{x \to +\infty} \frac{1}{x} = 0.$$

Our example leads us to a general definition.

**3.4.2
Definitions**

1. We say that

$$\lim_{x \to +\infty} f(x) = L$$

if we can get $f(x)$ as close to L as we like by choosing any x large enough positively.

2. We say that

$$\lim_{x \to -\infty} f(x) = L$$

if we can get $f(x)$ as close to L as we like by choosing any x large enough negatively.

EXAMPLE 3   Investigate

$$\lim_{x \to -\infty} \frac{1}{x^p} \quad \text{and} \quad \lim_{x \to +\infty} \frac{1}{x^p},$$

where p is a positive integer.

As in the case of $1/x$, whenever x is large enough positively or negatively, we see that $1/x^p$ is close to zero. Therefore, $\lim_{x \to -\infty} (1/x^p) = 0$ and $\lim_{x \to +\infty} (1/x^p) = 0$.

EXAMPLE 4   Evaluate

$$\lim_{x \to +\infty} \frac{x^2 - 3}{8x^3 + 2x - 5}.$$

We first divide the numerator and the denominator of the given fraction by x^3:

$$\frac{x^2 - 3}{8x^3 + 2x - 5} = \frac{1/x - 3/x^3}{8 + 2/x^2 - 5/x^3}.$$

Whenever x is large enough positively, we see that

$$\frac{1}{x} - \frac{3}{x^3} \text{ is close to zero}$$

and that

$$8 + \frac{2}{x^2} - \frac{5}{x^3} \text{ is close to 8.}$$

Hence

$$\frac{1/x - 3/x^3}{8 + 2/x^2 - 5/x^3} \text{ is close to } \frac{0}{8} = 0.$$

We can write

$$\lim_{x \to +\infty} \frac{x^2 - 3}{8x^3 + 2x - 5} = 0.$$

EXAMPLE 5   Evaluate

$$\lim_{x \to +\infty} \frac{6x^5 - 7x^2 + 3}{4x^5 + x^3 - 4x}.$$

Again, divide the numerator and the denominator by x^5 to obtain

$$\frac{6x^5 - 7x^2 + 3}{4x^5 + x^3 - 4x} = \frac{6 - 7/x^3 + 3/x^5}{4 + 1/x^2 - 4/x^4}.$$

Whenever x is large enough, we see that

$$6 - \frac{7}{x^3} + \frac{3}{x^5} \text{ is close to 6}$$

and that

$$4 + \frac{1}{x^2} - \frac{4}{x^4} \text{ is close to 4.}$$

Hence

$$\frac{6 - 7/x^3 + 3/x^5}{4 + 1/x^2 - 4/x^4} \text{ is close to } \frac{6}{4} = \frac{3}{2}.$$

Thus

$$\lim_{x \to +\infty} \frac{6x^5 - 7x^2 + 3}{4x^5 + x^3 - 4x} = \frac{3}{2}.$$

EXAMPLE 6   Consider the function

$$f(x) = \frac{2x - 1}{x - 1}.$$

By the methods of this section one can show that

$$\lim_{x \downarrow 1} \frac{2x - 1}{x - 1} = +\infty$$

$$\lim_{x \uparrow 1} \frac{2x - 1}{x - 1} = -\infty$$

$$\lim_{x \to +\infty} \frac{2x - 1}{x - 1} = 2$$

$$\lim_{x \to -\infty} \frac{2x - 1}{x - 1} = 2.$$

The graph of f appears in Figure 3.18. The lines $x = 1$ and $y = 2$ are called **vertical** and **horizontal asymptotes** respectively.

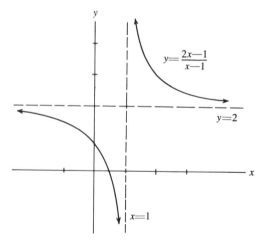

Figure 3.18

Using the symbols $+\infty$ and $-\infty$, it is possible to attach meaning to the following:

$$\lim_{x \to +\infty} f(x) = +\infty$$

$$\lim_{x \to +\infty} f(x) = -\infty$$

$$\lim_{x \to -\infty} f(x) = +\infty$$

$$\lim_{x \to -\infty} f(x) = -\infty.$$

We shall indicate the meaning of one of them.

3.4.3
Definition
We say that

$$\lim_{x \to +\infty} f(x) = +\infty$$

if we can get $f(x)$ as large positively as we like by choosing any x large enough positively.

EXAMPLE 7   Evaluate

$$\lim_{x \to -\infty} \frac{3x^2 + 5}{7x - 9}.$$

Divide the numerator and the denominator by x to obtain

$$\frac{3x^2 + 5}{7x - 9} = \frac{3x + 5/x}{7 - 9/x}.$$

Whenever x is large enough negatively, we see that

$$3x + \frac{5}{x} \text{ is large negatively}$$

and that

$$7 - \frac{9}{x} \text{ is close to 7.}$$

Hence

$$\frac{3x + 5/x}{7 - 9/x} \text{ is large negatively.}$$

Thus

$$\lim_{x \to -\infty} \frac{3x^2 + 5}{7x - 9} = -\infty.$$

EXERCISES

1. Evaluate the following limits.

(a) $\lim_{x\uparrow 4}[1/(x - 4)]$, $\lim_{x\downarrow 4}[1/(x - 4)]$.

(b) $\lim_{x\uparrow 4}[1/(x - 4)^2]$, $\lim_{x\downarrow 4}[1/(x - 4)^2]$.

(c) $\lim_{x\uparrow 4}[1/(x - 4)^p]$, $\lim_{x\downarrow 4}[1/(x - 4)^p]$, where p is an odd positive integer.

(d) $\lim_{x\uparrow 4}[1/(x - 4)^p]$, $\lim_{x\downarrow 4}[1/(x - 4)^p]$, where p is an even positive integer.

2. Compute the following limits.

(a) $\lim_{x\uparrow -1} [(x - 1)/(x + 1)]$, $\lim_{x\downarrow -1} [(x - 1)/(x + 1)]$.

(b) $\lim_{x\uparrow 2} [-3x^2/(x - 2)^2]$, $\lim_{x\downarrow 2} [-3x^2/(x - 2)^2]$.

3. (a) Determine $\lim_{x\downarrow 3} x^2 - 5x + 6$.

(b) Find $\lim_{x\downarrow 3} [1/(x^2 - 5x + 6)]$.

(c) More generally, if $\lim_{x\downarrow c} f(x) = 0$ and $f(x) > 0$ for $x > c$, what can you say

about $\lim_{x\downarrow c} 1/f(x)$?

4. Evaluate the following limits.

(a) $\lim_{x\to -\infty} [(2x^3 + x)/(x^4 + x^2 + 6)]$.

(b) $\lim_{x\to +\infty} [(6x^5 - 7x^2 + 3)/(4x^5 + x^3 - 4x)]$.

(c) $\lim_{x\to +\infty} [(6 - x - x^5)/(1 + x^4)]$.

(d) $\lim_{x\to -\infty} [-5x^2/(x - 4)^2]$.

5. Let $f(x) = (x - 1)/(x + 1)$.

(a) Determine each of the following.

(1) $\lim_{x\uparrow -1} f(x)$, $\lim_{x\downarrow -1} f(x)$.   (2) $\lim_{x\to +\infty} f(x)$, $\lim_{x\to -\infty} f(x)$.

(b) Use the information from part (a) to sketch the graph of f.

6. Let $f(x) = -3x^2/(x - 2)^2$.

(a) Determine the following.

(1) $\lim_{x\uparrow 2} f(x)$, $\lim_{x\downarrow 2} f(x)$.   (2) $\lim_{x\to +\infty} f(x)$, $\lim_{x\to -\infty} f(x)$.

(b) Use the information from part (a) to sketch the graph of f.

7. Let $f(x) = 4x/(2x^2 + x - 3)$. Determine each of the following.

(a) $\lim_{x\downarrow 1} f(x)$.   (b) $\lim_{x\uparrow 1} f(x)$.   (c) $\lim_{x\uparrow -3/2} f(x)$.

(d) $\lim_{x\downarrow -3/2} f(x)$.   (e) $\lim_{x\uparrow 0} f(x)$.   (f) $\lim_{x\downarrow 0} f(x)$.

(g) $\lim_{x\to 0} f(x)$.   (h) $\lim_{x\to +\infty} f(x)$.   (i) $\lim_{x\to -\infty} f(x)$.

8. Determine the following.

(a) $\lim_{x\uparrow 25} [x/(\sqrt{x} - 5)]$.   (b) $\lim_{x\to +\infty} (2x/\sqrt{9x^2 - 5})$.

3.5 A RESTATEMENT OF THE CONCEPT OF LIMIT (OPTIONAL)

Recall the concept of $\lim_{x \to c} f(x) = L$ as explained in Section 3.1. As acceptable as this description has been for our purposes, it is not as precise a description as one can give. In particular, the statement in 3.1.3 is not a workable one, for just how does one show conclusively that "we can get $f(x)$ as close to L as we like" when there are infinitely many numbers close to L? Any man-made list of values of f, as in our tables, is necessarily finite. It is our task in this section to recast the statement in 3.1.3 into a definition which is workable and precise.

The idea of *closeness* appears twice in 3.1.3. Mathematically, one measures the closeness of two numbers by computing the absolute value of the difference of the two numbers. Therefore, to say that "we can get $f(x)$ as close to L as we like" is the same as saying

"One can make $|f(x) - L|$ as small as one likes,"

which, in turn, is the same as saying

"Given any $\varepsilon > 0$ (no matter how small), one can make $|f(x) - L| < \varepsilon$."

Substituting this new phrase into 3.1.3, we have

"Given any $\varepsilon > 0$, one can make $|f(x) - L| < \varepsilon$ by choosing any x within a certain distance of c, $x \neq c$."

This is the same as saying

"Given any $\varepsilon > 0$, one can make $|f(x) - L| < \varepsilon$ by choosing any x such that $|x - c|$ is small enough, $x \neq c$,"

which, in turn, means

"Given any $\varepsilon > 0$, there exists a sufficiently small $\delta > 0$ such that $|f(x) - L| < \varepsilon$ for every x satisfying $|x - c| < \delta$, $x \neq c$."

Observing that $x \neq c$ is equivalent to $0 < |x - c|$, our statement reads

"Given any $\varepsilon > 0$, there exists a sufficiently small $\delta > 0$ such that $|f(x) - L| < \varepsilon$ for every x satisfying $0 < |x - c| < \delta$."

Note that the choice of δ depends on the given ε.

We formally restate our translation of 3.1.3.

3.5.1 Definition

We say
$$\lim_{x \to c} f(x) = L$$
if for each $\varepsilon > 0$ there exists a $\delta > 0$ such that
$$|f(x) - L| < \varepsilon$$
for every x satisfying
$$0 < |x - c| < \delta.$$

Definition 3.5.1 is nothing but a concise, precise, and economical mathematical expression of the notion of a limit expressed in 3.1.3. As abstract as it might be, it is a workable definition, for it allows us to prove all the limit statements previously accepted as true by 3.1.3.

EXAMPLE 1   (see Example 1, Section 3.1)

$$\lim_{x \to 5}(2x + 8) = 18.$$

Proof: Let $\varepsilon > 0$ be given. We must find a $\delta > 0$ such that

$$|(2x + 8) - 18| < \varepsilon$$

for any x satisfying

$$0 < |x - 5| < \delta.$$

Now note that

$$|(2x + 8) - 18| < \varepsilon \Leftrightarrow |2x - 10| < \varepsilon$$

$$\Leftrightarrow 2|x - 5| < \varepsilon$$

$$\Leftrightarrow |x - 5| < \frac{\varepsilon}{2}.$$

Therefore, if we choose $\delta = \varepsilon/2$, then for any x satisfying

$$0 < |x - 5| < \delta$$

it follows that

$$|(2x + 8) - 18| = |2x - 10| = 2|x - 5| < 2\delta = 2\left(\frac{\varepsilon}{2}\right) = \varepsilon.$$

EXAMPLE 2   (see Example 2, Section 3.1)

$$\lim_{x \to 3} \frac{x^2 - 5x + 6}{(x - 3)} = 1.$$

Proof: Let $\varepsilon > 0$ be given. We must find $\delta > 0$ such that

$$\left| \frac{x^2 - 5x + 6}{(x - 3)} - 1 \right| < \varepsilon$$

for any x satisfying

$$0 < |x - 3| < \delta.$$

For $0 < |x - 3|$ (that is, for $x \neq 3$)

$$\left| \frac{x^2 - 5x + 6}{x - 3} - 1 \right| < \varepsilon \Leftrightarrow \left| \frac{(x - 2)(x - 3)}{x - 3} - 1 \right| < \varepsilon$$

$$\Leftrightarrow |(x - 2) - 1| < \varepsilon$$

$$\Leftrightarrow |x - 3| < \varepsilon.$$

Therefore, if we choose $\delta = \varepsilon$, then for any x satisfying

$$0 < |x - 3| < \delta$$

it follows that

$$\left|\frac{x^2 - 5x + 6}{x - 3} - 1\right| = \left|\frac{(x - 2)(x - 3)}{x - 3} - 1\right| = |(x - 2) - 1|$$

$$= |x - 3| < \delta = \varepsilon.$$

EXAMPLE 3

$$\lim_{x \to 0}(2 - x^2) = 2.$$

Proof: Let $\varepsilon > 0$ be given. We must find a $\delta > 0$ such that

$$|(2 - x^2) - 2| < \varepsilon$$

for any x satisfying

$$0 < |x - 0| < \delta.$$

Now note that

$$|(2 - x^2) - 2| < \varepsilon \Leftrightarrow |-x^2| < \varepsilon$$

$$\Leftrightarrow |x|^2 < \varepsilon$$

$$\Leftrightarrow |x| < \sqrt{\varepsilon}$$

$$\Leftrightarrow |x - 0| < \sqrt{\varepsilon}.$$

Therefore, if we choose $\delta = \sqrt{\varepsilon}$, then for any x satisfying

$$0 < |x - 0| < \delta$$

it follows that

$$|(2 - x^2) - 2| = |-x^2| = |x|^2 = |x - 0|^2 < \delta^2 = (\sqrt{\varepsilon})^2 = \varepsilon.$$

EXAMPLE 4

$$\lim_{x \to -5}\frac{x - 3}{x - 1} = \frac{4}{3}.$$

Proof: Let $\varepsilon > 0$ be given. We must find $\delta > 0$ such that

$$\left|\frac{x - 3}{x - 1} - \frac{4}{3}\right| < \varepsilon$$

for any x satisfying

$$0 < |x - (-5)| < \delta.$$

Note that

$$\left|\frac{x - 3}{x - 1} - \frac{4}{3}\right| = \left|\frac{-x - 5}{3(x - 1)}\right|$$

$$= \frac{|x - (-5)|}{3|x - 1|}.$$

Let us restrict x such that $-6 < x < -4$; that is, x satisfies
$$|x - (-5)| < 1.$$
With this restriction,
$$|x - 1| \geq 5.$$
Therefore
$$3|x - 1| = 15$$
and hence
$$\frac{1}{3|x - 1|} < \frac{1}{15}.$$
Therefore
$$\left| \frac{x - 3}{x - 4} - \frac{4}{3} \right| = \frac{|x - (-5)|}{3|x - 1|} = \frac{|x - (-5)|}{15}.$$
Let us make the further restriction that
$$|x - (-5)| < \delta_1,$$
where δ_1 is yet to be chosen. Then
$$\left| \frac{x - 3}{x - 1} - \frac{4}{3} \right| = \frac{|x - (-5)|}{15} < \frac{\delta_1}{15}.$$
Since we desire
$$\left| \frac{x - 3}{x - 1} - \frac{4}{3} \right| < \varepsilon,$$
we choose $\delta_1/15 < \varepsilon$; that is,
$$\delta_1 = 15\varepsilon.$$
Therefore, if we choose
$$\delta = \min\{1, 15\varepsilon\},$$
then for any x satisfying
$$0 < |x - (-5)| < \delta$$
we have
$$\left| \frac{x - 3}{x - 1} - \frac{4}{3} \right| = \frac{|x - (-5)|}{3|x - 1|} < \frac{15\varepsilon}{15} = \varepsilon.$$

EXAMPLE 5   (see Example 3, Section 3.1). Consider the function h defined on **R** by
$$h(x) = \begin{cases} 2 - x^2, & x \leq 0, \\ 3 + x, & x > 0. \end{cases}$$
Then $\lim_{x \to 0} h(x)$ does not exist.

Proof: Suppose the contrary: that $\lim_{x \to 0} h(x) = b$ for some number b. By Definition 3.5.1, for each $\varepsilon > 0$ there must exist a $\delta > 0$ such that
$$|h(x) - b| < \varepsilon$$
for any x satisfying
$$0 < |x - 0| < \delta.$$

In particular, for $\varepsilon = \frac{1}{2}$ there must exist a δ_1 such that

$$|h(x) - b| < \tfrac{1}{2}$$

for any x satisfying

$$0 < |x - 0| < \delta_1.$$

Consider the points

$$x_1 = \frac{\delta_1}{2} \qquad \text{and} \qquad x_2 = -\frac{\delta_1}{2};$$

note that

$$h(x_1) = 3 + x_1 = 3 + \frac{\delta_1}{2}$$

and that

$$h(x_2) = 2 - x_2^2 = 2 - \frac{\delta_1^2}{4}.$$

Since

$$0 < |x_1 - 0| < \delta_1 \qquad \text{and} \qquad 0 < |x_2 - 0| < \delta_1,$$

then it must be that

$$|h(x_1) - b| < \tfrac{1}{2} \qquad \text{and} \qquad |h(x_2) - b| < \tfrac{1}{2};$$

that is,

$$\left|\left(3 + \frac{\delta_1}{2}\right) - b\right| < \frac{1}{2} \qquad \text{and} \qquad \left|\left(2 - \frac{\delta_1^2}{4}\right) - b\right| < \frac{1}{2};$$

that is,

$$\left|b - \left(3 + \frac{\delta_1}{2}\right)\right| < \frac{1}{2} \qquad \text{and} \qquad \left|b - \left(2 - \frac{\delta_1^2}{4}\right)\right| < \frac{1}{2}.$$

Now

$$\left|b - \left(3 + \frac{\delta_1}{2}\right)\right| < \frac{1}{2} \Leftrightarrow \frac{5}{2} + \frac{\delta_1}{2} < b < \frac{7}{2} + \frac{\delta_1}{2}.$$

Since $\delta_1 > 0$, it follows that

$$b > \tfrac{5}{2}.$$

On the other hand,

$$\left|b - \left(2 - \frac{\delta_1^2}{4}\right)\right| < \frac{1}{2} \Leftrightarrow \frac{3}{2} - \frac{\delta_1^2}{4} < b < \frac{5}{2} - \frac{\delta_1^2}{4}.$$

Since $\delta_1 > 0$, we have

$$b < \tfrac{5}{2}.$$

Obviously, there is no such number b. Thus

$$\lim_{x \to 0} h(x) \text{ does not exist.}$$

EXAMPLE 6   (proof of Fact 3.2.3 using Definition 3.5.1). Since $\lim_{x \to c} f(x) = L$, then for each $\varepsilon > 0$ there exists a $\delta > 0$ such that

$$|f(x) - L| < \varepsilon$$

for any x satisfying

$$0 < |x - c| < \delta.$$

In particular, for $\varepsilon = L/2 > 0$ there exists a $\delta_1 > 0$ such that

$$|f(x) - L| < \frac{L}{2}$$

for any x satisfying

$$0 < |x - c| < \delta_1.$$

But

$$|f(x) - L| < \frac{L}{2} \Leftrightarrow -\frac{L}{2} < f(x) - L < \frac{L}{2}$$

$$\Leftrightarrow \frac{L}{2} < f(x) < \frac{3L}{2}$$

and

$$|x - c| < \delta_1 \Leftrightarrow -\delta_1 < x - c < \delta_1$$

$$\Leftrightarrow c - \delta_1 < x < c + \delta_1.$$

Therefore

$$f(x) > \frac{L}{2} > 0$$

for any x in the interval

$$(c - \delta_1, c + \delta_1), \qquad x \neq c.$$

EXAMPLE 7   (proof of limit rule 3 using Definition 3.5.1). If $\lim_{x \to c} f(x) = L$ and $\lim_{x \to c} g(x) = M$, then we wish to show that

$$\lim_{x \to c} [f(x) + g(x)] = L + M.$$

Let $\varepsilon > 0$ be given. Then there exists a $\delta_1 > 0$ such that

$$|f(x) - L| < \frac{\varepsilon}{2}$$

for any x satisfying

$$0 < |x - c| < \delta_1.$$

Also, there exists a $\delta_2 > 0$ such that

$$|g(x) - M| < \frac{\varepsilon}{2}$$

for any x satisfying

$$0 < |x - c| < \delta_2.$$

Choose $\delta = \min\{\delta_1, \delta_2\}$. Then for any x satisfying

$$0 < |x - c| < \delta$$

it follows that

$$|[f(x) + g(x)] - [L + M]| = |(f(x) - L) + (g(x) - M)|$$
$$\leq |f(x) - L| + |g(x) - M|$$
$$< \frac{\varepsilon}{2} + \frac{\varepsilon}{2} = \varepsilon.$$

EXERCISES

1. Prove the following using 3.5.1.

(a) $\lim\limits_{x \to 2}(3x - 2) = 4$.

(b) $\lim\limits_{x \to 8}[(x^2 - 64)/(x - 8)] = 16$.

(c) $\lim\limits_{x \to 1}(x^2 + 5) = 6$.

(d) $\lim\limits_{x \to -1} x^3 = -1$.

(e) $\lim\limits_{x \to 5}(1/x) = \frac{1}{5}$.

(f) $\lim\limits_{x \to 3} \sqrt{x + 1} = 2$.

(g) $\lim\limits_{x \to 5}[(x - 3)/(x - 1)] = \frac{1}{2}$.

2. Let

$$f(x) = \begin{cases} 1, & (x \geq 0), \\ -1, & (x < 0). \end{cases}$$

Show that $\lim\limits_{x \to 0} f(x)$ does not exist.

3. Let $f(x)$ be the postage stamp function (see Example 6, Section 2.1). Show that $\lim\limits_{x \to n} f(x)$ does not exist for any integer n.

4. Give an ε-δ definition of "f is continuous at c."

5. Let $f(x) = |x|$. Using Exercise 4, show that f is continuous at any point c.

6. Using 3.5.1, verify $\lim\limits_{x \to c} kx^r = kc^r$, where k is a constant for

(a) $r = 2$.

(b) $r = -1$.

(c) $r = \frac{1}{2}$. [*Hint:* Consider the cases $c = 0$ and $c > 0$.]

7. Suppose that $\lim\limits_{x \to c} f(x) = L$. Show that there exist $M > 0$ and $\delta > 0$ such that

$$|f(x)| \leq M \text{ for all } x \text{ satisfying } 0 < |x - c| < \delta.$$

8. Using 3.5.1, prove limit rule 4. [*Hint:* Use

$$|f(x)g(x) - LM| = |(f(x) - L)g(x) + (g(x) - M)L|$$

and Exercise 7.]

9. Suppose that $\lim\limits_{x \to c} f(x) = L$, where $L \neq 0$. Show that there exist $M > 0$ and $\delta > 0$ such that $|f(x)| \geq M$ for all x satisfying $0 < |x - c| < \delta$.

10. Using 3.5.1, prove that if $\lim\limits_{x \to c} g(x) = M \neq 0$, then $\lim\limits_{x \to c} 1/g(x) = 1/M$.
[*Hint:* Use $|1/g(x) - 1/M| = |(M - g(x))/Mg(x)|$ and Exercise 9.]

11. Prove limit rule 5. [*Hint:* $f(x)/g(x) = f(x) \cdot 1/g(x)$.]

Chapter 4

THE DERIVATIVE
AND ITS APPLICATIONS

4.1 INTRODUCTION

In the branch of calculus known as **differential calculus**, we study the rate of change in the functional value $f(x)$ with respect to a change in the variable x. This rate of change will be expressed precisely as a limit (called the **derivative**). It turns out that this limit arises in a variety of situations including geometry, physics, and economics. For instance, in geometry the limit is the slope of a line (called the tangent line), in physics the limit can be interpreted as in- stantaneous velocity, and in economics the limit gives the rate at which the cost of a product is changing with respect to total output. Let us look at three examples.

Tangent Line

To indicate what is meant by a **tangent line to a curve at a point P_0**, we consider lines passing through P_0 and points P on the curve which are near P_0. These lines are called *secant lines* (see Figure 4.1).

It should be noted that the closer P is to P_0, the better the secant line approximates the curve near P. If, as P approaches P_0, the secant lines approach a fixed line L passing through P_0, then we call this limiting line the tangent line to the curve at P_0. The slope of line L is that number approached by the slopes of these secant lines.

Let us consider a specific curve

$$y = x^2$$

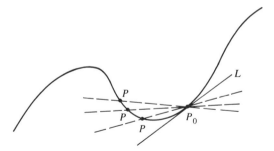

Figure 4.1

at the point (1, 1). To describe the tangent line at (1, 1) we need only indicate its slope since we know a point on the line. Therefore we shall compute the slopes of lines joining (1, 1) with points nearby; these points will be designated by $(1 + h, (1 + h)^2)$. We should consider both positive and negative values of h, but small in absolute value (see Figure 4.2).

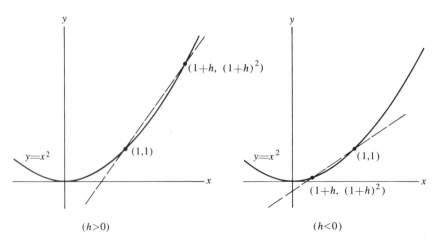

Figure 4.2

The line through the points (1, 1) and $(1 + h, (1 + h)^2)$ is a *secant line*; the slope of this line is

$$\frac{(1 + h)^2 - 1}{h}.$$

If we let $f(x) = x^2$, then the slope of the secant line can be written as

$$\frac{f(1 + h) - f(1)}{h}.$$

We shall now compute this number for various values of h. The results for $h < 0$ appear on the left and those for $h > 0$ are on the right.

h	$\dfrac{f(1 + h) - f(1)}{h}$	h	$\dfrac{f(1 + h) - f(1)}{h}$
-0.5	1.5	1	3
-0.2	1.8	0.3	2.3
-0.07	1.93	0.06	2.06
-0.03	1.97	0.01	2.01
-0.005	1.995	0.007	2.007
-0.001	1.999	0.002	2.0019999
-0.0006	1.9994	0.0004	2.0003999
-0.0002	1.9998	0.00009	2.00008999
-0.00004	1.99996	0.00001	2.000009999

We observe that

$$\lim_{h \to 0} \frac{f(1 + h) - f(1)}{h} = 2;$$

that is, the slopes of the secant lines approach 2 (see Figure 4.3). Therefore the tangent line T is the line passing through $(1, 1)$ with slope 2.

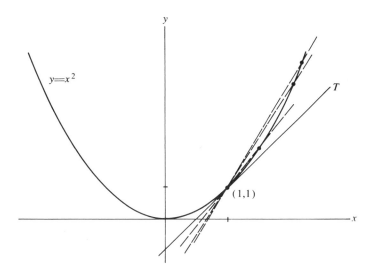

Figure 4.3

Instantaneous Velocity

Suppose that we take a 2-hour train trip and suppose that the function

$$s(t) = 20t^2 + 10t \qquad (0 \le t \le 2)$$

indicates the distance the train has traveled after t hours. For example, after 1 hour the train has gone $s(1) = 30$ miles, and after 2 hours, it has gone $s(2) = 100$ miles.

The average velocity in a certain time interval is defined to be the distance traveled in the interval divided by the length of the time interval. In our example, the average velocity between 1 and 2 hours is

$$\frac{100 - 30}{2 - 1} = 70 \text{ miles per hour.}$$

We now want to discuss the meaning of the speedometer reading on the train at the end of 1 hour; that is, what is the **instantaneous velocity** at the 1-hour mark?

We shall do something similar to the computation of the slopes of secant lines. In fact, we shall compute the average velocity in the time interval $[1, 1 + h]$ if $h > 0$ and in the time interval $[1 + h, 1]$ if $h < 0$. Again, we take values of h which are small in absolute value. The average velocity is given by

$$\frac{s(1 + h) - s(1)}{h}.$$

Our computations are listed below.

h	$\dfrac{s(1 + h) - s(1)}{h}$	h	$\dfrac{s(1 + h) - s(1)}{h}$
−0.9	32	1.5	80
−0.2	46	0.6	62
−0.05	49	0.1	52
−0.01	49.8	0.05	51
−0.007	49.86	0.009	50.179999
−0.003	49.94	0.002	50.039999
−0.0005	49.99	0.0008	50.0159999
−0.0001	49.998	0.0001	50.0019999
−0.00004	49.9992	0.00004	50.00079999
−0.000006	49.99988	0.000003	50.00005999

We see that

$$\lim_{h \to 0} \frac{s(1 + h) - s(1)}{h} = 50.$$

We call the number 50, which is approached by the average velocities on smaller and smaller time intervals about 1, the instantaneous velocity at $t = 1$. This is the speedometer reading at $t = 1$.

Marginal Cost

One of the things a manufacturer is interested in knowing is how his cost of production is changing with respect to the total output at a given instant. Such an indicator is called the **marginal cost**.

As an example, suppose that the cost of production is given by a cost function

$$C(x) = 90 + 3\sqrt[3]{x},$$

where x is the number of articles produced. We wish to compute the marginal cost when $x = 64$; that is, how is the cost changing with respect to total output when 64 items have been produced?

At this time, for the purposes of mathematical analysis, we assume that $C(x)$ is not just defined for the positive integers but for all values of $x \geq 0$ and thus represents a continuous function for $x \geq 0$.

We shall compute the average change in the cost of production per output when h additional items are produced and when h fewer items are produced. More precisely, we shall compute

$$\frac{C(64 + h) - C(64)}{h}$$

for both $h > 0$ and $h < 0$. These computations will be done successively on smaller and smaller production intervals $[64, 64 + h]$ if $h > 0$ and $[64 + h, 64]$ if $h < 0$, thus giving an indication of the change in production per output at 64.

h	$\dfrac{C(64 + h) - C(64)}{h}$	h	$\dfrac{C(64 + h) - C(64)}{h}$
-1	0.0628	2	0.0619
-0.5	0.0627	1	0.0622
-0.1	0.06253	0.4	0.0624
-0.07	0.06252	0.1	0.06247
-0.02	0.062507	0.05	0.06248
-0.008	0.062503	0.01	0.062497
-0.004	0.0625013	0.006	0.062498
-0.001	0.0625003	0.002	0.062499
-0.0005	0.06250016	0.0003	0.0624999
-0.00003	0.0625000097	0.00008	0.062499973

Therefore we observe that

$$\lim_{h \to 0} \frac{C(64 + h) - C(64)}{h} = 0.0625.$$

We say that the marginal cost is 0.0625 when $x = 64$.

As mentioned in Section I.1, Fermat (in 1629) was the first to discuss the essential idea of **differential calculus**. He was interested in determining where a function has maximum and minimum values. Fermat indicated that at such extreme points,

$$\lim_{h \to 0} \frac{f(x + h) - f(x)}{h} = 0.$$

Once again the quotient

$$\frac{f(x + h) - f(x)}{h}$$

appears in another type of problem, which we shall discuss in Section 4.6.

4.2 DEFINITION AND EXAMPLES

We have seen in the introduction to this chapter that the so-called **difference quotient** $[f(x + h) - f(x)]/h$ occurs in several different situations—slope of a tangent line, velocity, and marginal cost. Differential calculus is the study of the limit of this difference quotient and its applications.

4.2.1
Definition

Let f be defined in an open interval containing x and suppose that $h \neq 0$.
    If

$$\lim_{h \to 0} \frac{f(x + h) - f(x)}{h} \text{ exists,}$$

then f is said to be **differentiable** at x and the value of this limit is denoted by

$$f'(x).$$

It should be noted that in computing

$$\lim_{h \to 0} \frac{f(x + h) - f(x)}{h}$$

x is regarded as a constant as $h \to 0$. The value of the limit—$f'(x)$—is called the **derivative of f at x**.
    Leibniz introduced an alternative notation for the derivative. If

$$y = f(x),$$

let

$$h = \Delta x$$

and

$$f(x + h) - f(x) = \Delta y.$$

The Greek letter Δ indicates *change* in the variable; that is Δx indicates change in the domain values and Δy indicates change in the functional values which occur in the difference quotient. Therefore

$$\frac{f(x + h) - f(x)}{h} = \frac{\Delta y}{\Delta x}.$$

Since $h \to 0 \Leftrightarrow \Delta x \to 0$, the derivative at x is

$$\lim_{\Delta x \to 0} \frac{\Delta y}{\Delta x}.$$

Leibniz chose a symbol which appears as a quotient for the value of the preceding limit, namely

$$\frac{dy}{dx} = \lim_{\Delta x \to 0} \frac{\Delta y}{\Delta x}.$$

This notation for the derivative was of assistance to the mathematicians of continental Europe in their development of calculus.

We sometimes write

$$\frac{dy}{dx} = \frac{d}{dx}\{y\},$$

read "the derivative of y with respect to x." If $y = f(x)$, we write

$$\frac{dy}{dx} = f'(x) = \frac{d}{dx}\{f(x)\}.$$

Let us look at some very simple examples.

EXAMPLE 1   Let $f(x) = a$, where a is a constant. Then, if $h \neq 0$, we have

$$\frac{f(x + h) - f(x)}{h} = \frac{a - a}{h} = 0.$$

Therefore

$$\lim_{h \to 0} \frac{f(x + h) - f(x)}{h} = \lim_{h \to 0} 0 = 0,$$

and hence

$$f'(x) = 0,$$

or, equivalently,

$$\frac{d}{dx}\{a\} = 0.$$

EXAMPLE 2   Let $f(x) = ax$, where a is a constant. Then, if $h \neq 0$, we have

$$\frac{f(x + h) - f(x)}{h} = \frac{a(x + h) - ax}{h} = \frac{ah}{h} = a.$$

Therefore

$$\lim_{h \to 0} \frac{f(x + h) - f(x)}{h} = \lim_{h \to 0} a = a,$$

and hence

$$f'(x) = a$$

or

$$\frac{d}{dx} \{ax\} = a.$$

EXAMPLE 3   Let $f(x) = ax^2$, where a is a constant. Then, if $h \neq 0$,

$$\frac{f(x + h) - f(x)}{h} = \frac{a(x + h)^2 - ax^2}{h}$$

$$= \frac{a[x^2 + 2xh + h^2] - ax^2}{h}$$

$$= 2ax + ah.$$

Hence

$$\lim_{h \to 0} \frac{f(x + h) - f(x)}{h} = \lim_{h \to 0}(2ax + ah) = 2ax;$$

that is,

$$f'(x) = 2ax$$

or

$$\frac{d}{dx} \{ax^2\} = 2ax.$$

EXAMPLE 4   Let $f(x) = ax^{1/2}$, where a is a constant and $x > 0$. If $h \neq 0$,

$$\frac{f(x + h) - f(x)}{h} = \frac{a(x + h)^{1/2} - ax^{1/2}}{h}.$$

We shall rationalize the numerator of the right-hand side by multiplying numerator and denominator by $(x + h)^{1/2} + x^{1/2}$:

$$\frac{f(x + h) - f(x)}{h} = a \left[\frac{(x + h)^{1/2} - x^{1/2}}{h} \right] \left[\frac{(x + h)^{1/2} + x^{1/2}}{(x + h)^{1/2} + x^{1/2}} \right]$$

$$= \frac{ah}{h((x + h)^{1/2} + x^{1/2})}$$

$$= \frac{a}{(x + h)^{1/2} + x^{1/2}}.$$

Hence

$$\lim_{h \to 0} \frac{f(x + h) - f(x)}{h} = \lim_{h \to 0} \frac{a}{(x + h)^{1/2} + x^{1/2}}$$

$$= \frac{a}{2x^{1/2}} = \frac{1}{2} ax^{-1/2};$$

that is,

$$f'(x) = \frac{1}{2} ax^{-1/2}$$

or

$$\frac{d}{dx} \{ax^{1/2}\} = \frac{1}{2} ax^{-1/2}.$$

The preceding examples indicate a more general formula.

4.2.2
Power Rule

$$\frac{d}{dx} \{ax^r\} = arx^{(r-1)}.$$

The above rule is valid whenever x^{r-1} is defined.

EXAMPLE 5

$$\frac{d}{dx} \{1\} = 0 \qquad \frac{d}{dx} \{2x\} = 2$$

$$\frac{d}{dx} \left\{ \frac{5}{x^2} \right\} = \frac{d}{dx} \{5x^{-2}\} = (5)(-2)x^{-2-1} = -10x^{-3}$$

$$\frac{d}{dx} \{3x^{1/4}\} = 3 \left\{ \frac{1}{4} \right\} x^{1/4-1} = \frac{3}{4} x^{-3/4}$$

$$\frac{d}{dx} \{x^{2/3}\} = \frac{2}{3} x^{2/3-1} = \frac{2}{3} x^{-1/3}.$$

EXAMPLE 6   Let $f(x) = \frac{1}{6}x^4$ and find $f'(2)$. Since $f'(x) = \frac{2}{3}x^3, f'(2) = \frac{2}{3}(2)^3 = \frac{16}{3}$.

EXAMPLE 7   What is $f'(x)$ if $f(x) = 1/(x + 2)$? If $h \neq 0$,

$$\frac{f(x + h) - f(x)}{h} = \frac{[1/(x + h + 2)] - [1/(x + 2)]}{h}.$$

If we place the numerator of the right-hand side over a common denominator, we obtain

$$\frac{f(x + h) - f(x)}{h} = \frac{[(x + 2) - (x + h + 2)]/[(x + h + 2)(x + 2)]}{h}$$

$$= \frac{-h}{h(x + h + 2)(x + 2)}$$

$$= -\frac{1}{(x + h + 2)(x + 2)}.$$

Therefore

$$\lim_{h \to 0} \frac{f(x + h) - f(x)}{h} = \lim_{h \to 0} \left[-\frac{1}{(x + h + 2)(x + 2)} \right] = -\frac{1}{(x + 2)^2};$$

that is,

$$f'(x) = -\frac{1}{(x + 2)^2},$$

which is valid for $x \neq -2$.

Recall that to say that f is differentiable at x_0 is to say that

$$\lim_{h \to 0} \frac{f(x_0 + h) - f(x_0)}{h}$$

exists. In this situation we are assuming that f is defined in an open interval containing x_0, since the definition involves a two-sided limit.

However, if f is defined for $x \geq x_0$, then it makes sense to investigate the one-sided limit:

$$\lim_{x \downarrow 0} \frac{f(x_0 + h) - f(x_0)}{h}.$$

The value of such a limit, if it exists, is called the **right-hand derivative of f at x_0**. If f is defined for $x \leq x_0$, then

$$\lim_{h \uparrow 0} \frac{f(x_0 + h) - f(x_0)}{h},$$

if it exists, is called the **left-hand derivative of f at x_0**. Recalling 3.1.2, we note that to say the two-sided derivative at x_0 exists and equals b is the same as saying that each of the one-sided derivatives exists and equals b.

EXAMPLE 8   Let $f(x) = |x|$. We wish to investigate the differentiability of f at zero. Therefore if $h \neq 0$,

$$\frac{f(0 + h) - f(0)}{h} = \frac{|0 + h| - |0|}{h}$$

$$= \frac{|h|}{h} = \begin{cases} 1 & \text{if } h > 0, \\ -1 & \text{if } h < 0. \end{cases}$$

Hence

$$\lim_{h \downarrow 0} \frac{f(0 + h) - f(0)}{h} = \lim_{h \downarrow 0} 1 = 1;$$

that is, the right-hand derivative at zero is 1, and

$$\lim_{h \uparrow 0} \frac{f(0 + h) - f(0)}{h} = \lim_{h \uparrow 0}(-1) = -1;$$

that is, the left-hand derivative at zero is -1.

Even though f has a left-hand derivative at zero and a right-hand derivative at zero, these values do not coincide, and so f is not differentiable at zero.

The concept of one-sided derivatives allows us to define differentiability on an interval I. In particular, we say that f **is differentiable on** $[a, b]$ if

1. f is differentiable at each x_0, where $a < x_0 < b$.
2. f has a right-hand derivative at a.
3. f has a left-hand derivative at b.

Differentiability on other types of intervals is defined analogously, keeping in mind the existence of the appropriate one-sided limits at end points. If f is differentiable at each number, then we say that f **is differentiable on R**.

EXAMPLE 9   Let $f(x) = |x|$.
 Case 1: $x > 0$. Then $f(x) = x$ and $f'(x) = 1$. In addition, we saw that f has a right-hand derivative, namely 1, at $x = 0$. Hence it is proper to say that f is differentiable on the interval $[0, 2]$.
 Case 2: $x < 0$. Then $f(x) = -x$ and $f'(x) = -1$. Again, we note that f has a left-hand derivative, namely -1, at $x = 0$. Therefore we can say that f is differentiable on $[-2, 0]$.
 However, we cannot say that f is differentiable on $[-2, 2]$ since f is not differentiable at $x = 0$.
 It is interesting to note that at this point of nondifferentiability the graph of f has a corner (see Figure 4.4).

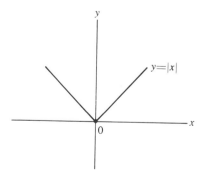

Figure 4.4

In our study of calculus in the following sections, we shall see many applications of the derivative. In some problems the derivative of f' adds some additional information in helping to solve the problem.

We call the derivative of f' the **second derivative of f** and denote it by

$$f''.$$

In other words,

$$f''(x) = \lim_{h \to 0} \frac{f'(x + h) - f'(x)}{h}.$$

If $y = f(x)$, we write

$$f''(x) = \frac{d^2 y}{dx^2}.$$

Also,

$$f''(x) = \frac{d^2}{dx^2} \{y\}$$

$$= \frac{d^2}{dx^2} \{f(x)\}.$$

Higher-order derivatives are denoted by $f'''(x), f^{(IV)}(x), f^{(V)}(x)$, etc.

EXAMPLE 10   Let $f(x) = x^{5/2}$. Then

$$f'(x) = \tfrac{5}{2} x^{3/2}$$
$$f''(x) = \tfrac{15}{4} x^{1/2}$$
$$f'''(x) = \tfrac{15}{8} x^{-1/2}$$
$$f^{(IV)}(x) = -\tfrac{15}{16} x^{-3/2}.$$

In particular,

$$f'''(4) = \frac{15}{8} 4^{-1/2} = \frac{15}{8} \frac{1}{\sqrt{4}} = \frac{15}{16}.$$

EXERCISES

1. Use Definition 4.2.1 to verify the following special case of the power rule:

$$\frac{d}{dx} \left\{ \frac{a}{x} \right\} = \frac{-a}{x^2}.$$

2. Use Definition 4.2.1 to verify the following special case of the power rule:

$$\frac{d}{dx} \left\{ \frac{a}{x^2} \right\} = \frac{-2a}{x^3}.$$

3. Use Definition 4.2.1 to verify the following special case of the power rule:

$$\frac{d}{dx} \left\{ \frac{a}{\sqrt{x}} \right\} = -\frac{a}{2x^{3/2}}.$$

4. For each of the following, use the power rule to compute dy/dx and d^2y/dx^2.

(a) $y = -7$.   (b) $y = \sqrt{2x}$.

(c) $y = x^2/13$.   (d) $y = 3^{2/3}$.

(e) $y = 3\sqrt{x}$.   (f) $y = -1/(3x)$.

(g) $y = 2x^2/9$.   (h) $y = -4/\sqrt{x}$.

(i) $y = x^{3/5}$.   (j) $y = 4x^{-10/7}$.

(k) $y = 3x^{42}$.

5. Let $f(x) = 2x^3$.

(a) Find $f'(x)$ and compute $f'(0)$, $f'(4)$.

(b) Find $f''(x)$ and compute $f''(-1)$, $f''(\frac{1}{2})$.

(c) Find $f'''(x)$ and compute $f'''(1)$, $f'''(-\frac{4}{3})$.

(d) Find $f^{(IV)}(x)$ and compute $f^{(IV)}(0)$, $f^{(IV)}(3)$.

6. Use Definition 4.2.1 to find $f'(x)$ if

(a) $f(x) = 3x + 4$.   (b) $f(x) = 6 - 2x + 7x^2$.

(c) $f(x) = x^4 + 2x - 1$.   (d) $f(x) = 1/(x^2 + 1)$.

(e) $f(x) = \sqrt{x + 1}$.   (f) $f(x) = x^3 + 2$.

(g) $f(x) = 1/x - 1$.

7. For each of the following functions f, evaluate the right- and left-hand derivatives of f at the specified point x_0, and determine if f is differentiable at x_0. Graph f in each case.

(a) $f(x) = |x - 3|$, $x_0 = 3$.   (b) $f(x) = -|x + 2|$, $x_0 = -2$.

(c) $f(x) = \begin{cases} 0, & x \le 0, \\ x^2, & x > 0, \end{cases}$ $x_0 = 0$.

(d) $f(x) = \begin{cases} x/1000, & x \le 0, \\ x^2, & x > 0, \end{cases}$ $x_0 = 0$.

(e) $f(x) = \begin{cases} x^3, & x \le 0, \\ x^2, & x > 0, \end{cases}$ $x_0 = 0$.

8. Let $f(x) = x|x|$. Find $f'(x)$ for all x.

9. (a) Let $f'(x) = f(x)$. What is $f''(x) - f(x)$?

(b) Let $f'(x) = -g(x)$ and $g'(x) = f(x)$. What is $f''(x) + f(x)$?

10. (a) Compute $\dfrac{d}{dx}\{7x^2\}$ and $\dfrac{d}{dx}\{-2x\}$.

(b) What is $\dfrac{d}{dx}\{7x^2\} + \dfrac{d}{dx}\{-2x\}$?

(c) Use Definition 4.2.1 to find $\dfrac{d}{dx}\{7x^2 - 2x\}$.

(d) Are the answers to parts (b) and (c) equal?

(e) Is it possible that the formula $\dfrac{d}{dx}\{f(x) + g(x)\} = \dfrac{d}{dx}\{f(x)\} + \dfrac{d}{dx}\{g(x)\}$

can be valid?

11. (a) Compute $\dfrac{d}{dx}\{7x^2\}$ and $\dfrac{d}{dx}\{-2x\}$.

(b) Compute the product $\left[\dfrac{d}{dx}\{7x^2\}\right]\left[\dfrac{d}{dx}\{-2x\}\right]$.

(c) Compute $\dfrac{d}{dx}\{(7x^2)(-2x)\} = \dfrac{d}{dx}\{-14x^3\}$.

(d) Are the answers to parts (b) and (c) equal?

(e) Is it possible that the formula $\left[\dfrac{d}{dx}\{f(x)\}\right]\left[\dfrac{d}{dx}\{g(x)\}\right] = \dfrac{d}{dx}\{f(x)g(x)\}$ can be valid?

12. (a) Compute $\dfrac{d}{dx}\{7x^2\}$ and $\dfrac{d}{dx}\{-2x\}$.

(b) Compute the quotient $\dfrac{\dfrac{d}{dx}\{7x^2\}}{\dfrac{d}{dx}\{-2x\}}$.

(c) Compute $\dfrac{d}{dx}\left\{\dfrac{7x^2}{-2x}\right\} = \dfrac{d}{dx}\left\{-\dfrac{7}{2}x\right\}$.

(d) Are the answers to (b) and (c) equal?

(e) Is it possible that the formula $\dfrac{d}{dx}\left\{\dfrac{f(x)}{g(x)}\right\} = \dfrac{\dfrac{d}{dx}\{f(x)\}}{\dfrac{d}{dx}\{g(x)\}}$ can be valid?

13. For each function f defined in Exercise 7, answer the following three questions.

(a) Is f differentiable on $[0, 5]$?

(b) Is f differentiable on $[-5, 0]$?

(c) Is f differentiable on $[-5, 5]$?

4.3 GEOMETRIC INTERPRETATION OF THE DERIVATIVE

It is our intention in this section to give geometric interpretations of the statements

"f is differentiable at the point x_0"

and

"f is differentiable on the interval I."

As we shall see, the former implies the existence of a tangent line to the curve

$$y = f(x)$$

at the point $(x_0, f(x_0))$, while the latter can be interpreted in terms of the notion of a *smooth* curve on interval I.

To begin, recall the discussion in Section 4.1 of the line tangent to the graph of $y = x^2$ at the point $(1, 1)$. We now wish to generalize on that development. Therefore, consider the curve

$$y = f(x)$$

at the point $(x_0, f(x_0))$, and consider the secant line joining the points $(x_0, f(x_0))$ and $(x_0 + h, f(x_0 + h))$, as in Figure 4.5. Its slope is

$$\frac{f(x_0 + h) - f(x_0)}{h}.$$

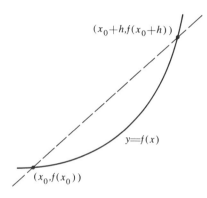

Figure 4.5

If we change h, we, of course, get a different secant line with a different slope. We observe that as $h \to 0$, the secant lines approach a line T which touches the curve at the point $(x_0, f(x_0))$ and is close to the curve near the point (see Figure 4.6). The slope of this limiting line T is the limit of the slopes of the secant lines, namely

$$\lim_{h \to 0} \frac{f(x_0 + h) - f(x_0)}{h},$$

if it exists.

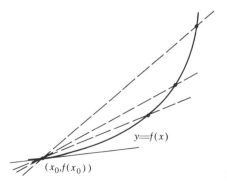

Figure 4.6

Therefore the curve $y = f(x)$ has a tangent line at the point $(x_0, f(x_0))$ if and only if f is differentiable at x_0.

4.3.1

Equation of the Tangent Line

Let f have a derivative at x_0 and let $y_0 = f(x_0)$.
The tangent line to the curve $y = f(x)$ at the point (x_0, y_0) has slope

$$f'(x_0)$$

and its equation is

$$y - y_0 = f'(x_0)(x - x_0).$$

For instance, from Examples 3 and 8 of Section 4.2, we can conclude that

There is a tangent line T at each point (a, ax_0^2) on the curve $y = ax^2$, and the slope of T is $2ax_0$,

and

There is no tangent line to the curve $y = |x|$ at the point $(0, 0)$. (see Figure 4.7).

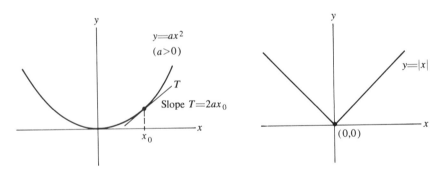

Figure 4.7

EXAMPLE 1   What is the equation of the tangent line to the curve $y = x^3$ at the point $(2, 8)$?

Let $f(x) = x^3$; then

$$f'(x) = 3x^2$$

and hence

$$f'(2) = 12.$$

Therefore the slope of the tangent line is 12 and its equation is

$$y - 8 = 12(x - 2).$$

The graph of the curve and this tangent line are given in Figure 4.8.

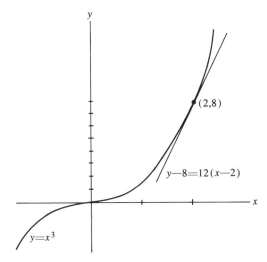

Figure 4.8

The next example introduces the idea of a vertical tangent at a point.

EXAMPLE 2   Let us consider the continuous function defined by

$$f(x) = x^{2/3} \qquad \text{(see Figure 4.9)}.$$

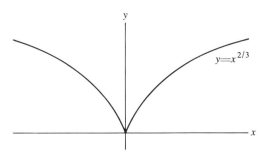

Figure 4.9

We observe the following one-sided limits:

$$\lim_{h \downarrow 0} \frac{f(0 + h) - f(0)}{h} = \lim_{h \downarrow 0} \frac{h^{2/3}}{h}$$

$$= \lim_{h \downarrow 0} \frac{1}{h^{1/3}}$$

$$= +\infty$$

and

$$\lim_{h \uparrow 0} \frac{f(0 + h) - f(0)}{h} = \lim_{h \uparrow 0} \frac{1}{h^{1/3}}$$

$$= -\infty.$$

Therefore, f is not differentiable at zero. However, is there some geometric interpretation we can give to the above limits?

If $h > 0$, then

$$\frac{f(0 + h) - f(0)}{h}$$

represents slopes of secant lines for various values of h (see Figure 4.10).

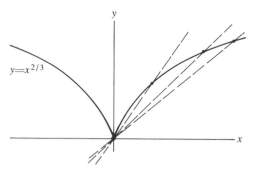

Figure 4.10

The fact that

$$\lim_{h \downarrow 0} \frac{f(0 + h) - f(0)}{h} = +\infty$$

says that the slopes of these secant lines are getting large positively, which we interpret geometrically to mean that the secant lines are getting steeper and steeper. In fact, they are approaching a vertical line, namely $x = 0$.

A similar interpretation can be given to

$$\lim_{h \uparrow 0} \frac{f(0 + h) - f(0)}{h} = -\infty.$$

It then seems reasonable to say that the curve $y = x^{2/3}$ has a **vertical tangent** at $(0, 0)$, namely $x = 0$.

In the preceding example, we note that the function is continuous on **R**; that is, there are no breaks in the curve on **R**. However, it is not differentiable at $x = 0$, where we observed that the graph has a corner. This raises an interesting question concerning the relationship between continuity and differentiability. It turns out that a differentiable function must be continuous, but, as the above example shows, a continuous function need not be differentiable.

4.3.2
Theorem   If f is differentiable at x_0, then f is continuous at x_0.

Proof: If $x \neq x_0$, then

$$f(x) - f(x_0) = \left[\frac{f(x) - f(x_0)}{x - x_0}\right](x - x_0).$$

Therefore

$$\lim_{x \to x_0} [f(x) - f(x_0)] = \lim_{x \to x_0} \left\{\left[\frac{f(x) - f(x_0)}{x - x_0}\right](x - x_0)\right\}$$

$$= \lim_{x \to x_0} \left[\frac{f(x) - f(x_0)}{x - x_0}\right] \lim_{x \to x_0} (x - x_0) \qquad \text{(limit rule 4)}$$

$$= f'(x_0) \cdot 0 = 0.$$

Therefore

$$\lim_{x \to x_0} f(x) = f(x_0),$$

which means that f is continuous at x_0.

Let us summarize our observations.

In Chapter 3 we saw that the graph of a continuous function is an unbroken curve. There is a possibility that the graph has corners where there is no nonvertical tangent line to the graph (for instance, Example 2 of this section and Example 8 of Section 4.2). On the other hand, we have shown that differentiability at a point indicates the existence of a nonvertical tangent line and hence there is no corner on the graph at that point. Let us call an unbroken curve which has no corners a **smooth curve**. *To say that f is differentiable on I (and hence continuous) is the same as saying that the graph of f is smooth (and hence unbroken).*

EXAMPLE 3   Let f be the function on $[-5, 4]$ whose graph is given in Figure 4.11. Then f is continuous everywhere on $[-5, 4]$ except at $x = 0$ and $x = 1$, and f is differentiable everywhere on $[-5, 4]$ except at $x = -3$, $x = 0$, and $x = 1$.

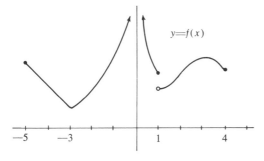

Figure 4.11

EXERCISES

1. For each of the following functions f, determine the slope of the tangent line to the graph of f at the specified point $(x_0, f(x_0))$.

(a) $f(x) = x^2$, $(-1, 1)$ (b) $f(x) = 2x^3$, $(1, 2)$

(c) $f(x) = \sqrt{x}$, $(9, 3)$ (d) $f(x) = 1/\sqrt[3]{x}$, $(8, \frac{1}{2})$

(e) $f(x) = -4/x$, $(-\frac{1}{3}, 12)$

2. For each part in Exercise 1, find the equation of the tangent line to the graph of f at the specified point.

3. The **normal line to the curve** $y = f(x)$ **at the point** $(x_0, f(x_0))$ is the line passing through $(x_0, f(x_0))$ and perpendicular to the tangent line at $(x_0, f(x_0))$.

(a) Let f have a derivative at x_0 [assume that $f'(x_0) \neq 0$] and let $y_0 = f(x_0)$. Verify that the normal line to the curve $y = f(x)$ at the point (x_0, y_0) has slope $- [1/f'(x_0)]$ and that its equation is $y - y_0 = - [1/f'(x_0)](x - x_0)$.

(b) For each part in Exercise 1, find the equation of the normal line to the graph of f at the specified point.

4. Sketch the graph of each of the following functions f. Without computing derivatives, determine from the graph points of discontinuity and points where no derivative exists.

(a) $f(x) = |x - 7|$.   (b) $f(x) = x/x$.

(c) $f(x) = x|x|$.   (d) $f(x) = \begin{cases} 0, & x \leq 0, \\ x, & x > 0. \end{cases}$

(e) $f(x) = \begin{cases} 0, & x \leq 0, \\ x^2, & x > 0. \end{cases}$ (f) $f(x) = \begin{cases} 2x + 4, & x \leq -1, \\ x^2, & -1 < x < 1, \\ x, & x \geq 1. \end{cases}$

5. A pair of intersecting curves is said to be **orthogonal** if at each point of intersection the respective tangent lines are perpendicular. Determine whether or not each of the following pairs of curves is orthogonal.

(a) $y = x^3$, $y = 1/x$.   (b) $y = (\frac{1}{3})x^2$, $y = 1/x$.

(c) $y = x$, $y = 1/x$.   (d) $y = 2x$, $y = 2/x$.

6. Let $f(x) = x^{1/3}$. Investigate

$$\lim_{h \uparrow 0} \frac{f(0 + h) - f(0)}{h} \quad \text{and} \quad \lim_{h \downarrow 0} \frac{f(0 + h) - f(0)}{h}.$$

Graph f and interpret these results geometrically.

7. $f(x) = \sqrt{x - 1}$. Investigate

$$\lim_{h \downarrow 0} \frac{f(1 + h) - f(1)}{h}.$$

Graph f and interpret this result geometrically.

8. Does $y = |x|$ have a vertical tangent at $x = 0$? Explain.

Supplement (optional)

It is our purpose in this supplement to show that the tangent line is—in some sense—a good approximation to the curve $y = f(x)$ at the point $(x_0, f(x_0))$.

Let f have a derivative at x_0; then

$$f'(x_0) = \lim_{h \to 0} \frac{f(x_0 + h) - f(x_0)}{h}.$$

If $x = x_0 + h$, then

$$x \to x_0 \Leftrightarrow h \to 0.$$

Therefore with this substitution

$$f'(x_0) = \lim_{x \to x_0} \frac{f(x) - f(x_0)}{x - x_0},$$

which may be rewritten as

$$\lim_{x \to x_0} \left\{ \frac{f(x) - f(x_0)}{x - x_0} - f'(x_0) \right\} = 0.$$

Now let

$$\frac{f(x) - f(x_0)}{x - x_0} - f'(x_0) = \varepsilon(x),$$

so that

$$\lim_{x \to x_0} \varepsilon(x) = 0.$$

Then

(4.3.3)
$$f(x) - f(x_0) - f'(x_0)(x - x_0) = \varepsilon(x)(x - x_0)$$
$$f(x) = f(x_0) + f'(x_0)(x - x_0) + \varepsilon(x)(x - x_0),$$

where

$$\lim_{x \to x_0} \varepsilon(x) = 0.$$

Since

$$\lim_{x \to x_0} \varepsilon(x)(x - x_0) = 0,$$

the term $(x - x_0)\varepsilon(x)$ is small in absolute value whenever x is close enough to x_0.

Hence, from Equation 4.3.3 we see that whenever x is close enough to x_0, $f(x)$ is approximately equal to

$$f(x_0) + f'(x_0)(x - x_0);$$

this we write as

(4.3.4)
$$f(x) \approx f(x_0) + f'(x_0)(x - x_0).$$

In other words, the curve $y = f(x)$ and its tangent line $y = f(x_0) + f'(x_0)(x - x_0)$ at $(x_0, f(x_0))$ are close whenever x is close enough to x_0.

We close this supplement by applying the approximation formula 4.3.4 in a numerical problem.

EXAMPLE 4   Approximate the value of $\sqrt{17}$.

Let $f(x) = \sqrt{x}$ and $x_0 = 16$; using 4.3.4,

$$f(x) \approx f(x_0) + f'(x_0)(x - x_0).$$

Since $f'(x) = \dfrac{1}{2\sqrt{x}}$ and $f'(x_0) = \frac{1}{8}$, then

$$\sqrt{x} \approx 4 + \tfrac{1}{8}(x - 16).$$

In particular,

$$\sqrt{17} \approx 4 + \tfrac{1}{8}(17 - 16) = \tfrac{33}{8}.$$

4.4 RULES FOR DIFFERENTIATION

Sometimes a given function can be expressed in terms of simpler functions. If so, can we compute the derivative of the given function knowing the derivatives of the simpler functions? A reason for asking the question, as the Exercises in the preceding sections suggest, is that the computation of derivatives using only the definition is often a long and tedious process. Here we want to discuss some rules which will facilitate our computation of derivatives, much like the limit rules in Section 3.2, which aid us in the computation of limits.

4.4.1
Differentiation
Rules

(1) $\dfrac{d}{dx}\{cf(x)\} = c\,\dfrac{d}{dx}\{f(x)\}.$

(2) $\dfrac{d}{dx}\{f(x) + g(x)\} = \dfrac{d}{dx}\{f(x)\} + \dfrac{d}{dx}\{g(x)\}.$

(3) $\dfrac{d}{dx}\{f(x)g(x)\} = \dfrac{d}{dx}\{f(x)\}g(x) + f(x)\,\dfrac{d}{dx}\{g(x)\}.$

(4) $\dfrac{d}{dx}\left\{\dfrac{f(x)}{g(x)}\right\} = \dfrac{g(x)\,\dfrac{d}{dx}\{f(x)\} - f(x)\,\dfrac{d}{dx}\{g(x)\}}{g^2(x)},$

where $g(x) \neq 0$.

Rule 4.4.1(3) is referred to as the **product rule** for derivatives and rule 4.4.1(4) is referred to as the **quotient rule** for derivatives.

Each of these rules can be verified by using Definition 4.2.1. For 4.4.1(2), let

$$\phi(x) = f(x) + g(x)$$

and observe that

$$\frac{\phi(x + h) - \phi(x)}{h} = \frac{f(x + h) - f(x)}{h} + \frac{g(x + h) - g(x)}{h}.$$

However,

$$\lim_{h \to 0} \frac{\phi(x + h) - \phi(x)}{h} = \lim_{h \to 0} \frac{f(x + h) - f(x)}{h} + \lim_{h \to 0} \frac{g(x + h) - g(x)}{h};$$

that is,

$$\phi'(x) = f'(x) + g'(x).$$

For 4.4.1(3), let

$$\psi(x) = f(x)g(x).$$

Then

$$\frac{\psi(x + h) - \psi(x)}{h} = \frac{f(x + h)g(x + h) - f(x)g(x)}{h}.$$

By adding and subtracting $[f(x)g(x + h)]/h$ and then factoring, we obtain

$$\frac{\psi(x + h) - \psi(x)}{h} = \left[\frac{f(x + h) - f(x)}{h}\right] g(x + h)$$

$$+ f(x) \left[\frac{g(x + h) - g(x)}{h}\right].$$

Therefore

$$\lim_{h \to 0} \frac{\psi(x + h) - \psi(x)}{h} = \lim_{h \to 0} \left\{ \left[\frac{f(x + h) - f(x)}{h}\right] g(x + h) \right\}$$

$$+ \lim_{h \to 0} \left\{ f(x) \left[\frac{g(x + h) - g(x)}{h}\right] \right\}.$$

Since $\lim_{h \to 0} g(x + h) = g(x)$ by 4.3.2, it follows that

$$\psi'(x) = f'(x)g(x) + f(x)g'(x).$$

The other rules can be verified in a similar fashion.

EXAMPLE 1   Let $y = x^3 - 2x + 1$. What is dy/dx?

$$\frac{dy}{dx} = \frac{d}{dx} \{x^3 - 2x + 1\}$$

$$= \frac{d}{dx} \{x^3\} + \frac{d}{dx} \{-2x\} + \frac{d}{dx} \{1\} \qquad [4.4.1(2)]$$

$$= \frac{d}{dx} \{x^3\} - 2 \frac{d}{dx} \{x\} + \frac{d}{dx} \{1\} \qquad [4.4.1(1)]$$

$$= 3x^2 - 2.$$

EXAMPLE 2   If $y = 3x^{32} + 25x^2 - 12/\sqrt{x} + 1/x^5$, then determine dy/dx.

$$\frac{dy}{dx} = \frac{d}{dx} \{3x^{32} + 25x^2 - 12x^{-1/2} + x^{-5}\}$$

$$= \frac{d}{dx} \{3x^{32}\} + \frac{d}{dx} \{25x^2\} + \frac{d}{dx} \{-12x^{-1/2}\} + \frac{d}{dx} \{x^{-5}\}$$

$$= 3\frac{d}{dx} \{x^{32}\} + 25\frac{d}{dx} \{x^2\} - 12\frac{d}{dx} \{x^{-1/2}\} + \frac{d}{dx} \{x^{-5}\}$$

$$= 96x^{31} + 50x + 6x^{-3/2} - 5x^{-6}.$$

EXAMPLE 3   What is $\dfrac{d}{dx} \{(x + 2)(\sqrt{x} + 1)\}$?

Method 1: By multiplying, we obtain

$$(x + 2)(\sqrt{x} + 1) = x^{3/2} + 2x^{1/2} + x + 2.$$

Therefore

$$\frac{d}{dx} \{(x + 2)(\sqrt{x} + 1)\} = \frac{d}{dx} \{x^{3/2}\} + \frac{d}{dx} \{2x^{1/2}\} + \frac{d}{dx} \{x\} + \frac{d}{dx} \{2\}$$

$$= \frac{d}{dx} \{x^{3/2}\} + 2\frac{d}{dx} \{x^{1/2}\} + \frac{d}{dx} \{x\} + \frac{d}{dx} \{2\}$$

$$= \tfrac{3}{2}x^{1/2} + x^{-1/2} + 1.$$

Method 2: By using the product rule, we have

$$\frac{d}{dx} \{(x + 2)(\sqrt{x} + 1)\} = \frac{d}{dx} \{x + 2\}(\sqrt{x} + 1) + (x + 2)\frac{d}{dx} \{\sqrt{x} + 1\}$$

$$= (1)(\sqrt{x} + 1) + (x + 2)(\tfrac{1}{2}x^{-1/2})$$

$$= \tfrac{3}{2}x^{1/2} + x^{-1/2} + 1.$$

EXAMPLE 4   What is $\dfrac{d}{dx} \left\{\dfrac{2x + 3}{x^2 + 2}\right\}$?

By the quotient rule,

$$\frac{d}{dx} \left\{\frac{2x + 3}{x^2 + 2}\right\} = \frac{(x^2 + 2)\dfrac{d}{dx} \{2x + 3\} - (2x + 3)\dfrac{d}{dx} \{x^2 + 2\}}{(x^2 + 2)^2}$$

$$= \frac{(x^2 + 2)(2) - (2x + 3)(2x)}{(x^2 + 2)^2}$$

$$= \frac{-2x^2 - 6x + 4}{(x^2 + 2)^2}.$$

EXAMPLE 5 Let $f(x) = 3x^3 + 7/x^2 + 5\sqrt{x}$, and find f', f'', f'''.

Since $f(x) = 3x^3 + 7x^{-2} + 5x^{1/2}$,

$$f'(x) = 9x^2 - 14x^{-3} + \tfrac{5}{2}x^{-1/2}$$
$$f''(x) = 18x + 42x^{-4} - \tfrac{5}{4}x^{-3/2}$$
$$f'''(x) = 18 - 168x^{-5} + \tfrac{15}{8}x^{-5/2}.$$

EXERCISES

1. In each case, compute dy/dx.

(a) $y = 5x^3 - 6x^2 + 2x - 7$.   (b) $y = 9 - 3x^2 + 4x^5$.

(c) $y = x^3 + 2\sqrt{x} - 3/x^{2/3} - 2$, $x > 0$.

(d) $y = x - 1/x$, $x \neq 0$.   (e) $y = 2/\sqrt{x} - \sqrt[3]{x}$, $x > 0$.

(f) $y = (2x - 1)^2$.   (g) $y = (\sqrt{x} - 1/\sqrt{x})^2$, $x > 0$.

(h) $y = (2x + 5)(3x^2 + 2x - 4)$.   (i) $y = (x + 1/x)(\sqrt{x} - 1)$, $x > 0$.

(j) $y = \sqrt[4]{x}(3x^5 - 7)$, $x > 0$.   (k) $y = (7x - 1)/(x^2 + 4)$.

(l) $y = (3x^2 + 5x - 2)/(2x - 1)$, $x \neq \tfrac{1}{2}$.

(m) $y = \sqrt{x}/(x + 1)$, $x \neq -1$.

(n) $y = (1 + x^2)/(\sqrt{x} - 1)$, $x > 0$, $x \neq 1$.

2. In each part of Exercise 1, compute d^2y/dx^2.

3. At the specified point, find the equation of the tangent line and the equation of the normal line. Sketch the graph.

(a) $y = 2x^2 - 1$, $(2, 7)$.   (b) $y = x^2 + 6x - 2$, $(1, 5)$.

(c) $y = 7x - 3$, $(2, 11)$.   (d) $y = x^{2/3}$, $(-8, 4)$.

(e) $y = x^{2/3} + 1$, $(8, 5)$.   (f) $y = x^{2/3} + 1$, $(0, 1)$.

4. Use the differentiation rules to find formulas for the following.

(a) $\dfrac{d}{dx}\{[f(x)]^2\}$.   (b) $\dfrac{d}{dx}\left\{\dfrac{1}{g(x)}\right\}$, $g(x) \neq 0$.

(c) $\dfrac{d}{dx}\{c_1 f(x) + c_2 g(x) + c_3 h(x)\}$, c_1, c_2, c_3 constants.

(d) $\dfrac{d}{dx}\{f(x)g(x)h(x)\}$.

5. (a) If $\dfrac{d}{dx}\{f(x)\} = f(x)$, what are $\dfrac{d}{dx}\{(f(x))^2\}$ and $\dfrac{d^2}{dx^2}\{f(x)\}$?

(b) If $\dfrac{d}{dx}\{f(x)\} = -g(x)$ and $\dfrac{d}{dx}\{g(x)\} = f(x)$, what is $\dfrac{d}{dx}\{f(x)g(x)\}$?

6. Use Definition 4.2.1 to verify differentiation rule 4.4.1(1).

7. Use Definition 4.2.1 to verify differentiation rule 4.4.1(4). [*Hint:* Derive a formula for $\dfrac{d}{dx}\{1/g(x)\}$, $g(x) \neq 0$, and observe that $f(x)/g(x) = f(x) \cdot [1/g(x)]$, and then use 4.4.1(3).]

4.5 THE CHAIN RULE

In the preceding section we discussed rules which make it much easier to compute derivatives of certain functions which are expressed in terms of simpler functions. Using only the rules in Section 4.4, it would be difficult to compute $f'(x)$ if

$$f(x) = (x^2 + 1)^{74},$$

for we would have to expand $(x^2 + 1)^{74}$ as the sum of 75 terms and then differentiate each term. This is obviously not a practical method. We now want to discuss a rule which will make the computation of $f'(x)$ much easier.

Let us consider the following question: Suppose that f has a derivative at x

$$f'(x)$$

and g has a derivative at $f(x)$

$$g'(f(x)).$$

If

$$\phi(x) = g(f(x)),$$

then what is

$$\phi'(x)?$$

The answer is, in fact, the product of the above-mentioned derivatives; that is,

$$\phi'(x) = g'(f(x))f'(x).$$

4.5.1

**Theorem
(The Chain Rule)**

Let $\phi(x) = g(f(x))$.

If f has a derivative at x and g has a derivative at $f(x)$, then ϕ has a derivative at x and, in fact,

$$\phi'(x) = g'(f(x))f'(x).$$

A proof of this theorem is given in the optional supplement following this section.

Once more, the appropriateness of Leibniz's fractional notation for the derivative is evident, for if one lets $y = f(x)$, then the conclusion of the chain rule may be written as

$$\frac{d}{dx}\{g(y)\} = \frac{d}{dy}\{g(y)\}\frac{dy}{dx},$$

or, more briefly,

$$\frac{dg}{dx} = \frac{dg}{dy}\frac{dy}{dx}.$$

If we apply the chain rule to the particular case

$$g(y) = y^r \qquad (r \text{ rational}),$$

then

$$\phi(x) = [f(x)]^r.$$

Since $g'(y) = ry^{r-1}$ whenever y^{r-1} is defined, then

$$\phi'(x) = r[f(x)]^{r-1}f'(x)$$

whenever $[f(x)]^{r-1}$ is defined. Stated concisely, we have the following.

4.5.2
Fact

$$\frac{d}{dx}\{[f(x)]^r\} = r[f(x)]^{r-1}\frac{d}{dx}\{f(x)\}.$$

If we let $y = f(x)$, then this result can be written as

$$\frac{d}{dx}\{y^r\} = \frac{d}{dy}\{y^r\}\frac{dy}{dx}$$

$$= ry^{r-1}\frac{dy}{dx}.$$

Let us indicate the validity of 4.5.2 in some special cases.
  If $r = 2$, then

$$\frac{d}{dx}\{[f(x)]^2\} = \frac{d}{dx}\{f(x)f(x)\}$$

$$= \frac{d}{dx}\{f(x)\}f(x) + f(x)\frac{d}{dx}\{f(x)\}$$

$$= 2f(x)\frac{d}{dx}\{f(x)\}.$$

For $r = 3$, we have

$$\frac{d}{dx}\{[f(x)]^3\} = \frac{d}{dx}\{[f(x)]^2 f(x)\}$$

$$= \frac{d}{dx}\{[f(x)]^2\}f(x) + [f(x)]^2\frac{d}{dx}\{f(x)\}$$

$$= 2f(x)\frac{d}{dx}\{f(x)\}f(x) + [f(x)]^2\frac{d}{dx}\{f(x)\}$$

$$= 3[f(x)]^2\frac{d}{dx}\{f(x)\}.$$

Finally, if $r = 4$, then

$$\frac{d}{dx}\{[f(x)]^4\} = \frac{d}{dx}\{[f(x)]^3 f(x)\}$$

$$= \frac{d}{dx}\{[f(x)]^3\}f(x) + [f(x)]^3 \frac{d}{dx}\{f(x)\}$$

$$= 3[f(x)]^2 \frac{d}{dx}\{f(x)\}f(x) + [f(x)]^3 \frac{d}{dx}\{f(x)\}$$

$$= 4[f(x)]^3 \frac{d}{dx}\{f(x)\}.$$

EXAMPLE 1   If $y = (4x^5 + 2x + 3)^2$, find dy/dx.

Method 1: If we square $(4x^5 + 2x + 3)$, we obtain

$$(4x^5 + 2x + 3)^2 = 16x^{10} + 16x^6 + 24x^5 + 4x^2 + 12x + 9.$$

Therefore

$$\frac{d}{dx}\{(4x^5 + 2x + 3)^2\} = 160x^9 + 96x^5 + 120x^4 + 8x + 12.$$

Method 2: Using 4.5.2, we have

$$\frac{d}{dx}\{(4x^5 + 2x + 3)^2\} = 2(4x^5 + 2x + 3)^1 \frac{d}{dx}\{4x^5 + 2x + 3\}$$

$$= 2(4x^5 + 2x + 3)(20x^4 + 2).$$

EXAMPLE 2   Let $\phi(x) = \sqrt{x^2 + 1}$; what is $\phi'(x)$?

We first note that $f(x) = x^2 + 1$ is always positive, and so $\phi(x)$ is defined for all x. Then,

$$\frac{d}{dx}\{\sqrt{x^2 + 1}\} = \frac{d}{dx}\{(x^2 + 1)^{1/2}\}$$

$$= \frac{1}{2}(x^2 + 1)^{-1/2} \frac{d}{dx}\{x^2 + 1\}$$

$$= \frac{1}{2}(x^2 + 1)^{-1/2}(2x)$$

$$= \frac{x}{\sqrt{x^2 + 1}}.$$

EXAMPLE 3   What is $\dfrac{d}{dx}\{\sqrt{(2x+1)^2+3)}\}$?

Using 4.5.2, we have

$\dfrac{d}{dx}\{\sqrt{(2x+1)^2+3}\}$

$$= \frac{1}{2}[(2x+1)^2+3]^{-1/2}\frac{d}{dx}\{(2x+1)^2+3\}$$

$$= \frac{1}{2}[(2x+1)^2+3]^{-1/2}\left[\frac{d}{dx}\{(2x+1)^2\}+\frac{d}{dx}\{3\}\right]$$

$$= \frac{1}{2}[(2x+1)^2+3]^{-1/2}\left[2(2x+1)^1\frac{d}{dx}\{2x+1\}\right]$$

$$= \frac{1}{2}[(2x+1)^2+3]^{-1/2}[2(2x+1)\cdot 2]$$

$$= \frac{2(2x+1)}{\sqrt{(2x+1)^2+3}}\,.$$

EXAMPLE 4   What is the equation of the tangent line to the circle $x^2+y^2=9$ at $(3/\sqrt{2},\ -3/\sqrt{2})$?

Previously, we have discussed only the equation of the tangent line to the curve $y=f(x)$ at $(x_0,\ y_0)$, where the function f is differentiable at point x_0. We can determine a function of interest in the given problem by solving the given equation for y,

$$y = \pm\sqrt{9-x^2},$$

and observing that the point $(3/\sqrt{2},\ -3/\sqrt{2})$ lies on the lower semicircle

$$y = -\sqrt{9-x^2}.$$

Therefore we seek the equation of the tangent line to the curve

$$y = f(x) = -\sqrt{9-x^2}$$

at the point $(3/\sqrt{2},\ -3/\sqrt{2})$. Now

$$f'(x) = -\tfrac{1}{2}(9-x^2)^{-1/2}(-2x)$$

$$= \frac{x}{\sqrt{9-x^2}}$$

and

$$f'\left(\frac{3}{\sqrt{2}}\right) = \frac{3/\sqrt{2}}{\sqrt{9-\frac{9}{2}}} = 1.$$

Therefore the desired tangent line has the equation

$$y - \left(\frac{-3}{\sqrt{2}}\right) = 1\left(x - \frac{3}{\sqrt{2}}\right)$$

or

$$y + \frac{3}{\sqrt{2}} = x - \frac{3}{\sqrt{2}}.$$

Let us reflect a moment on Example 4. Since the key to the problem is to find the slope of the desired tangent line, a question arises immediately. Namely, is it possible to determine $f'(3/\sqrt{2})$ without knowing f explicitly as a function of x? The answer is yes, for using the given equation

$$x^2 + y^2 = 9,$$

we assume that there is a differentiable function f such that

$$f\left(\frac{3}{\sqrt{2}}\right) = -\frac{3}{\sqrt{2}}$$

and

(4.5.3) $x^2 + [f(x)]^2 = 9.$

In this case we say that function f **is defined implicitly** by the equation

$$x^2 + y^2 = 9.$$

The graph of f is then a portion of the graph of $x^2 + y^2 = 9$, and it contains the point $(3/\sqrt{2}, -3/\sqrt{2})$.

If we differentiate both sides of Equation (4.5.3), using Fact 4.5.2 on the term $[f(x)]^2$, we obtain

$$2x + 2f(x)f'(x) = 0.$$

Therefore

$$f'(x) = \frac{-x}{f(x)},$$

where $f(x) \neq 0$, and

$$f'(3/\sqrt{2}) = \frac{-3/\sqrt{2}}{f(3/\sqrt{2})} = \frac{-3/\sqrt{2}}{-3/\sqrt{2}} = 1.$$

This procedure can be generalized. Suppose that f is a differentiable function defined implicitly by an equation involving x and y* [that is, $y = f(x)$ satisfies the equation]. Replace y by $f(x)$ in the given equation and differentiate both sides of the equation; then solve algebraically for $f'(x)$. This process is referred to as **implicit differentiation**.

---

* There is a question of whether there exists a function defined by a given equation. For example, there is no function which satisfies $x^2 + y^2 = -1$. It is possible (although we shall not do it here) to give conditions when a differentiable function is defined implicitly by some equation.

EXAMPLE 5   Assume that there is a differentiable function f defined by
$$x^3 + x^2y^5 + 2y = 12.$$
What is $f'(x)$?

Let $y = f(x)$; then
$$x^3 + x^2[f(x)]^5 + 2f(x) = 12.$$
Differentiating and using 4.5.2, we get
$$3x^2 + 2x[f(x)]^5 + x^2(5[f(x)]^4f'(x)) + 2f'(x) = 0$$
or
$$f'(x)[5x^2[f(x)]^4 + 2] = -3x^2 - 2x[f(x)]^5$$
or
$$f'(x) = \frac{-3x^2 - 2x[f(x)]^5}{5x^2[f(x)]^4 + 2}.$$

In the process of implicit differentiation one may wish to save some writing by not replacing y by $f(x)$—but always remembering that y is a function of x and that dy/dx should be written in place of $f'(x)$.

EXAMPLE 6   In Example 5, we had
$$x^3 + x^2y^5 + 2y = 12.$$
Differentiating both sides yields
$$3x^2 + 2xy^5 + x^2\left(5y^4\frac{dy}{dx}\right) + 2\frac{dy}{dx} = 0.$$
Solving for dy/dx, we have
$$\frac{dy}{dx} = \frac{-3x^2 - 2xy^5}{5x^2y^4 + 2}.$$

EXERCISES

1. Let $\phi(x) = (1 - 5x^2)^3$.

   (a) Compute $\phi'(x)$ by expanding $(1 - 5x^2)^3$ and using the rules of Section 4.4.

   (b) Compute $\phi'(x)$ by using the chain rule.

2. In each of the following, determine $\phi'(x)$.

   (a) $\phi(x) = (4x^2 - 2)^4$.   (b) $\phi(x) = (2 - x + x^2)^3$.

   (c) $\phi(x) = (3x^2 + 1)^{20}$.   (d) $\phi(x) = (6x - 5)^{-5}$, $x \neq \frac{5}{6}$.

   (e) $\phi(x) = (x - 1/x)^3$, $x \neq 0$.

   (f) $\phi(x) = [(3x^2 + 1)/(x + 1)]^2$, $x \neq -1$.

   (g) $\phi(x) = \sqrt{x^2 + 5}$.   (h) $\phi(x) = (1 - x)\sqrt{3x^2 + 1}$.

   (i) $\phi(x) = (2x^2 + 7)^{1/4}$.   (j) $\phi(x) = 1/\sqrt{x^2 + 1}$.

   (k) $\phi(x) = [(3x - 2)^5 + 8]^2$.   (l) $\phi(x) = \sqrt[3]{x + (5 - x)^4}$.

3. For each part of Exercise 2, determine $\phi''(x)$.

4. Let $f(x) = 5x + 1$ and $g(x) = (x - 3)^4$.

(a) Compute $\phi(x) = f(g(x))$ and $\psi(x) = g(f(x))$.

(b) Find $\phi'(x)$ and $\psi'(x)$.

(c) Find $\phi'(2)$ and $\psi'(1)$.

5. Assume that each of the following equations defines y as a function of x. Find dy/dx wherever it exists.

(a) $x^2 + y^2 = 25$.   (b) $x^3 - y^3 = 1$.

(c) $x^3 + 2xy - y^2 - 6 = 0$.   (d) $x^4 = 13 - xy - y^5$.

(e) $y^6 = 4x^9 y + x$.

6. At the specified point, find the equation of the tangent and normal lines to the given curve.

(a) $y^2 = x$, $(4, -2)$.   (b) $x^2 y^4 - 2xy + 1 = 0$, $(1, -1)$.

(c) $x^2 + y^2 - 2y = 0$, $(0, 2)$.

7. Find $\phi'(x)$ in terms of $f(x)$ and $f'(x)$ in each of the following.

(a) $\phi(x) = f(x + 7)$.   (b) $\phi(x) = f(2x)$.

(c) $\phi(x) = \sqrt{f(x)}$.   (d) $\phi(x) = f(x)/\sqrt{1 + f(x)}$.

(e) $\phi(x) = [f(-2x)]^3$.   (f) $\phi(x) = \sqrt{f(5x)}$.

8. (a) Let $f'(x) = f(x)$. Show that for any positive integer n, $\dfrac{d}{dx}\{[f(x)]^n\} = n[f(x)^n]$.

(b) Let $f'(x) = g(x)$ and $g'(x) = -f(x)$. Show that $\dfrac{d}{dx}\{f^2(x) + g^2(x)\} = 0$.

Supplement (optional)

Our purpose here is to prove the chain rule (4.5.1).

Note first that if h is differentiable at t_0, then

$$\frac{h(t) - h(t_0)}{t - t_0} - h'(t_0) = \varepsilon(t),$$

where $\varepsilon(t) \to 0$ as $t \to t_0$. In other words,

$$h(t) - h(t_0) = (t - t_0)h'(t) + \varepsilon(t)(t - t_0),$$

where $\varepsilon(t) \to 0$ as $t \to t_0$.

Now let

$$\phi(x) = g(f(x)).$$

We wish to prove that if f has a derivative at x_0 and g has a derivative at $f(x_0)$, then ϕ has a derivative at x_0 and

$$\phi'(x_0) = g'(f(x_0))f'(x_0).$$

Since f has a derivative at x_0,

(4.5.4) $f(x) - f(x_0) = (x - x_0)[f'(x_0) + \varepsilon_1(x)]$,

where $\varepsilon_1(x) \to 0$ as $x \to x_0$.

If g has a derivative at y_0, then

(4.5.5) $g(y) - g(y_0) = (y - y_0)[g'(y_0) + \varepsilon_2(y)]$,

where $\varepsilon_2(y) \to 0$ as $y \to y_0$.

In Equation 4.5.5 let $y_0 = f(x_0)$ and $y = f(x)$. Then

$$g(f(x)) - g(f(x_0)) = [f(x) - f(x_0)][g'(f(x_0)) + \varepsilon_2(y)];$$

that is,

(4.5.6) $\phi(x) - \phi(x_0) = [f(x) - f(x_0)][g'(f(x_0)) + \varepsilon_2(y)]$.

Substituting Equation 4.5.4 into the right-hand side of 4.5.6, we obtain

$$\phi(x) - \phi(x_0) = (x - x_0)[f'(x_0) + \varepsilon_1(x)][g'(f(x_0)) + \varepsilon_2(y)]$$

or

$$\frac{\phi(x) - \phi(x_0)}{x - x_0} = [f'(x_0) + \varepsilon_1(x)][g'(f(x_0)) + \varepsilon_2(y)].$$

Recall that as $x \to x_0$, $\varepsilon_1(x) \to 0$. Moreover, as $x \to x_0$, $y \to y_0$ (since f is continuous at x_0) and hence $\varepsilon_2(y) \to 0$. Therefore

$$\lim_{x \to x_0} \frac{\phi(x) - \phi(x_0)}{x - x_0} = \lim_{x \to x_0} [f'(x_0) + \varepsilon_1(x)][g'(f(x_0)) + \varepsilon_2(y)];$$

that is,

$$\phi'(x_0) = f'(x_0)g'(f(x_0)).$$

4.6 MAXIMA AND MINIMA

In this section we shall illustrate a use of the derivative in solving the following basic problem:

Given a continuous function on $[a, b]$, find the maximum and minimum values of f on $[a, b]$.

We recall that $f(c)$ is **the maximum value of f on $[a, b]$** if

$$f(x) \le f(c)$$

for every x in $[a, b]$, and that $f(d)$ is **the minimum value of f on $[a, b]$** if

$$f(x) \ge f(d)$$

for every x in $[a, b]$. Recall also that for a continuous function on $[a, b]$ these maximum and minimum values of f always exist [3.3.4(I)].

Before discussing a method for finding these values, we want to discuss a related concept, namely that of a local maximum and a local minimum. A local maximum or local minimum is sometimes referred to as a **local extreme value**.

4.6.1
Definition

Let f be defined in an open interval containing x_0.
   1. f is said to have a **local maximum at x_0** if

$$f(x) \leq f(x_0)$$

   for all x in some open interval containing x_0.
   2. f is said to have a **local minimum at x_0** if

$$f(x) \geq f(x_0)$$

   for all x in some open interval containing x_0.

In Figure 4.12, f has local maxima at points x_1 and x_2, and g has local minima at points x_3 and x_4. We want to show that the graphs in Figure 4.12 are typical; that is, at a local extreme value, either there is no (nonvertical) tangent line, or, if there is, it is horizontal.

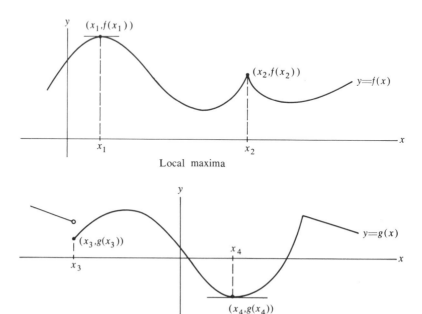

Figure 4.12

4.6.2

Theorem

If f has a local maximum (or local minimum) at x_0 and if f is differentiable x_0, then

$$f'(x_0) = 0.$$

Proof: We shall consider the case of a local maximum. Since f has a local maximum at x_0, then

$$f(x) \leq f(x_0)$$

in some open interval containing x_0. Therefore

$$f(x_0 + h) \leq f(x_0)$$

for h in some open interval I containing zero; that is,

$$f(x_0 + h) - f(x_0) \leq 0$$

for h in I.

Now if h is in I and $h > 0$, then

$$\frac{f(x_0 + h) - f(x_0)}{h} \leq 0;$$

hence by Fact 3.2.4

$$\lim_{h \downarrow 0} \frac{f(x_0 + h) - f(x_0)}{h} \leq 0.$$

On the other hand, if h is in I and $h < 0$, then

$$\frac{f(x_0 + h) - f(x_0)}{h} \geq 0;$$

so again by Fact 3.2.4,

$$\lim_{h \uparrow 0} \frac{f(x_0 + h) - f(x_0)}{h} \geq 0.$$

By hypothesis, f is differentiable at x_0, and so the left-hand and right-hand derivatives at x_0 are both equal to $f'(x_0)$. Therefore we have

$$f'(x_0) \leq 0 \quad \text{and} \quad f'(x_0) \geq 0,$$

and so

$$f'(x_0) = 0.$$

The importance of Theorem 4.6.2 lies in the fact that it gives a way of determining those points at which f might have a local extreme value. *The theorem says that local extreme values occur either at points where f is not differentiable or at points where $f'(x) = 0$.*

EXAMPLE 1   Let $f(x) = -x^2$; then $f'(x) = -2x$ and $f'(0) = 0$. Therefore, $f(0)$ is a possible local extreme value. In fact, f has a local maximum at zero, as its graph in Figure 4.13 indicates.

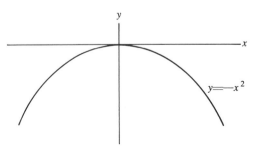

Figure 4.13

EXAMPLE 2   Let $f(x) = |x|$. We have already seen that f does not have a derivative at zero. Hence, $f(0)$ is a possible local extreme value; in fact, f has a local minimum at zero (see Figure 4.14).

EXAMPLE 3   Let $f(x) = x^3$; then $f'(x) = 3x^2$ and $f'(0) = 0$. Therefore, $f(0)$ is a possible extreme value, but by observing its graph in Figure 4.15, we see that f has neither a local maximum nor local minimum at $x = 0$.

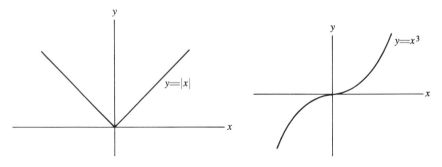

Figure 4.14   Figure 4.15

In Section 4.8 we shall state two tests for determining whether f has a local extreme value at a point without a knowledge of the graph of the function.

At this point, we want to turn our attention to solving the problem presented at the beginning of the section. Namely, we have a function continuous on $[a, b]$ and we seek the maximum and minimum values of f on $[a, b]$.

What are the candidates for the maximum and minimum values? If

$$a < x_0 < b$$

and $f(x_0)$ is a maximum or minimum value, it then follows that $f(x_0)$ is a local extreme value. In this case, either

$$f \text{ is not differentiable at } x_0$$

or

$$f'(x_0) = 0.$$

In addition, the values at the end points of interval $[a, b]$, namely

$$f(a) \quad \text{and} \quad f(b),$$

are potential maximum or minimum values.

Therefore our method for finding the maximum and minimum values can be stated as follows: Check the values of f at those points x ($a < x < b$) where f is not differentiable or where $f'(x) = 0$; then check the values of f at points a and b (the end points). For example, say that f fails to have a derivative at

$$x_1 \quad \text{and} \quad x_2$$

and that $f'(x) = 0$ at

$$x_3, \quad x_4, \quad \text{and} \quad x_5.$$

Thus the numbers we are considering are

$$f(a), \quad f(b), \quad f(x_1), \quad f(x_2), \quad f(x_3), \quad f(x_4), \quad \text{and} \quad f(x_5).$$

These are, in fact, then *only* candidates for maximum and minimum values of f. Hence the largest of these numbers is the maximum value of f and the smallest is the minimum value of f.

EXAMPLE 4   Find the maximum and minimum values of $f(x) = \frac{5}{2}x^{4/5} - x$ on $[-1, 243]$.
Since

$$f'(x) = 2x^{-1/5} - 1$$
$$= x^{-1/5}[2 - x^{1/5}],$$

then $f'(32) = 0$ and f' does not exist at $x = 0$. Checking the functional values at $-1, 0, 32,$ and 243,

x	-1	0	32	243
$f(x)$	$\frac{7}{2}$	0	8	$-\frac{81}{2}$

we see that the maximum value is 8 and that the minimum value is $-\frac{81}{2}$.

EXAMPLE 5   Find the maximum and minimum values of $f(x) = 3x^4 - 4x^3 - 12x^2 + 2$ on $[-2, 3]$.
Note that for each x,

$$f'(x) = 12x^3 - 12x^2 - 24x$$
$$= 12x(x^2 - x - 2)$$
$$= 12x(x + 1)(x - 2).$$

Therefore, $f'(x) = 0$ if $x = 0, -1, 2$. All three values lie in $[-2, 3]$. We now list the candidates for maximum and minimum values.

x	-2	-1	0	2	3
$f(x)$	44	7	12	-20	39

Therefore the maximum value is 44 and the minimum value is -20.

EXAMPLE 6 (with 1940 prices) An amusement park operator over a long period of time has observed that 400 people will attend daily if the admission is 30 cents and that the attendance will decrease by 20 for each nickel added to the admission charge. What admission price will yield the greatest gross receipts?

Let x be the number of nickels added to 30 cents. Then the admission price is
$$30 + 5x.$$
The attendance is
$$400 - 20x.$$
Since the gross receipts equal the admission price times the attendance, we consider the function
$$G(x) = (30 + 5x)(400 - 20x)$$
$$= 12{,}000 + 1400x - 100x^2.$$

Furthermore, since adding $1 (20 nickels) to the 30-cent price will make the attendance zero (and hence gross receipts zero), we seek the value of x in $[0, 20]$ for which $G(x)$ is a maximum. Therefore we proceed as usual. For any x,
$$G'(x) = 1400 - 200x$$
and
$$G'(x) = 0 \quad \text{only when } x = 7.$$

The only candidates for our table are 0, 7, 20.

x	0	7	20
$G(x)$	12,000	16,900	0

Hence an admission charge of 65 cents will yield the greatest gross receipts (namely, $169). The graph of the gross receipt function is given in Figure 4.16.

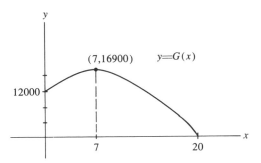

Figure 4.16

EXAMPLE 7   Find the rectangle of maximum area that can be inscribed in a semicircular region.

Because of the symmetry of a semicircular region, we shall inscribe the rectangle symmetrically (see Figure 4.17).

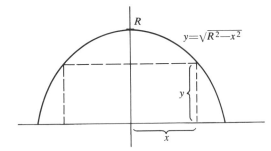

Figure 4.17

The area of the rectangle is $2xy$. Since the rectangle is inscribed in a semicircular arc $y = \sqrt{R^2 - x^2}$, we have the area

$$A(x) = 2x\sqrt{R^2 - x^2} \qquad (0 \le x \le R).$$

We seek the value of x in $[0, R]$ for which $A(x)$ is a maximum. Now

$$A'(x) = 2(R^2 - x^2)^{1/2} + 2x(\tfrac{1}{2}(R^2 - x^2)^{-1/2})(-2x)$$

$$= 2(R^2 - x^2)^{-1/2}[(R^2 - x^2) - x^2]$$

$$= \frac{2(R^2 - 2x^2)}{\sqrt{R^2 - x^2}}.$$

Thus, $A'(x) = 0$ if

$$R^2 - 2x^2 = 0$$

or

$$x = \frac{R}{\sqrt{2}}.$$

Noting that $A(0) = 0$ and $A(R) = 0$, we see that $A(R/\sqrt{2})$ is the maximum value. Therefore the dimensions of the desired rectangle are

$$\frac{2R}{\sqrt{2}} \quad \text{by} \quad \frac{R}{\sqrt{2}}.$$

EXERCISES

1. In each of the following, find the maximum and minimum values of the given continuous function f on the given closed interval.

   (a) $f(x) = 3x - 7$, $[-5, 5]$.   (b) $f(x) = x^2 + 5$, $[-2, 3]$.

   (c) $f(x) = 9 - x^2$, $[-1, 1]$.   (d) $f(x) = 3x^2 + 3x + 2$, $[1, 4]$.

   (e) $f(x) = 4x^3 + 2x^2 - x + 3$, $[-2, 2]$.

   (f) $f(x) = \sqrt{x - 4}$, $[4, 13]$.   (g) $f(x) = 10 - x^{2/3}$, $[-1, 1]$.

   (h) $f(x) = |2 - x|$, $[0, 3]$.

   (i) $f(x) = x^4 + \frac{2}{3}x^3 - 3x^2 + 1$, $[-2, 2]$.

   (j) $f(x) = x/(1 + x^2)$, $[-2, 3]$.

2. Find the maximum value of $f(x) = 1/(x^2 - 4x + 13)$ on $[-2, 2]$.

3. Find the two numbers whose sum is 24 and whose product is as large as possible.

4. Find the two numbers whose product is 24 and whose sum is as large as possible.

5. Determine the shortest distance from the point $(1, 0)$ to the graph of $y = \sqrt{x}$.

6. Determine the point on the graph of $y = x^2$ which is nearest the point $(3, 0)$.

7. Find the rectangle of maximum perimeter that can be inscribed in a semicircular region.

8. Find the rectangle of maximum area which can be inscribed in the region shaded in Figure 4.18.

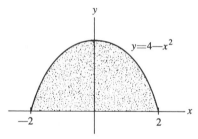

Figure 4.18

9. Find the right circular cylinder of greatest volume that can be inscribed in a right circular cone of radius r and height h.

10. Of all the right circular cylinders with a given volume V, determine the dimensions of the one with minimal surface area.

11. Of all the triangles with a given base and a given perimeter, show that the one with the largest area is an isosceles triangle.

12. Of all the rectangles with a given perimeter, show that the one with the largest area is a square.

13. One hundred yards of fence are to be used to bound a child's rectangular playground. Find the dimensions which will give the maximum playing area.

14. One hundred yards of fence are to be used to bound three sides of a child's rectangular playground; the fourth side is a wall. Find the dimensions which will give the maximum playing area.

15. An aquarium (no top) with a square base and a capacity of 4 cubic feet is to be built. The glass for the bottom costs $2 per square foot and the glass for the sides costs $1 per square foot. Find the dimensions of the aquarium so that the cost will be the least.

16. A piece of wire 1 yard long is to be cut into two pieces. If one piece is to be bent into the shape of a square and the second piece into the shape of a circle, determine where the cut should be made so that the sum of the areas is greatest.

17. A merchant has determined how the price p of his product relates to demand d, and how the demand relates to manufacturing and selling costs c. His findings are $d = 75 - \frac{3}{5}p$ and $c = d^2 + 15d + 500$. Determine the price of the product which will minimize selling costs, and the price of the product which will give the maximum profit. [*Hint:* Profit = (demand × price) − costs.]

18. A certain fraternal organization has planned a special vacation trip for its members, and it has arranged with an airline for a reduced group rate. If 300 or fewer members go, the round-trip plane ticket will cost $250 per person. If more than 300 members go, then the ticket price for everybody will be reduced by 50 cents for each person over 300. Find the number of passengers which will give the airline maximum gross receipts.

4.7 TWO CLASSICAL THEOREMS

Previously, it was noted that

$$f(x) = c \text{ on } I \Rightarrow f'(x) = 0 \text{ on } I$$

and that

$$f(x) - g(x) = c \text{ on } I \Rightarrow f'(x) = g'(x) \text{ on } I,$$

where I is an interval and c is a constant.

One objective of this section is to indicate that the converses of these results are also true; that is,

$$f'(x) = 0 \text{ on } I \Rightarrow f(x) = c \text{ on } I$$

and

$$f'(x) = g'(x) \text{ on } I \Rightarrow f(x) - g(x) = c \text{ on } I.$$

To begin, suppose that the curve $y = f(x)$ is unbroken on $[a, b]$ and smooth on (a, b) and that $f(a) = f(b)$. If f is not a constant on $[a, b]$, then f has values different from $f(a)$. Consider the case in which f attains values greater than $f(a)$. Then the graph of $y = f(x)$ begins at $(a, f(a))$, has points above the line $y = f(a)$, and has ends at $(b, f(b))$. Since the graph of f is unbroken and $f(a) = f(b)$, then f attains its maximum value at a point x_0 in the interval (a, b). At that point f has a local maximum value (see Figure 4.19). The smoothness of f on (a, b) ensures that a tangent line can be drawn at each point. Therefore at the extreme point the tangent line is horizontal; that is, the derivative at x_0 is zero.

We have thus justified the following statement, a complete proof of which is given in the optional supplement at the end of this section.

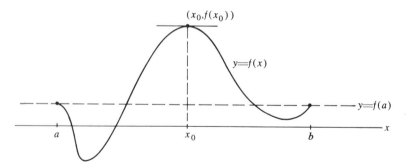

Figure 4.19

4.7.1
Rolle's Theorem

If f is continuous on $[a, b]$ and differentiable on (a, b) and $f(a) = f(b)$, then there exists a point x_0 in (a, b) such that

$$f'(x_0) = 0.$$

EXAMPLE 1   Let $f(x) = x^2 - 3x - 4$ on $[-1, 4]$.

Since f is continuous on $[-1, 4]$, differentiable on $(-1, 4)$, and

$$f(-1) = f(4) = 0,$$

we can conclude from Rolle's theorem that

$$f'(x_0) = 0,$$

where $-1 < x_0 < 4$. In fact,

$$f'(x) = 2x - 3,$$

and so

$$f'(\tfrac{3}{2}) = 0.$$

Here, $x = \tfrac{3}{2}$ is the only point at which f' is zero (see Figure 4.20).

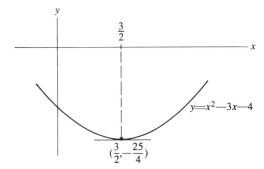

Figure 4.20

Consider now a curve $y = f(x)$ which is unbroken on $[a, b]$ and smooth on (a, b). The line joining the two points $(a, f(a))$ and $(b, f(b))$ has slope

$$\frac{f(b) - f(a)}{b - a}.$$

The smoothness of the curve ensures that among all the tangent lines to the curve $y = f(x)$, there is at least one whose slope is the same as $[f(b) - f(a)]/(b - a)$. That is, there is a tangent line which is parallel to the line joining $(a, f(a))$ and $(b, f(b))$ (see Figure 4.21).

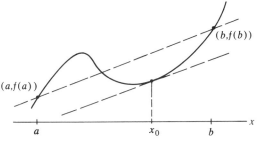

Figure 4.21

We express this fact formally as follows.

4.7.2
Mean Value
Theorem

If f is continuous on $[a, b]$ and differentiable on (a, b), then there exists a point x_0 in (a, b) such that

$$\frac{f(b) - f(a)}{b - a} = f'(x_0).$$

It is interesting to note that if the graph of f in Figure 4.21 were rotated so that the dotted lines were horizontal, we would have the graph of a function which satisfies the hypothesis of Rolle's Theorem. This idea is the heart of the proof of the Mean Value Theorem; this proof is also given in the supplement to this section.

EXAMPLE 2

Let $f(x) = x^2 + x$ on $[-1, 1]$. Find a point x_0 as given by the Mean Value Theorem. Note that

$$\frac{f(1) - f(-1)}{(1) - (-1)} = \frac{2 - 0}{2} = 1$$

and that

$$f'(x) = 2x + 1.$$

Thus

$$2x + 1 = 1 \Leftrightarrow x = 0;$$

our point is

$$x_0 = 0 \quad \text{(see Figure 4.22)}.$$

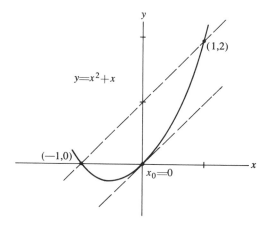

Figure 4.22

The Mean Value Theorem allows us to prove certain properties about a function on an interval knowing something about the derivative of the function at each point of the interval.

4.7.3
Consequence

If $f'(x) = 0$ at each point x in the interval I, then

$$f(x) = c,$$

where c is a fixed constant.

Proof: We first observe that $f(x) = c$ on I means that f has the same value for any two points x_1 and x_2 in I.

Therefore let $x_1 < x_2$ and let us apply the Mean Value Theorem to f on $[x_1, x_2]$. Then

$$\frac{f(x_2) - f(x_1)}{x_2 - x_1} = f'(x_0),$$

where $x_1 < x_0 < x_2$. Since $f'(x_0) = 0$, it follows that

$$\frac{f(x_2) - f(x_1)}{x_2 - x_1} = 0;$$

that is,

$$f(x_1) = f(x_2).$$

4.7.4
Consequence

If $f'(x) = g'(x)$ at each point x in the interval I, then

$$f(x) = g(x) + c,$$

where c is a fixed constant.

Proof: Let $\phi(x) = f(x) - g(x)$; then

$$\phi'(x) = f'(x) - g'(x) = 0.$$

By 4.7.3,

$$f(x) - g(x) = c;$$

that is,

$$f(x) = g(x) + c.$$

EXAMPLE 3　　Find a function f given that $f'(x) = 4x^3 + 2x$ and $f(0) = 2$.
It is easy to see that

$$\frac{d}{dx}(x^4 + x^2) = 4x^3 + 2x.$$

Therefore, by 4.7.4,

$$f(x) = (x^4 + x^2) + c.$$

Since $f(0) = 2$, then $2 = c$. Therefore,

$$f(x) = x^4 + x^2 + 2.$$

EXERCISES

1. For each of the following functions f, find a number x_0 in the given interval, as guaranteed by the Mean Value Theorem.

   (a) $f(x) = x^2$, $[2, 3]$. 　　　　　(b) $f(x) = 2x^2 - x$, $[1, 2]$.
   (c) $f(x) = \sqrt{x}$, $[1, 4]$. 　　　　　(d) $f(x) = x^3$, $[-2, 3]$.

2. For each of the following functions f, verify that the hypotheses of Rolle's Theorem are satisfied on the given interval. Then find a number x_0 in this interval as guaranteed by the theorem.

   (a) $f(x) = x^2 + x$, $[-\frac{3}{4}, -\frac{1}{4}]$. 　　　(b) $f(x) = \sqrt{9 - x^2}$, $[-3, 3]$.
   (c) $f(x) = \frac{1}{3}x^3 - x$, $[0, 3]$.

3. Let $f(x) = 1 - x^{2/3}$. Therefore, $f(-1) = 0 = f(1)$. Can you conclude that $f'(x_0) = 0$ for some point x_0 in $[-1, 1]$?

4. Consider the function h defined on $[-2, 2]$ by

$$h(x) = \begin{cases} -1, & -2 \le x \le 0, \\ 1, & 0 < x \le 2. \end{cases}$$

   Does h contradict 4.7.2? If not, why not?

5. Suppose that f is differentiable on the interval I, that x_1 and x_2 are points of I, and that x_1 and x_2 are roots of the equation $f'(x) = 0$. Verify that if $f'(x) = 0$ has no roots in (x_1, x_2), then $f(x) = 0$ has at most one root in $[x_1, x_2]$.

6. Use Exercise 5 to do each of the following.

   (a) Determine intervals in which there exist at most one root of $f(x) = 0$ if $f'(x) = x^3 - 3x^2 + 2x$.

   (b) Show that there is at most one root of $x^4/4 - 2x^2 + 1 = 0$ in $[0, 1]$.

   (c) Show that zero is the only root of $x^3 + x^2 - 5x = 0$ in $[-1, 1]$.

7. Suppose that f and g are continuous on $[a, b]$ and differentiable on (a, b) and that $f(a) = g(a)$, $f(b) = g(b)$. Show that there exists an x_0 in (a, b) such that $f'(x_0) = g'(x_0)$ and interpret this result geometrically.

8. Suppose that $f(0) = 0$ and $f'(x) \le 1$ for all $x > 0$. Show that $f(x) \le x$ for $x > 0$. [*Hint:* $[f(x) - f(0)]/(x - 0) = f'(z), 0 < z < x$.]

Supplement (optional)

Our purpose here is to prove Rolle's Theorem (4.7.1) and the Mean Value Theorem (4.7.2).

Proof of Rolle's Theorem:

Case 1: $f(x) = c$, where c is a constant. Then $f'(x) = 0$ for all x in (a, b).

Case 2: $f(x)$ is not constant on $[a, b]$. Let us say that f assumes values greater than $f(a) = f(b)$. Since f is continuous on $[a, b]$, by Fact 3.3.4(1) f has a maximum value on $[a, b]$, say $f(x_0)$. Since $f(a)$ and $f(b)$ cannot be the maximum values, we have

$$a < x_0 < b.$$

Therefore, f has a local maximum at x_0. By 4.6.2

$$f'(x_0) = 0.$$

The proof is similar in the case that f assumes values less than $f(a) = f(b)$.

Proof of the Mean Value Theorem: First, we note that

$$y = f(a) + \left[\frac{f(b) - f(a)}{b - a} \right] (x - a)$$

is the equation of the line joining the two points $(a, f(a))$ and $(b, f(b))$.

We define a function ϕ as follows:

$$\phi(x) = \left\{ f(a) + \left[\frac{f(b) - f(a)}{b - a} \right] (x - a) \right\} - f(x).$$

The value of the function ϕ at x can be interpreted as the vertical distance (disregarding the sign) between the curve $y = f(x)$ and the line $y = f(a) + \{[f(b) - f(a)]/(b - a)\}(x - a)$ (see Figure 4.23).

It should be noted that the point x_0 we seek is a point where ϕ has a local extreme value. To see that this is indeed the situation, we note first that ϕ satisfies the hypothesis of Rolle's Theorem. Since f is continuous on $[a, b]$ and differentiable on (a, b), ϕ is also continuous on $[a, b]$ and differentiable on (a, b). Moreover,

$$\phi(a) = \phi(b) = 0.$$

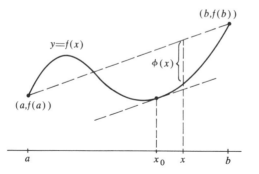

Figure 4.23

By 4.7.1, there is a point x_0 in (a, b) such that

$$\phi'(x_0) = 0.$$

Since $\phi'(x) = \{[f(b) - f(a)]/(b - a)\} - f'(x)$, then

$$\phi'(x_0) = \frac{f(b) - f(a)}{b - a} - f'(x_0) = 0$$

or

$$\frac{f(b) - f(a)}{b - a} = f'(x_0).$$

4.8 GEOMETRIC IMPLICATIONS OF THE FIRST AND SECOND DERIVATIVES

If one wishes to examine a given function carefully, it is frequently helpful to analyze the graph of that function. The differential calculus gives us additional techniques for sketching graphs accurately.

To draw graphs accurately, it is useful to determine those intervals where the graph of the function appears as one of the arcs shown in Figure 4.24. In this section we shall see that the signs of the first and second derivatives tell us which of the pictures in Figure 4.24 is correct. Once this is known for various intervals, then by joining together these intervals, one can determine whether a certain extreme point is a local maximum or a local minimum (see Figure 4.25).

In discussing the slope m of a (nonvertical) straight line, we distinguish

Figure 4.24

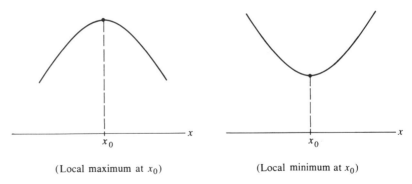

(Local maximum at x_0)   (Local minimum at x_0)

Figure 4.25

between the case when the line *rises* from left to right ($m > 0$) and the case when the line *falls* from left to right ($m < 0$). We now want to make similar remarks concerning a more general curve.

4.8.1
Definitions

1. f is **increasing on an interval I** if

$$f(x_1) \leq f(x_2)$$

whenever x_1 and x_2 are in I and $x_1 < x_2$.

2. f is **decreasing on an interval I** if

$$f(x_1) \geq f(x_2)$$

whenever x_1 and x_2 are in I and $x_1 < x_2$.

Geometrically, the situation might appear as in Figure 4.26.

For differentiable functions, there is a simple test to determine whether a function is increasing or decreasing.

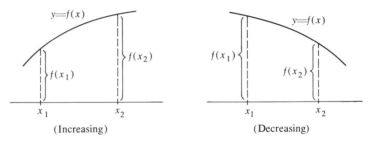

(Increasing)   (Decreasing)

Figure 4.26

4.8.2
Increasing-Decreasing Test

1. If $f'(x) \geq 0$ on I, then f is increasing on I.
2. If $f'(x) \leq 0$ on I, then f is decreasing on I.

These results follow from the equation

$$\frac{f(x_2) - f(x_1)}{x_2 - x_1} = f'(x_0) \qquad (x_1 < x_0 < x_2),$$

which holds by the Mean Value Theorem for any x_1, x_2 in I; for

$$f'(x) \geq 0 \text{ on } I \Rightarrow [f(x_2) - f(x_1)]/(x_2 - x_1) \geq 0.$$

Since $x_1 < x_2$, then $f(x_2) \geq f(x_1)$; that is, f is increasing on I. A similar argument can be given to show that $f'(x) \leq 0$ implies that f is decreasing.

If we recall that continuity on an interval means geometrically that the graph is unbroken on that interval, then the following statement is evident.

(4.8.3) Suppose that f is continuous on (a, b) and $a < x_0 < b$.

1. If f is increasing on (a, x_0) and decreasing on (x_0, b), then f has a local maximum at x_0.
2. If f is decreasing on (a, x_0) and increasing on (x_0, b), then f has a local minimum at x_0 (see Figure 4.27).

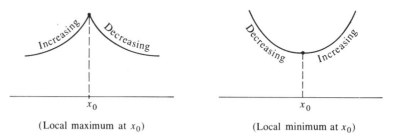

(Local maximum at x_0)   (Local minimum at x_0)

Figure 4.27

Now if f is differentiable on (a, b), except possibly at x_0 where it is continuous, then one may determine whether f is increasing (decreasing) on the intervals (a, x_0), (x_0, b) by looking at the sign of the first derivative (4.8.2). When this method is used, we refer to statement 4.8.3 as the **first derivative test**.

EXAMPLE 1   Let $f(x) = x^{2/3}$. Then $f'(x) = \frac{2}{3}x^{-1/3}$.

Case 1: x < 0. Then $f'(x) < 0$ and f is decreasing.
Case 2: x > 0. Then $f'(x) > 0$ and f is increasing.

By the first derivative test, f has a local minimum at zero (see Figure 4.28).

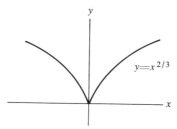

Figure 4.28

As we have just observed, the knowledge that f is increasing on an interval I gives us some indication of the appearance of its graph. Two examples are given in Figure 4.29. The two functions shown are both

Figure 4.29

increasing on I, but there is obviously something different about their appearance. From the graph on the right in Figure 4.29 we see that the slopes of tangent lines to the curve increase as we move from left to right (see Figure 4.30).

Figure 4.30

In this case we say that f is concave up on I. The situation at the left in Figure 4.29 is just the opposite. More precisely, note the following.

4.8.4
Definitions

1. **f is concave up on I** if f' is increasing on I.
2. **f is concave down on I** if f' is decreasing on I.

If f is twice differentiable, then we have a test to indicate concavity.

4.8.5
Concavity Test

1. If $f''(x) \geq 0$ on I, then f is concave up on I.
2. If $f''(x) \leq 0$ on I, then f concave down on I.

If $f''(x) \geq 0$ on I, then f' is increasing on I [by 4.8.2(1)] and so f is concave up on I. Also, if $f''(x) \leq 0$ on I, then f' is decreasing on I [by 4.8.2(2)] and so f is concave down on I.

EXAMPLE 2   Let $f(x) = x^3$. Then $f'(x) = 3x^2$ and $f''(x) = 6x$. Since $f'(0) = 0$ and $f''(0) = 0$, we examine two cases.

Case 1: $x \geq 0$. Then $f''(x) \geq 0$, and hence f is concave up.

Case 2: $x \leq 0$. Then $f''(x) \leq 0$, and hence f is concave down (see Figure 4.31).

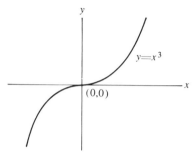

Figure 4.31

The point $(0, 0)$ is a point of special interest in the preceding example. In general, the following holds.

4.8.6

Definition   We say that f has a **point of inflection** at x_0 if (x_0, y_0) is a point on the graph of f and f is concave in one sense on some interval $(a, x_0]$ and is concave in the opposite sense on some interval $[x_0, b)$.

Therefore, if f has a point of inflection at x_0 and f is twice differentiable at x_0, then $f''(x_0) = 0$. It follows that the only candidates for points of inflection are those points for which $f''(x) = 0$ and those points where $f''(x)$ fails to exist.

EXAMPLE 3   Let $f(x) = x^{1/3}$. Then $f'(x) = \frac{1}{3}x^{-2/3}$ and $f''(x) = -\frac{2}{9}x^{-5/3}$. Since f' and f'' do not exist at zero, we have two cases.

Case 1: $x > 0$. Then $f'(x) > 0$ and $f''(x) < 0$, and hence f is increasing and concave down.

Case 2: $x < 0$. Then $f'(x) > 0$ and $f''(x) > 0$, and hence f is increasing and concave up. Note that f has a point of inflection at zero (see Figure 4.32).

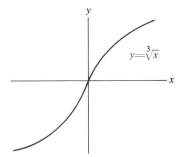

$y=\sqrt[3]{x}$

Figure 4.32

x	$x < 0$	$x > 0$
$f'(x)$	$+$	$+$
$f''(x)$	$+$	$-$
f	Increasing Concave up	Increasing Concave down

In the case that $f'(x_0) = 0$ and $f''(x_0)$ exists but $f''(x_0) \neq 0$, there is an alternative way to determine whether f has a local maximum or a local minimum at x_0.

Suppose that $f'(x_0) = 0$ and that $f''(x_0)$ exists.

1. If $f''(x_0) < 0$, then f has a local maximum at x_0.
2. If $f''(x_0) > 0$, then f has a local minimum at x_0 (see Figure 4.33).

This statement is referred to as the **second derivative test** (see Exercise 10).

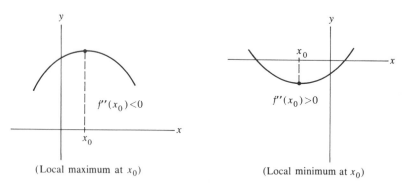

(Local maximum at x_0)   (Local minimum at x_0)

Figure 4.33

EXAMPLE 4

Let $f(x) = x^3 - 2x^2 + 1$. Then $f'(x) = 3x^2 - 4x = x(3x - 4)$. Since $f'(0) = 0$ and $f'(\frac{4}{3}) = 0$, we examine three cases.

Case 1: $x < 0$. Then $x < 0$ and $3x - 4 < 0$; therefore, $f'(x) > 0$, and hence f is increasing.

Case 2: $0 < x < \frac{4}{3}$. Then $x > 0$ and $3x - 4 < 0$; therefore $f'(x) < 0$, and hence f is decreasing.

Case 3: $x > \frac{4}{3}$. Then $x > 0$ and $3x - 4 > 0$; therefore, $f'(x) > 0$, and hence f is increasing.

For the second derivative we have

$$f''(x) = 6x - 4 = 2(3x - 2).$$

Since $f''(\frac{2}{3}) = 0$, we need examine only two cases.

Case 1: $x < \frac{2}{3}$. Then $2(3x - 2) < 0$ and so $f''(x) < 0$. Hence f is concave down.

Case 2: $x > \frac{2}{3}$. Then $2(3x - 2) > 0$ and so $f''(x) > 0$. Hence f is concave up.

x	$x < 0$	$0 < x < \frac{2}{3}$	$\frac{2}{3} < x < \frac{4}{3}$	$x > \frac{4}{3}$
f'	$+$	$-$	$-$	$+$
f''	$-$	$-$	$+$	$+$
f	Increasing Concave down	Decreasing Concave down	Decreasing Concave up	Increasing Concave up

Analyzing the f' row of the chart, the first derivative test allows us to conclude that

$$f \text{ has a local maximum at zero,}$$

and that

$$f \text{ has a local minimum at } \tfrac{4}{3}.$$

Alternatively, since

$$f''(0) = -4 < 0 \quad \text{and} \quad f''(\tfrac{4}{3}) = 4 > 0,$$

the second derivative test gives us those same conclusions.

The f'' row of the chart and the concavity test lead us to conclude that f has a point of inflection at $\tfrac{2}{3}$ (see Figure 4.34).

Figure 4.34

EXERCISES

1. In each of the following, graph the function f and determine from the graph those intervals where f is increasing, decreasing, concave up, concave down; also, determine points of inflection.

   (a) $f(x) = 1/x$.   (b) $f(x) = 1/|x|$,
   (c) $f(x) = x^{3/4} + 1$.   (d) $f(x) = \sqrt{9 - x^2}$.
   (e) $f(x) = -|x + 5|$.   (f) $f(x) = |x| - x$.

   (g) $f(x) = [x] - x$.   (h) $f(x) = \begin{cases} 1, & x \text{ rational,} \\ -1, & x \text{ irrational.} \end{cases}$

   (i) $f(x) = x|x|$.   (j) $f(x) = x^2 + 7$.

2. For each of the following functions f, use tests 4.8.2 and 4.8.5 to determine intervals where f is increasing, decreasing, concave up, concave down; also, determine points of inflection. (Do *not* be concerned with graphing f at this time.)

   (a) $f(x) = (1 + x)^2$.   (b) $f(x) = 3x - x^3$.
   (c) $f(x) = x^2 - x + 2$.   (d) $f(x) = x^3 - 3x + 1$.
   (e) $f(x) = x - 1/x$.   (f) $f(x) = x^2(1 + x)$.
   (g) $f(x) = (x + 1)/(x - 1)$.   (h) $f(x) = x^2/(1 + x)$.
   (i) $f(x) = x^{2/3}(x + 5)$.

3. Graph each of the following functions and from the graph determine those points, if any, at which f has a local minimum and those points, if any, at which f has a local maximum.

(a) $f(x) = 16 - x^2$.

(b) $f(x) = |16 - x^2|$.

(c) $f(x) = |x^3|$.

(d) $f(x) = x^{1/3}$.

(e) $f(x) = |x^{1/3}|$.

(f) $f(x) = -\sqrt{25 - x^2}$.

(g) $f(x) = 5 - x^3$.

(h) $f(x) = \begin{cases} 0, & x \le 0, \\ x^2, & 0 < x < 1, \\ 2 - x, & x \ge 1. \end{cases}$

(i) $f(x) = ||2 - x| - 3|$.

4. For each of the following functions use the first derivative test to determine those points, if any, at which f has a local maximum and those points, if any, at which f has a local minimum. (Do *not* be concerned about graphing f at this time.)

(a) $f(x) = x^2 - 8x + 10$.

(b) $f(x) = 6x - x^2 - 5$.

(c) $f(x) = 27x - x^3$.

(d) $f(x) = x^4 - 8x^2 + 10$.

(e) $f(x) = x^2 - 1/x$.

(f) $f(x) = 1 - x^{1/7}$.

(g) $f(x) = x/(x + 1)$.

(h) $f(x) = x/(x^2 + 1)$.

(i) $f(x) = x^{3/4} - 3x$.

(j) $f(x) = (2x - x^2)^{1/5}$.

5. Consider $f(x) = mx + b$. Show that

(a) f is increasing everywhere if $m > 0$.

(b) f is decreasing everywhere if $m < 0$.

6. Show that the equation $3x^4 - 4x^3 - 36x^2 + 199 = 0$ has no root greater than 3. [*Hint:* Let $f(x) = 3x^4 - 4x^3 - 36x^2 + 199$. Evaluate $f(3)$ and investigate $f'(x)$ for $x > 3$.]

7. Suppose that ϕ is increasing and let $f(x) = [\phi(x)]^2$, $g(x) = -\phi(x)$, and $h(x) = 1/\phi(x)$. Determine whether f, g, h are increasing.

8. Suppose that ϕ is increasing and that ψ is decreasing. Let $f(x) = \phi(x) + \psi(x)$ and $g(x) = \phi(x) - \psi(x)$. What can you say about f and g?

9. f is said to be **strictly increasing on** I if $x_1 < x_2$ implies that $f(x_1) < f(x_2)$.
(a) Show that $f'(x) > 0$ on I implies that f is strictly increasing on I.
(b) Show that $f(x) = x^3$ is a counterexample to the converse of part (a); that is, "f is strictly increasing implies that $f'(x) > 0$" is false.

10. Verify the second derivative test. [*Hint:*

$$f''(x_0) = \lim_{h \to 0} \frac{f'(x_0 + h) - f'(x_0)}{h} = \lim_{h \to 0} \frac{f'(x_0 + h)}{h}.$$

Use Fact 3.2.3 and the first derivative test.]

4.9 CURVE SKETCHING

In this section we shall employ the methods of the preceding section in describing an effective procedure which will allow us to graph many types of functions.

To begin, let us list some steps one may wish to perform in sketching a curve. Just how many of these steps one performs depends on how much information one wants from the graph.

1. Determine those points where $f(x)$ is not defined.

2. Compute $f'(x)$ where it exists. Determine those points where $f'(x) = 0$ or $f'(x)$ fails to exist. (These are the only points at which f can have a local maximum or local minimum.)

3. Compute $f''(x)$ where it exists. Determine those points where $f''(x) = 0$ or $f''(x)$ fails to exist. (The possible points of inflection are among these.)

4. Make a table with four rows labeled

$$x, \quad f(x), \quad f'(x), \quad \text{and} \quad f''(x).$$

The x row should span the entire domain of f, but with separate columns for those x values gathered in steps (1)–(3). Place the respective functional values in the $f(x)$ row. Do not place functional values in the $f'(x)$ and $f''(x)$ rows, but rather place appropriately the signs plus or minus for positive or negative, or zero, or does not exist. One need only glance at the chart now to apply the first derivative test, the second derivative test, the increasing-decreasing test, and the concavity test.

5. Determine the points of discontinuity of f. If c is such a point, then investigate the one-sided limits

$$\lim_{x \downarrow c} f(x) \quad \text{and} \quad \lim_{x \uparrow c} f(x).$$

If either of these limits is $+\infty$ or $-\infty$, then the vertical line $x = c$ is a vertical asymptote.

6. Investigate the two limits

$$\lim_{x \to +\infty} f(x) \quad \text{and} \quad \lim_{x \to -\infty} f(x).$$

If either of these limits is a number, say L, then the horizontal line $y = L$ is a horizontal asymptote.

7. If possible, determine the x- and y-intercepts of the curve $y = f(x)$.
   (a) The y-intercept is determined by setting $x = 0$, and so it is $f(0)$.

(b) The x-intercept is determined by setting $y = 0$, and so we have to solve the equation $f(x) = 0$; it might be quite difficult to solve or a method of solution might not be available.

8. Plot some additional points (x, y), where $y = f(x)$.

EXAMPLE 1   Sketch the graph of $y = f(x) = \dfrac{-x^3}{3} + \dfrac{x^2}{2} + 1$.

Both $f(x)$ and $f'(x)$ exist for all x, and

$$f'(x) = -x^2 + x = -x(x - 1),$$

and so

$$f'(0) = 0 \quad \text{and} \quad f'(1) = 0.$$

Also, $f''(x)$ exists for all x and

$$f''(x) = -2x + 1;$$

hence, $f''(\tfrac{1}{2}) = 0$.

x	$x < 0$	0	$0 < x < \tfrac{1}{2}$	$\tfrac{1}{2}$	$\tfrac{1}{2} < x < 1$	1	$x > 1$
$f(x)$		1		$\tfrac{13}{12}$		$\tfrac{7}{6}$	
$f'(x)$	$-$	0	$+$	$+$	$+$	0	$-$
$f''(x)$	$+$	$+$	$+$	0	$-$	$-$	$-$

By the first derivative test (or the second derivative test),

f has a local minimum at 0

and

f has a local maximum at 1.

By the increasing-decreasing test,

f is decreasing for all $x < 0$ and $x > 1$

and

f is increasing on $[0, 1]$.

By the concavity test,

f is concave up for $x < \tfrac{1}{2}$

and

f is concave down for $x > \tfrac{1}{2}$.

Hence, f has a point of inflection at $\tfrac{1}{2}$. The graph of $y = f(x)$ is drawn in Figure 4.35.

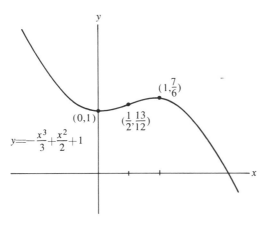

$(1, \frac{7}{6})$

$(0,1)$

$(\frac{1}{2}, \frac{13}{12})$

$y = -\frac{x^3}{3} + \frac{x^2}{2} + 1$

Figure 4.35

EXAMPLE 2 Sketch the graph of $y = f(x) = \dfrac{x}{x - 2}$.

We observe that f, f', and f'' exist for all $x \neq 2$. In fact,

$$f'(x) = \frac{-2}{(x - 2)^2}$$

and

$$f''(x) = \frac{4}{(x - 2)^3}.$$

x	$x < 2$	2	$x > 2$
$f(x)$		n.d.	
$f'(x)$	$-$	n.d.	$-$
$f''(x)$	$-$	n.d.	$+$

Note that $f'(x) < 0$ for all x, and hence, f is decreasing where it is defined and f has no local extreme points. Also, f is concave down for $x < 2$ and concave up for $x > 2$. Since

$$\lim_{x \downarrow 2} \frac{x}{x - 2} = +\infty \quad \text{and} \quad \lim_{x \uparrow 2} \frac{x}{x - 2} = -\infty,$$

the line $x = 2$ is a vertical asymptote. Furthermore, since

$$\lim_{x \to +\infty} \frac{x}{x - 2} = 1 \quad \text{and} \quad \lim_{x \to -\infty} \frac{x}{x - 2} = 1,$$

the line $y = 1$ is a horizontal asymptote.

Finally, we note that the only intercept is the origin $(0, 0)$ (see Figure 4.36).

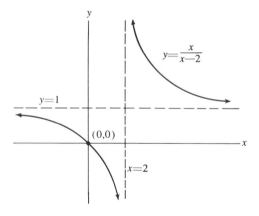

Figure 4.36

EXAMPLE 3   Sketch the graph of $y = f(x) = \frac{3}{2}x^{2/3} - x$.
First, we see that

$$f'(x) = x^{-1/3} - 1 = x^{-1/3}(1 - x^{1/3}).$$

Therefore, $f'(x) = 0$ if $x = 1$ and f' does not exist at $x = 0$. Also,

$$f''(x) = -\tfrac{1}{3}x^{-4/3}.$$

Hence the second derivative does not exist at $x = 0$.

x	$x < 0$	0	$0 < x < 1$	1	$x > 1$
$f(x)$		0		$\frac{1}{2}$	
$f'(x)$	$-$	n.d.	$+$	0	$-$
$f''(x)$	$-$	n.d.	$-$	$-$	$-$

By the first derivative test, f has a local minimum at $x = 0$ and a local maximum at $x = 1$.

By the increasing-decreasing test, f is decreasing for all $x < 0$ and $x > 1$ and f is increasing for $0 < x < 1$.

By the concavity test, f is always concave down. Therefore, f has no point of inflection.

The graph of $y = f(x)$ is given in Figure 4.37.

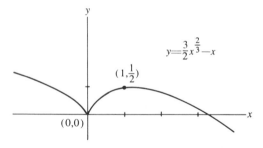

Figure 4.37

EXERCISES

Sketch the graph of the following equations.

1. $y = x^2 - 8x + 10$.   2. $y = 6x - x^2 - 5$.
3. $y = 27x - x^3$.   4. $y = x^4 - 8x^2 + 10$.
5. $y = x^2 - 1/x$.   6. $y = x + 1/x$.
7. $y = x^2 + 1/x$.   8. $y = x - 1/x$.
9. $y = (x + 1)/(x - 1)$.   10. $y = x^2/(1 + x)$.
11. $y = 1/[(x - 1)(x + 2)]$.   12. $y = 1 - x^7$.
13. $y = x^{3/4} - 3x$.   14. $y = x/(x^2 + 1)$.
15. $y = |x|/(x^2 + 1)$.

4.10 RATES OF CHANGE

The notions of tangent line, instantaneous velocity, and marginal cost were introduced at the beginning of this chapter. The tangent line idea was examined more carefully in Section 4.3. In this section and the one following we shall, among other things, study in greater detail the concepts of instantaneous velocity and marginal cost. In doing so we shall employ some of the techniques developed in Sections 4.8 and 4.9.

Let us begin with a projectile problem.

EXAMPLE 1   A projectile is shot upward and its height (in feet) $s(t)$ after t seconds is given by

$$s(t) = -16t^2 + 48t + 64.$$

The fact that $s(0) = 64$ indicates that the projectile was shot upward from 64 feet above ground and $s(4) = 0$ indicates that it hit the ground after 4 seconds. There are several questions one can ask here. What is the maximum height traveled by the projectile? With what velocity does the projectile hit the ground?

We shall return to the problem shortly to analyze the motion of this projectile.

EXAMPLE 2   Let $x(t) = t^2/2 - t + 1$ be the position of a point moving on a coordinatized line. Then

$$x(0) = 1, \qquad x(\tfrac{1}{2}) = \tfrac{5}{8}, \qquad x(1) = \tfrac{1}{2}, \qquad \text{and} \qquad x(\tfrac{3}{2}) = \tfrac{5}{8}.$$

Here we see that the point started at $x = 1$ and then apparently moved to the left. However, sometime between $t = \tfrac{1}{2}$ and $t = \tfrac{3}{2}$ the point has changed direction and moved to the right. One might ask: What is the total distance traveled by the point in 4 seconds?

We shall return to the problem shortly to analyze the motion of this point.

4.10.1
Definition

Let $x(t)$ be the position of a point after t seconds. The **average velocity** of the point in the time interval $a \leq t \leq b$ is defined by

$$\frac{x(b) - x(a)}{b - a}.$$

As in the introduction to this chapter, we fix t and compute average velocities

$$\frac{x(t + h) - x(t)}{h}$$

in the time intervals $[t, t + h]$, $h > 0$, and $[t + h, h]$, $h < 0$, for smaller and smaller h. Then we get close to a number which we call the instantaneous velocity at t.

4.10.2
Definition

Let $x(t)$ be the position of a point after t seconds. The **instantaneous velocity** $v(t)$ of the point at time t is defined by

$$v(t) = x'(t) = \lim_{h \to 0} \frac{x(t + h) - x(t)}{h}.$$

$v(t) = x'(t)$ is also called the (instantaneous) rate of change of x with respect to t.

Let us make several observations here.

1. If $v(t_0) = 0$, then the point has stopped at $t = t_0$.
2. If $v(t_0) > 0$ on the time interval (a, b) [that is $x'(t) > 0$ on (a, b)], then the function $x(t)$ is increasing in the time interval (a, b) (that is, the point is moving to the right on the line).
3. If $v(t) < 0$ on (a, b), then $x(t)$ is decreasing on (a, b) (that is, the point is moving to the left).

4.10.3
Definition

Let $x(t)$ be the position of a point after t seconds. The **speed** of the point is defined to be

$$|v(t)| = |x'(t)|.$$

We see that the velocity is concerned with direction as well as magnitude, whereas the speed is concerned only with the magnitude of the rate of change of $x(t)$.

**4.10.4
Definition**

Let $v(t)$ be the velocity of a point after t seconds. The **average acceleration** of the point in the time interval $a \leq t \leq b$ is defined by

$$\frac{v(b) - v(a)}{b - a}.$$

If we compute average accelerations

$$\frac{v(t + h) - v(t)}{h}$$

in the time intervals $[t, t + h]$, $h > 0$, and $[t + h, t]$, $h < 0$, for smaller and smaller h, then we get close to a number which we call the instantaneous acceleration at t.

**4.10.5
Definition**

Let $v(t)$ be the velocity of a point after t seconds. The **instantaneous acceleration** at time t is given by

$$a(t) = v'(t) = \lim_{h \to 0} \frac{v(t + h) - v(t)}{h}.$$

Observe that the acceleration is the rate of change of the velocity and that

$$a(t) = x''(t).$$

EXAMPLE 3

(Continuation of Example 1). A projectile is shot upward and its height (in feet) $s(t)$ after t seconds is given by

$$s(t) = -16t^2 + 48t + 64.$$

What is the maximum height traveled by the projectile? With what velocity does the projectile hit the ground?

Since the velocity is zero at the point of maximum height, we see that

$$v(t) = s'(t) = -32t + 48 = 0$$

when $t = \frac{4}{3}$ seconds. Therefore the maximum height is

$$s(\tfrac{4}{3}) = -16(\tfrac{4}{3})^2 + 48(\tfrac{4}{3}) + 64 = \tfrac{896}{9} \text{ feet.}$$

Since $s(4) = 0$, the projectile has hit the ground after 4 seconds. Hence the velocity upon hitting the ground is

$$v(4) = (-32)(4) + 48 = -80 \text{ feet per second.}$$

EXAMPLE 4   (Continuation of Example 2). Let $x(t) = t^2/2 - t + 1$; then $x'(t) = t - 1$.

Case 1: $0 < t < 1$. Then $x'(t) < 0$, and hence the point is moving to the left.

Case 2: $t = 1$. Then $x'(t) = 0$, and hence the point has stopped.

Case 3: $t > 0$. Then $x'(t) > 0$, and so the point is moving to the right. Since $v(t) = t - 1$,

$$a(t) = 1;$$

thus the point moves with constant acceleration.

t	0	$0 < t < 1$	1	$t > 1$
$x(t)$	1		$\frac{1}{2}$	
$v(t)$	$-$	$-$	0	$+$
$a(t)$	$+$	$+$	$+$	$+$

The motion of the point is indicated in Figure 4.38.

Figure 4.38

EXAMPLE 5   If the motion of a point is described as in Example 4, then what is the total distance traveled after 4 seconds?

The point starts at $x(0) = 0$ and then moves to the left and gets to $x(1) = \frac{1}{2}$, where it stops. Therefore it has gone $\frac{1}{2}$ units in 1 second.

Now it moves to the right and gets to $x(4) = 5$. Thus between 1 second and 4 seconds it has gone $4\frac{1}{2}$ units.

The total distance traveled is

$$\tfrac{1}{2} + 4\tfrac{1}{2} = 5 \text{ units.}$$

EXAMPLE 6   Analyze the motion of the point which moves along a coordinatized line
according to the rule

$$x(t) = t^3 - 9t^2 + 15t + 1.$$

Since the position of the point at time t is $x(t) = t^3 - 9t^2 + 15t + 1$,

$$v(t) = x'(t) = 3t^2 - 18t + 15$$
$$= 3(t - 5)(t - 1)$$

and

$$a(t) = v'(t) = 6t - 18.$$

Observe that

$$v(t) = 0 \Leftrightarrow (t - 5)(t - 1) = 0$$
$$\Leftrightarrow t = 5, \quad t = 1$$

and that

$$a(t) = 0 \Leftrightarrow 6t - 18 = 0$$
$$\Leftrightarrow t = 3.$$

t	0	$0 < t < 1$	1	$1 < t < 3$	3	$3 < t < 5$	5	$t > 5$
$x(t)$	1		8		-8		-24	
$v(t)$	$+$	$+$	0	$-$	$-$	$-$	0	$+$
$a(t)$	$-$	$-$	$-$	$-$	0	$+$	$+$	$+$

Thus the point moves to the right for $0 < t < 1$, stops at $t = 1$, moves left
for $1 < t < 5$, stops at $t = 5$, and moves to the right for $t > 5$ (see Figure
4.39).

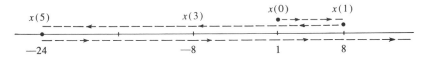

Figure 4.39

In addition to problems involving linear motion, there are instances
when we speak of rates of change of other functions with respect to time.
Often rates of change are related to each other.

EXAMPLE 7   A balloon is being inflated at a rate of 10 cubic inches per second. How fast
is the radius changing at the end of 5 seconds?

Let $V(t)$ be the volume after t seconds. Then the rate of change of V with respect to time t is given by

$$V'(t) = 10.$$

Moreover,

$$V(t) = \tfrac{4}{3}\pi[r(t)]^3,$$

where $r(t)$ is the radius after t seconds. Now

$$V'(t) = \tfrac{4}{3}\pi(3[r(t)]^2 r'(t)) \qquad (4.5.2)$$
$$= 4\pi[r(t)]^2 r'(t).$$

Therefore

$$V'(5) = 4\pi[r(5)]^2 r'(5),$$

and so

$$r'(5) = \frac{10}{4\pi[r(5)]^2}.$$

This gives us the rate of change of the radius after 5 seconds.

EXAMPLE 8   The height of a cylinder increases at the rate of 10 inches per minute. If the volume of the cylinder is always a constant, at what rate must the radius be diminished at each instant?

The volume V of a cylinder of radius r and height h is

$$V = \pi r^2 h.$$

Here V is a constant and r and h are functions of t. Therefore we shall write

$$V = \pi[r(t)]^2 h(t).$$

Differentiating both sides, we have

$$0 = \pi 2r(t)r'(t)h(t) + \pi[r(t)]^2 h'(t),$$

and so

$$r'(t) = \frac{-[r(t)]^2 h'(t)}{2r(t)h(t)}.$$

Since $h'(t) = 10$,

$$r'(t) = \frac{-10[r(t)]^2}{2r(t)h(t)};$$

that is,

$$r'(t) = \frac{-5r(t)}{h(t)}.$$

EXAMPLE 9   A point is moving along the curve $y = x^2$. When the point has coordinates $(2, 4)$, the rate of change of the y-coordinate is 3 units per second. What is the rate of change of the x-coordinate at this point?

We regard x and y as functions of t. Therefore

$$y(t) = [x(t)]^2;$$

then

$$y'(t) = 2x(t)x'(t).$$

Since $y'(t) = 3$ when $x(t) = 2$, we have

$$3 = (2)(2)x'(t);$$

that is,

$$x'(t) = \tfrac{3}{4} \text{ units per second.}$$

EXERCISES

1. A particle moves on the x-axis according to the law $x(t) = t^3 - 9$. Find the following:

    (a) Average velocity on $[2, 4]$.   (b) Velocity at $t = 2$.

    (c) Velocity at $t = 4$.

    (d) Average acceleration on time interval $[2, 4]$.

    (e) Acceleration at $t = 2$.   (f) Acceleration at $t = 4$.

2. For each of the following position functions, determine the velocity at any time t and the acceleration at any time t, and draw a picture of the motion.

    (a) $x(t) = 2t^2 - 6t, \, t \geq 0$.   (b) $x(t) = 3 + 16t - 4t^2, \, t \geq 0$.

    (c) $x(t) = t^3 - 9t^2 + 24t - 100, \, t \geq 0$.

    (d) $x(t) = 4t^4 - 16t^3, \, t \geq 0$.

    (e) $x(t) = t + 9/t, \, t > 0$.   (f) $x(t) = (t - 1)^{2/3}, \, t > 0$.

3. A particle moves on a coordinatized line according to the rule $x(t) = 10 + 36t + 3t^2 - 2t^3$. In the time interval $[0, 4]$ determine the following.

    (a) That instant at which the particle is at rest.

    (b) That instant at which the velocity is a maximum.

4. Describe in words (for example, moving to the right and slowing down) what is happening to a particle moving in time interval $[a, b]$ when for all t in $[a, b]$

    (a) $v(t) > 0$ and $a(t) > 0$.   (b) $v(t) > 0$ and $a(t) < 0$.

    (c) $v(t) < 0$ and $a(t) > 0$.   (d) $v(t) < 0$ and $a(t) < 0$.

5. The radius of a sphere is increasing at a rate of 6 inches per second. How fast is the volume changing when its radius is 3 inches?

6. The edge of a square is decreasing at a rate of 0.5 inches per second. Find the rate of change in its area when its edge is 6 inches long.

7. A conical tank (vertex downward) of height 4 feet and radius of base 3 feet is being filled with water at a constant rate of 2 cubic feet per minute. How fast is the water rising in the tank at any instant?

8. A point moves along the ellipse $x^2/4 + y^2/9 = 1$ so that its x-coordinate changes at the constant rate of 8 units per second. What is the rate of change of its y-coordinate at the point $(1, 3\sqrt{3}/2)$?

9. At a certain instant, cars A and B leave from the same spot, with car A traveling east at 50 miles per hour and car B traveling north at 60 miles per hour. How fast are they moving apart after 30 minutes?

10. Forced with a declining market in demand for his product, a merchant reduces the price p of his product (as time t increases) according to the rule $p(t) = 250 - t/100$. If his profit P is given by $P = 50 - \sqrt[7]{p^3}/2$, how fast is his profit changing with respect to time when $t = 100$?

11. A projectile is shot upward from the surface of the earth and its distance s from the earth at any time t is given by $s(t) = 320t - 16t^2$. At what time does the projectile reach its maximum height? What is the maximum height? How long does it take for the projectile to return to earth? With what velocity does the projectile hit the ground?

12. Suppose that the velocity v of a body is given as a function of its position x. Show that the acceleration can be written as $a = v'(x)v(x)$. [*Hint:* $a = \dfrac{d}{dt}(v(x(t)))$.]

13. Suppose that a particle moves in the plane and its position is given by $x(t) = 2t$ and $y(t) = t^2 - 1$.

   (a) Find the rate of change of the x-coordinate and the rate of change of the y-coordinate of the particle at any time t.

   (b) Find an equation whose graph is the path followed by this particle. [*Hint:* Eliminate t and obtain an equation in x and y.]

4.11　MARGINAL COST, MARGINAL REVENUE, MAXIMIZATION OF PROFITS

Now we shall consider a problem concerning the maximization of profit in the theory of economics.

Let C be a **cost function**, so that $C(x)$ is the manufacturer's cost of producing x items of a particular product. In Section 4.1, we briefly described the idea of marginal cost—the change in cost of production with respect to output.

4.11.1
Definition

Let $C(x)$ be a cost function. Then the **marginal cost** is defined as

$$\frac{d}{dx}\{C(x)\}.$$

A **revenue function** $R(x)$ is the revenue received for selling x items of a particular product. Analogously, the marginal revenue is the change in revenue received with respect to sales.

Let $R(x)$ be a revenue function. Then the **marginal revenue** is defined as

$$\frac{d}{dx} \{R(x)\}.$$

A **profit function** P is defined as $P(x) = R(x) - C(x)$. Of course, a manufacturer is interested in determining that level of production at which profits will be maximized. Let x_0 be the level of production at which profits are maximum. Then, by 4.6.2, it follows that

$$P'(x_0) = 0.$$

Since

$$\frac{d}{dx} \{P(x)\} = \frac{d}{dx} \{R(x)\} - \frac{d}{dx} \{C(x)\},$$

we see that

$$R'(x_0) - C'(x_0) = 0$$

or

$$R'(x_0) = C'(x_0);$$

that is, the marginal revenue must equal the marginal cost at the maximum level of production.

Since P is a maximum at x_0, we have

$$P'(x) \geq 0 \qquad \text{for } x \leq x_0$$

and

$$P'(x) \leq 0 \qquad \text{for } x \geq x_0;$$

that is,

$$R'(x) - C'(x) \geq 0 \qquad (x \leq x_0)$$

and

$$R'(x) - C'(x) \leq 0 \qquad (x \geq x_0),$$

or

$$R'(x) \geq C'(x) \qquad (x \leq x_0)$$

and

$$R'(x) \leq C'(x) \qquad (x \geq x_0).$$

The result $R'(x) \geq C'(x)$ for $x \leq x_0$ indicates that production should be increased if marginal cost is less than marginal revenue, while $R'(x) \leq C'(x)$ for $x \geq x_0$ indicates that production should be decreased if marginal cost is greater than marginal revenue.

The graphs of the equations

$$y = R'(x) \quad \text{and} \quad y = C'(x)$$

are called the **marginal curves** (see Figure 4.40). The point of intersection of the marginal curves gives the level of production x_0 for maximum profits. Moreover, the slope of the marginal revenue curve is smaller than the slope of the marginal cost curve at this point of intersection, for, at such a point,

$$P''(x_0) < 0;$$

that is

$$R''(x_0) - C''(x_0) < 0$$

or

$$R''(x_0) < C''(x_0).$$

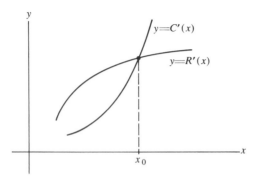

Figure 4.40

EXAMPLE 1   If an article sells for \$10 and the cost of producing x articles is given by the cost function

$$C(x) = 150 + 3.5x + 0.005x^2,$$

what is the profit function and at what level of production will it be a maximum?

It is clear that

$$R(x) = 10x,$$

and so

$$\begin{aligned} P(x) &= R(x) - C(x) \\ &= 10x - [150 + 3.5x + 0.005x^2] \\ &= -150 + 6.5x - 0.005x^2. \end{aligned}$$

Observe that

$$P'(x) = 6.5 - 0.010x,$$

and so $P'(x) = 0$ whenever

$$0.010x = 6.5$$

or

$$x = 650.$$

Therefore the level of production which will maximize profits is 650.

Sometimes the price of an item can be lowered if more items are sold and raised if fewer items are sold. In this case the price $P(x)$ may be given as a function of the number of items sold. Hence the revenue function is

$$R(x) = xP(x).$$

EXAMPLE 2 Suppose that a manufacturer can sell x items at a price

$$P(x) = 500 - 0.025x \text{ cents.}$$

His cost of producing x items is given by

$$C(x) = 75x + 3500.$$

What is the production level in order to maximize profits and what price should be charged for the items?

We see that

$$R(x) = xP(x)$$
$$= 500x - 0.025x^2,$$

and so

$$P(x) = R(x) - C(x)$$
$$= (500x - 0.025x^2) - (75x + 3500)$$
$$= -3500 + 425x - 0.025x^2.$$

Therefore

$$P'(x) = 425 - 0.050x$$

and $P'(x) = 0$ whenever

$$x = 8500;$$

that is, the maximum production level is 8500.

Since $f(8500) = 287.50$ cents, then the price should be \$2.88 for each item.

EXAMPLE 3 Suppose that a manufacturer sells an article for \$28 and that his marginal cost of production is given by

$$4 + 0.06x.$$

What is the output in order to maximize profits?

We know that

$$C'(x) = 4 + 0.06x$$

and that

$$R'(x) = 28 \qquad [\text{since } R(x) = 28x].$$

Since maximum profits occur at the point of intersection of the marginal curves, we see that

$$C'(x) = R'(x)$$

whenever

$$4 + 0.06x = 28;$$

that is,

$$x = 400.$$

EXERCISES

1. The cost function for a product is given by $C(x) = 750 + 15x + 0.005x^2$. What is the marginal cost? Sketch the graph of the cost function and the graph of the marginal cost function.

2. The cost function for a product is given by $C(x) = 2500 - 20x + 0.04x^2 + 0.002x^3$. What is the marginal cost of production?

3. The revenue function for a product is given by $R(x) = 165x - 0.01x^2$. What is the marginal revenue? Sketch the graph of the revenue function and the graph of the marginal revenue function.

4. Given the cost function $C(x) = 750 + 15x + 0.005x^2$ and the revenue function $R(x) = 165x - 0.01x^2$, determine the following.

   (a) Profit function P.

   (b) The level of production where profits are a maximum.

   (c) Graph of P.

5. A businessman can sell x items at a price $p(x) = 250 - 0.015x$ cents, and his cost of producing x items is $C(x) = 5000 + 16x$ cents. What price should be charged so as to maximize profits?

6. A businessman sells an item for \$25 and his marginal cost of production is given by $15 + 0.010x$. What is the output of production that maximizes profits?

7. In marketing a product it is determined that the marginal cost is $8 + 0.03x$ and that the marginal revenue is $50 - 0.04x$, where x is the number of items marketed. Determine the value of x which will maximize profits.

8. Let $p(x)$ be the price of x items of a commodity. Show the following.

   (a) If $p(x) = k$ for all x (k constant), then $R'(x) = k$.

   (b) In general $R'(x) = xp'(x) + p(x)$.

9. Suppose that the price $p(x)$ of x items of a commodity is given by $p(x) = 1/(3 + x^2)$. What is the marginal revenue?

10. The market demand for a product depends on the price p and is given by $x = (1 + p^2)/2p^2$, where x denotes the quantity demanded and p is its price. What is the marginal revenue function?

11. The market demand function for a product is given by $x = 150 - 30\sqrt[3]{p}$, where x denotes the quantity demanded and p is its price. What is the marginal revenue function?

12. Given the demand function $x = (480 - 12p)/2$ and the cost function $C(x) = 250 + 10x + 3x^2$, what level of production yields maximum profits?

4.12 ANTIDERIVATIVES

In the preceding two sections we dealt with the problems of finding the velocity of a moving body given its position function and of finding marginal cost given a cost function. We now turn our attention to a different type of problem where we seek a position function of a moving body given its velocity and a cost function given its marginal cost. The functions we seek are called antiderivatives of the given functions.

4.12.1 Definition

F is called an **antiderivative of f** if

$$F'(x) = f(x).$$

The concept of antiderivative will be used quite frequently in Chapter 5 in solving a variety of new problems.

EXAMPLE 1 Let $F(x) = x^3/3$ and $f(x) = x^2$; then F is an antiderivative of f since

$$F'(x) = f(x).$$

Not all functions have antiderivatives. For example,

$$f(x) = \begin{cases} -1, & x < 0, \\ 1, & x \geq 0, \end{cases}$$

does not have an antiderivative on $[-1, 1]$ (see Exercise 16). However, we shall see in Section 5.3 that each function continuous on $[a, b]$ has an antiderivative.

EXAMPLE 2 Let $f(x) = 4x + 5$ and let

$$F_1(x) = 2x^2 + 5x + 2 \qquad \text{and} \qquad F_2(x) = 2x^2 + 5x - 7.$$

Then

$$F_1'(x) = 4x + 5 \qquad \text{and} \qquad F_2'(x) = 4x + 5.$$

Therefore, F_1 and F_2 are antiderivatives of f.

This example shows that antiderivatives are not unique. In fact, if F_1 and F_2 are antiderivatives of f, that is, $F_1'(x) = f(x)$ and $F_2'(x) = f(x)$, then by 4.7.4,

(4.12.2) $$F_1(x) = F_2(x) + C.$$

4.12.3
Fact   For a rational number r, $r \neq -1$,

$$F(x) = \frac{x^{r+1}}{r+1}$$

is an antiderivative of

$$f(x) = x^r.$$

This is true because

$$\frac{d}{dx}\left\{\frac{x^{r+1}}{r+1}\right\} = x^r, \qquad r \neq -1.$$

EXAMPLE 3   What is an antiderivative of $f(x) = 1/x^2$?
Since $1/x^2 = x^{-2}$, it follows that

$$\frac{x^{-2+1}}{-2+1} = \frac{x^{-1}}{-1} = \frac{-1}{x}$$

is an antiderivative of

$$\frac{1}{x^2}.$$

4.12.4
Fact   For a rational number r, $r \neq -1$,

$$F(x) = \frac{[u(x)]^{r+1}}{r+1}$$

is an antiderivative of

$$f(x) = [u(x)]^r u'(x).$$

By the chain rule,

$$\frac{d}{dx}\left\{\frac{[u(x)]^{r+1}}{r+1}\right\} = [u(x)]^r u'(x).$$

EXAMPLE 4   What is an antiderivative of $f(x) = 2x(x^2+1)^3$?
If we let

$$u(x) = x^2 + 1,$$

then $u'(x) = 2x$. Therefore

$$f(x) = [u(x)]^3 u'(x).$$

By 4.12.4,

$$\frac{(x^2 + 1)^{3+1}}{3 + 1} = \frac{(x^2 + 1)^4}{4}$$

is an antiderivative of

$$2x(x^2 + 1)^3.$$

4.12.5

Fact   If $F(x)$ is an antiderivative of $f(x)$, then

$$cF(x)$$

is an antiderivative of

$$cf(x).$$

This follows from the fact that

$$\frac{d}{dx} \{cF(x)\} = c \frac{d}{dx} \{F(x)\} = cf(x).$$

EXAMPLE 5   What is an antiderivative of $2x^{1/7}$?
Since

$$\frac{x^{1/7+1}}{\frac{1}{7} + 1} = \frac{x^{8/7}}{\frac{8}{7}} = \tfrac{7}{8}x^{8/7}$$

is an antiderivative of

$$x^{1/7},$$

it follows that

$$2(\tfrac{7}{8}x^{8/7}) = \tfrac{7}{4}x^{8/7}$$

is an antiderivative of

$$2x^{1/7}.$$

EXAMPLE 6   What is an antiderivative of $(3x + 1)^3$?
If we let $u(x) = 3x + 1$, then

$$u'(x) = 3$$

and so

$$(3x + 1)^3 = \tfrac{1}{3}(3x + 1)^3 \cdot 3 = \tfrac{1}{3}[u(x)]^3 \cdot u'(x).$$

Since

$$\frac{(3x + 1)^4}{4}$$

is an antiderivative of

$$(3x + 1)^3 \cdot 3,$$

it follows that

$$\frac{1}{3}\frac{(3x + 1)^4}{4} = \frac{(3x + 1)^4}{12}$$

is an antiderivative of

$$(3x + 1)^3.$$

EXAMPLE 7   Determine an antiderivative of $x^2/\sqrt{1 + 2x^3}$.
Let $u(x) = 1 + 2x^3$; then

$$u'(x) = 6x^2.$$

Therefore

$$\frac{x^2}{\sqrt{1 + 2x^3}} = \frac{1}{6}\frac{6x^2}{\sqrt{1 + 2x^3}}$$

$$= \frac{1}{6}(1 + 2x^3)^{-1/2}(6x^2) = \frac{1}{6}[u(x)]^{-1/2}u'(x).$$

Since

$$\frac{(1 + 2x^3)^{1/2}}{\frac{1}{2}} = 2(1 + 2x^3)^{1/2},$$

is an antiderivative of

$$(1 + 2x^3)^{-1/2}(6x^2),$$

it follows that

$$\tfrac{1}{6}[2(1 + 2x^3)^{1/2}] = \tfrac{1}{3}(1 + 2x^3)^{1/2}$$

is an antiderivative of

$$(1 + 2x^3)^{-1/2}(x^2).$$

4.12.6
Fact   If $F(x)$ is an antiderivative of $f(x)$ and $G(x)$ is an antiderivative of $g(x)$, then

$$F(x) + G(x)$$

is an antiderivative of

$$f(x) + g(x).$$

This is true because

$$\frac{d}{dx}\{F(x) + G(x)\} = \frac{d}{dx}\{F(x)\} + \frac{d}{dx}\{G(x)\} = f(x) + g(x).$$

EXAMPLE 8   Determine an antiderivative of $3x^5 + 2/x^2 + x$.

Note that $3x^5 + 2/x^2 + x = 3x^5 + 2x^{-2} + x$. Since $x^6/2$ is an antiderivative of $3x^5$, $-2x^{-1}$ is an antiderivative of $2x^{-2}$, and $x^2/2$ is an antiderivative of x, it follows that

$$\frac{x^6}{2} - \frac{2}{x} + \frac{x^2}{2}$$

is an antiderivative of

$$3x^5 + \frac{2}{x^2} + x.$$

EXAMPLE 9   Find a function f satisfying

$$\begin{cases} f'(x) = 3x^5 + \dfrac{2}{x^2} + x \\[2mm] f(1) = 4. \end{cases}$$

In Example 8, we determined that

$$\frac{x^6}{2} - \frac{2}{x} + \frac{x^2}{2}$$

is an antiderivative of

$$3x^5 + \frac{2}{x^2} + x.$$

Thus

$$f(x) = \frac{x^6}{2} - \frac{2}{x} + \frac{x^2}{2} + k,$$

where k is a constant.

Since $f(1) = 4$,

$$4 = f(1) = \tfrac{1}{2} - 2 + \tfrac{1}{2} + k \qquad \text{and} \qquad k = 5.$$

Therefore

$$f(x) = \frac{x^6}{2} - \frac{2}{x} + \frac{x^2}{2} + 5.$$

EXAMPLE 10   A projectile 800 feet above ground is shot upward with an initial velocity of 100 feet per second. If only the constant acceleration of gravity (32 feet per second squared downward) acts on it, determine the position of the projectile at any time t.

Let $s(t)$ be the height above ground after t seconds. Since the projectile is initially 800 feet above ground,

$$s(0) = 800,$$

and since it is shot up with initial velocity 100 feet per second,

$$s'(0) = 100.$$

Therefore we seek the solution of the problem

$$s''(t) = -32$$

$$s(0) = 800$$

$$s'(0) = 100.$$

Since $-32t$ is an antiderivative of -32,

$$s'(t) = -32t + k_1,$$

where k_1 is a constant. However,

$$100 = s'(0) = 0 + k_1;$$

that is,

$$k_1 = 100.$$

Therefore

$$s'(t) = -32t + 100.$$

Now

$$-16t^2 + 100t$$

is an antiderivative of

$$-32t + 100.$$

Therefore

$$s(t) = -16t^2 + 100t + k_2,$$

where k_2 is a constant. However,

$$800 = s(0) = 0 + 0 + k_2;$$

that is,

$$k_2 = 800.$$

The position function at any time t is

$$s(t) = -16t^2 + 100t + 800.$$

EXAMPLE 11   Suppose that a manufacturer's total fixed cost in producing some product is $5000 and suppose that the marginal cost is given by

$$0.06x^2 - 0.24x + 3.$$

What is the cost of producing 1000 items?
Let $C(x)$ be the cost function; then

$$C'(x) = 0.06x^2 - 0.24x + 3$$

and $C(0) = 5000$.
Therefore, $C(x)$ is an antiderivative of

$$0.06x^2 - 0.24x + 3.$$

Therefore
$$C(x) = 0.02x^3 - 0.12x^2 + 3x + k,$$

where k is some constant. However,
$$C(0) = k = 5000;$$

hence
$$C(x) = 0.02x^3 - 0.12x^2 + 3x + 5000$$

and
$$C(1000) = 18{,}808{,}000.$$

EXAMPLE 12   Suppose that a manufacturer sells an article for $28 and that his marginal cost of production is given by
$$4 + 0.06x.$$

What is the maximum profit possible if his fixed cost is $200?
    In Example 3 of Section 4.11 we determined that maximum profit occurs when the output is 400.
    Moreover, $C(x)$ is an antiderivative of
$$4 + 0.06x.$$

Therefore
$$C(x) = 4x + 0.03x^2 + k,$$

where k is some constant. However,
$$C(0) = 200 = k;$$

thus
$$C(x) = 200 + 4x + 0.03x^2.$$

Furthermore,
$$P(x) = R(x) - C(x)$$
$$= (28x) - (200 + 4x + 0.03x^2)$$
$$= -200 + 24x - 0.03x^2.$$

Therefore the profit at the level of production $x = 400$ is given by
$$P(400) = \$4600.$$

EXERCISES

1. Find an antiderivative of each of the following.

(a) $3x^5$.   (b) 0.

(c) $2\sqrt[3]{x}$.   (d) 2.

(e) $2x - 7$.   (f) $6x^4 - 3x^3 + 2x^2 - x + 1$.

(g) $x^3 - 3/x^3$.   (h) $x^{2/5} - x$.

(i) $(6x - 3)^2$.

(j) $x\sqrt[3]{x^2 + 1}$.

(k) $(x^2 + 1)(x^3 + 3x - 6)^4$.

(l) $1/(3x - 2)^2$.

(m) $1/(1 - 5x)^{1/4}$.

(n) $2x^4/\sqrt{7 + x^5}$.

(o) $x\sqrt[5]{x}$.

(p) $(x^2 + 1)/x^2$.

(q) $(4\sqrt[3]{x} + 3\sqrt[4]{x})/\sqrt{x}$.

(r) $x/(1 + x)^3$. [*Hint:* $x/(1 + x)^3 = [(1 + x) - 1]/(1 + x)^3$.]

2. In each part of Exercise 1, find the antiderivative F with the property that $F(1) = 0$.

3. Let
$$F(x) = \begin{cases} -\frac{1}{2}x^2, & x < 0, \\ \frac{1}{2}x^2, & x \geq 0. \end{cases}$$

Show that $F(x)$ is an antiderivative of $|x|$. [Be sure to check $F'(0)$.]

4. Find a function f satisfying $f'(x) = 2x^2 - 3\sqrt{x}$ and $f(9) = 1$.

5. Find a function f satisfying $f''(x) = 3x^2 + 2x + 5, f(0) = 1$, and $f'(0) = -2$.

6. The acceleration of a point moving in a straight line is given by $a(t) = 2t - 3$. Determine the position $x(t)$ of the point if initially its velocity is 2 units per second and its position is zero.

7. The acceleration of a point moving in a straight line is given by $a(t) = 12t^2 + 6t$. If initially its position is 1 and 1 second later its position is 3, determine its position $x(t)$ at any time t.

8. A projectile y_0 feet above the ground is shot upward with an initial velocity v_0. If only a constant acceleration k acts downward on it, determine the position of the projectile at any time t.

9. A body falls from rest from a height of 3000 feet with an acceleration of 32 feet per second squared downward.
   (a) How long does it take for the body to hit the ground?
   (b) What is the velocity of the body when it hits the ground?

10. Let f and g be differentiable functions. Find an antiderivative of the following.
    (a) $f(x)g'(x) + f'(x)g(x)$.
    (b) $[g(x)f'(x) - f(x)g'(x)]/[g(x)]^2$.

11. The marginal cost of a product is given by $-20 + 0.08x + 0.006x^2$. The total fixed cost is \$2500. What is the cost function?

12. The marginal revenue of a product is given by $48 - 0.0024x^2$. What is the revenue function? (Assume that the revenue is zero when $x = 0$.)

13. The marginal cost of producing x items of a certain product is $150 + 0.03x^2$. Determine the difference in cost between producing 100 items and 200 items.

14. Let \$7000 be the fixed cost of producing a product. If the marginal cost of production is given by $15 + 0.010x$ and the marginal revenue is given by 25, what is the cost of production at the level where profits are a maximum? What is the revenue received at this level?

15. When requested, give a reason in the following proof of the **Intermediate Value Theorem for Derivatives**: If F is differentiable on $[a, b]$ and $F'(a) < c < F'(b)$ then there exists x_0 in (a, b) such that $F'(x_0) = c$. Proof:

(a) Let $G(x) = F(x) - cx$.

(b) Then $G'(x) = F'(x) - c$.

(c) $G'(a) < 0$ and $G'(b) > 0$. Why?

(d) G has a minimum value, say $G(x_0)$, where $a \leq x_0 \leq b$. Why?

(e) $\lim_{h \downarrow 0} \{[G(a + h) - G(a)]/h\} < 0$. Why?

(f) $G(a + h) < G(a)$ for $h > 0$ and h near 0. Why?

(g) Therefore, $x_0 \neq a$. Why?

(h) $\lim_{h \uparrow 0} \{[G(b + h) - G(b)]/h\} > 0$. Why?

(i) $G(b + h) < G(b)$ for $h < 0$ and h near 0. Why?

(j) Therefore, $x_0 \neq b$. Why?

(k) Therefore, $G'(x_0) = 0$. Why?

(l) Therefore, $F'(x_0) = c$. Why?

16. Show that f defined by

$$f(x) = \begin{cases} -1, & x < 0, \\ 1, & x \geq 0, \end{cases}$$

has no antiderivative on $[-1, 1]$. [*Hint:* Assume that there exists F defined on $[-1, 1]$ such that $F'(x) = f(x)$ and use Exercise 15.]

Chapter 5

THE INTEGRAL
AND ITS APPLICATIONS

5.1   INTRODUCTION

All of Chapter 4 was devoted to the study of the limit of the difference
quotient, called the derivative of the function. As we have seen, this particular limit has many and varied uses. Now we shall direct our attention to
another special limit having diverse application; this limit will be called the
integral of a function.

Recall that the geometric concept of tangent line was used in Chapter 4 to
motivate the definition of the derivative. Historically, the study of the integral,
and the process of integration, evolved from mathematicians grappling with
the problem of determining the **area of a plane region**.

At first, ancient geometers used the notion of *square units* to measure
certain areas. It was observed that a rectangle with l units of length and h
units of width contains lh square units. For example, in the accompanying
diagram (at the top of the following page) there are 44 square units. It was
then somewhat natural to assign the number

$$lh$$

as the area of a rectangle of base l and width h.

To assign a number as the area of a triangle of base b and height h,
consider the rectangle in Figure 5.1 containing the four triangles T_1, T_2, T_3, T_4.
It is easily seen that

$$\text{area}(T_1) + \text{area}(T_2) + \text{area}(T_3) + \text{area}(T_4) = bh.$$

224

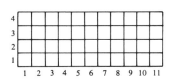

However,

$$\text{area}(T_1) = \text{area}(T_2) \quad \text{and} \quad \text{area}(T_3) = \text{area}(T_4),$$

and so we have

$$2\ \text{area}(T_2) + 2\ \text{area}(T_3) = bh$$

$$\text{area}(T_2) + \text{area}(T_3) = \tfrac{1}{2}bh;$$

that is,

$$\text{area}\ \triangle ABC = \tfrac{1}{2}bh.$$

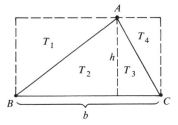

Figure 5.1

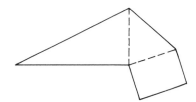

Figure 5.2

In addition, since any plane region bounded by straight lines can be sub-divided into rectangles and triangles, the area of the given region can be found by adding the areas of these constituent parts (see Figure 5.2).

But what about the areas of plane regions with curved boundaries? Approximate techniques of area measurement for such regions were intro-duced by Greek mathematicians as early as the fourth century B.C. This era could rightly be referred to as the birth of the integral calculus. Let us illustrate one of these techniques (sometimes called **methods of exhaustion**) by considering the plane region R in Figure 5.3. By drawing a grid of squares in R, one can estimate the area of R by counting the number of squares entirely within R and multiplying by the common area of these squares. For instance, if in Figure 5.3 each square has area 4, then the area of R is some-what larger than

$$4 \cdot 45 = 180.$$

Greater accuracy in estimating the area of R can be obtained by drawing a finer grid, that is, a grid with smaller squares.

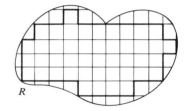

Figure 5.3

From a theoretical point of view, one could use the method of exhaustion described above to actually define the area of a plane region with a curved boundary. Again, consider the region R from above. First, draw a grid of squares in R where the side of each square has length s, and let A_1 be the sum of the areas of those squares lying entirely within R. Then draw a grid of squares in R where the side of each square has length $s/2$; let A_2 be the sum of the areas of those squares lying entirely within R (see Figure 5.4). Continue this process indefinitely with squares of sides

$$\frac{s}{4}, \quad \frac{s}{8}, \quad \frac{s}{16}, \quad \frac{s}{32}, \quad \cdots$$

Then one obtains a sequence of numbers

$$A_1 < A_2 < A_3 < \cdots.$$

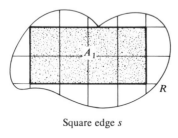
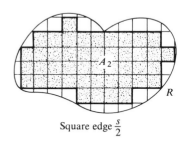

Square edge s Square edge $\frac{s}{2}$

Figure 5.4

Since each A_i in this sequence cannot exceed a certain fixed number (the area of any rectangle containing R will do), it must be that this increasing sequence of numbers closes in on (or converges to) a fixed number A. It is this number A which we call the area of region R.

With the advent of analytic geometry came a significant change in the statement of the area problem and subsequently a change in the method of attack, for one could then attempt to calculate the area of a region from the equation of the curve which bounded the region. It was noted that one could

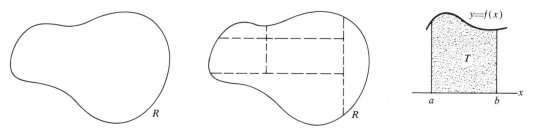

Figure 5.5

find the area of a region R if one could find areas of simpler regions, such as region T in Figure 5.5. The area of the region T is then approximated by areas of rectangles as follows: First, the interval $[a, b]$ is subdivided into intervals,

$$[a, x_1], [x_1, x_2], \ldots, [x_{n-1}, b],$$

and a number t_i is chosen in $[x_{i-1}, x_i]$. Then the sum

(5.1.1) $f(t_1)(x_1 - a) + f(t_2)(x_2 - x_1) + \cdots + f(t_n)(b - x_{n-1})$

is the sum of areas of rectangles, and this sum approximates the area of T (see Figure 5.6).

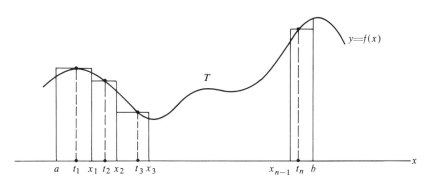

Figure 5.6

If we subdivide $[a, b]$ into smaller subintervals, then the sum 5.1.1 is a better approximation of the area of T. Therefore one is interested in finding the limit of 5.1.1 as the subintervals shrink in size.

An important development in mathematics occurred when the area under the graph of a function f was considered as a function defined on the interval $[a, b]$. For example, in Figure 5.7, $A(t)$ is the area of the shaded region. If $A(t)$ can be determined, then $A(b)$ is the area of the region T. It was discovered that

$$\frac{d}{dt} \{A(t)\} = f(t),$$

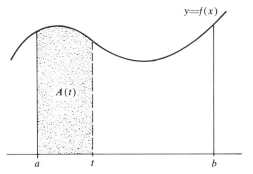

Figure 5.7

indicating that the area function we seek has a derivative equal to the given function. This discovery, which yields the **Fundamental Theorem of the Calculus,** precipitated the full development of the calculus by Isaac Newton and Gottfried Leibniz in the seventeenth century. As we shall see later in this chapter, in many instances one can use the Fundamental Theorem instead of methods of exhaustion to compute areas.

In an effort to generalize the procedure, mathematicians have determined that expression 5.1.1 converges to a number for a large class of functions f and that this convergence is independent of any interpretation in terms of area. It is this result which allows us to use the integral in the solution of a wide variety of problems.

5.2 APPROXIMATION OF AREA BY THE VERTICAL RECTANGLE METHOD

First, let us turn to the problem of computing the areas of certain types of plane regions.

Consider the region R of the plane described analytically by

$$0 \le y \le f(x) \qquad \text{and} \qquad a \le x \le b,$$

where f is a continuous, nonnegative function on $[a, b]$ (see Figure 5.8). It is our purpose in this section to describe a procedure, the **method of vertical rectangles,** for computing the area A of region R. We begin by using vertical rectangles to approximate certain areas. (Another method, using trapezoids, is outlined in Exercise 5 at the end of this section.)

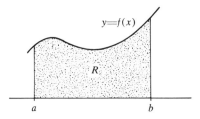

Figure 5.8

EXAMPLE 1   Approximate the area A of the region R bounded by the curves

$$y = 3\sqrt{x}, \qquad x = 0,$$

$$y = 0, \qquad\qquad x = 4 \qquad \text{(see Figure 5.9)}.$$

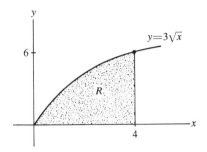

Figure 5.9

Divide the interval $[0, 4]$ by the points

$$0, \quad 1, \quad 2, \quad 3, \quad 4$$

into the four subintervals

$$[0, 1], \qquad [1, 2], \qquad [2, 3], \qquad [3, 4]$$

as shown in Figure 5.10. Such a collection of points is called a **partition of the interval** $[0, 4]$. The length of the longest subinterval is called the **norm of the partition**. In this case, each subinterval has length 1, and so the norm is 1.

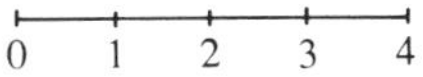

Figure 5.10

In Figure 5.11 we inscribe rectangles above these four subintervals, and the area of this shaded region is

$$3\sqrt{1}(1) + 3\sqrt{2}(1) + 3\sqrt{3}(1) = 12.44.$$

Or, if we circumscribe rectangles above these subintervals, as in Figure 5.12, then the area of this shaded region is

$$3\sqrt{1}(1) + 3\sqrt{2}(1) + 3\sqrt{3}(1) + 3\sqrt{4}(1) = 18.44.$$

Therefore we can conclude that

$$12.44 \le A \le 18.44.$$

Later in this chapter we shall be able to conclude that A is exactly 16.

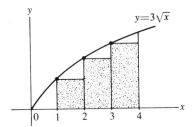

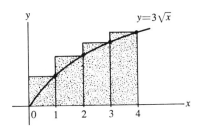

Figure 5.11

Figure 5.12

EXAMPLE 2   Approximate the area A of the region R bounded by the curves

$$y = \frac{1}{x}, \qquad x = 1$$

$$y = 0, \qquad x = 2 \qquad \text{(see Figure 5.13)}.$$

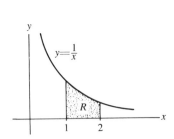

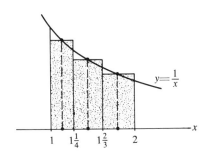

Figure 5.13

Figure 5.14

For the sake of variety, let us take a partition of $[1, 2]$ whose subintervals have varying lengths—for instance, the partition

$$1, \qquad 1\tfrac{1}{4}, \qquad 1\tfrac{2}{3}, \qquad 3;$$

the norm of this partition is $\tfrac{4}{3}$. Consider the midpoints of the three subintervals formed. The area of the shaded region in Figure 5.14 is

$$\frac{1}{(\tfrac{5}{4} + 1)/2}\,(\tfrac{5}{4} - 1) + \frac{1}{(\tfrac{5}{3} + \tfrac{5}{4})/2}\,(\tfrac{5}{3} - \tfrac{5}{4}) + \frac{1}{(3 + \tfrac{5}{3})/2}\,(3 - \tfrac{5}{3}) = 1.0794.$$

For an even better estimate of A, consider the partition

$$1, \qquad 1\tfrac{1}{8}, \qquad 1\tfrac{1}{4}, \qquad 1\tfrac{1}{2}, \qquad 1\tfrac{2}{3}, \qquad 1\tfrac{4}{5}, \qquad 2$$

of $[1, 2]$; the norm of this partition is $\tfrac{1}{4}$.

If we once again choose the midpoint of each subinterval,

$$\frac{1}{\frac{1}{2}(\frac{9}{8}+1)}(\frac{9}{8}-1) + \frac{1}{\frac{1}{2}(\frac{5}{4}+\frac{9}{8})}(\frac{5}{4}-\frac{9}{8}) + \frac{1}{\frac{1}{2}(\frac{3}{2}+\frac{5}{4})}(\frac{3}{2}-\frac{5}{4})$$

$$+ \frac{1}{\frac{1}{2}(\frac{5}{3}+\frac{3}{2})}(\frac{5}{3}-\frac{3}{2}) + \frac{1}{\frac{1}{2}(\frac{9}{5}+\frac{5}{3})}(\frac{9}{5}-\frac{5}{3}) + \frac{1}{\frac{1}{2}(2+\frac{9}{5})}(2-\frac{9}{5}) = 0.6922$$

is another approximation of A. In Chapter 6 we shall be able to compute A exactly.

In the next lengthy example we repeat the above procedure for partitions of smaller and smaller norms. This is done so as to indicate the exact area of a given region. As we shall see, it makes no difference whether the rectangles are inscribed, circumscribed, or "in between" (see Figure 5.15). The narrower the subinterval, the closer the area of any of these rectangles comes to the area above that subinterval.

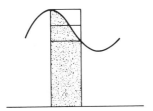

Figure 5.15

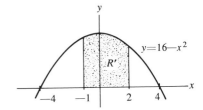

Figure 5.16

EXAMPLE 3 Consider the region R' bounded by the curves

$$x = -1, \quad x = 2, \quad y = 0, \quad y = 16 - x^2$$

(see Figure 5.16). We wish to determine the number A' which is the area of this region.

Let us begin by approximating A' in four different ways. Divide the interval $[-1, 2]$ by the points

$$-1, \quad -\tfrac{1}{2}, \quad \tfrac{1}{2}, \quad \tfrac{9}{10}, \quad \tfrac{4}{3}, \quad 2$$

into the five subintervals

$$[-1, -\tfrac{1}{2}], \quad [-\tfrac{1}{2}, \tfrac{1}{2}], \quad [\tfrac{1}{2}, \tfrac{9}{10}], \quad [\tfrac{9}{10}, \tfrac{4}{3}], \quad [\tfrac{4}{3}, 2],$$

as shown in Figure 5.17. Our four approximations to A' are obtained as shown on the following two pages.

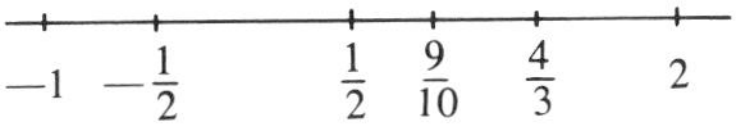

$$-1 \quad -\frac{1}{2} \qquad \frac{1}{2} \quad \frac{9}{10} \quad \frac{4}{3} \qquad 2$$

Figure 5.17

1. Choose the left end point in each subinterval. In this case,

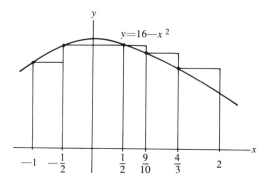

Figure 5.18

$$S_5(l) = f(-1)[-\tfrac{1}{2} - (-1)] + f(-\tfrac{1}{2})[\tfrac{1}{2} - (-\tfrac{1}{2})] + f(\tfrac{1}{2})[\tfrac{9}{10} - \tfrac{1}{2}]$$
$$+ f(\tfrac{9}{10})[\tfrac{4}{3} - \tfrac{9}{10}] + f(\tfrac{4}{3})[2 - \tfrac{4}{3}]$$
$$= (16 - 1)\tfrac{1}{2} + (16 - \tfrac{1}{4})1 + (16 - \tfrac{1}{4})\tfrac{2}{5}$$
$$+ (16 - \tfrac{81}{100})\tfrac{13}{30} + (16 - \tfrac{16}{9})\tfrac{2}{3}$$
$$= 45.61.$$

Since each of the five addends in $S_5(l)$ equals the area of one of the five rectangles shown in Figure 5.18, $S_5(l)$ approximates A'.

2. Choose the right end point in each subinterval. In this case,

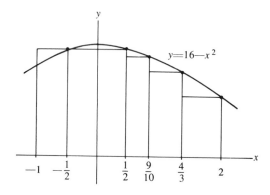

Figure 5.19

$$S_5(r) = f(-\tfrac{1}{2})[-\tfrac{1}{2} - (-1)] + f(\tfrac{1}{2})[\tfrac{1}{2} - (-\tfrac{1}{2})] + f(\tfrac{9}{10})[\tfrac{9}{10} - \tfrac{1}{2}]$$
$$+ f(\tfrac{4}{3})[\tfrac{4}{3} - \tfrac{9}{10}] + f(2)[2 - \tfrac{4}{3}]$$
$$= (16 - \tfrac{1}{4})\tfrac{1}{2} + (16 - \tfrac{1}{4})1 + (16 - \tfrac{91}{100})\tfrac{2}{5}$$
$$+ (16 - \tfrac{16}{9})\tfrac{13}{30} + (16 - 4)\tfrac{2}{3}$$
$$= 43.82.$$

Since each of the five addends in $S_5(r)$ equals the area of one of the five rectangles shown in Figure 5.19, then $S_5(r)$ approximates A'.

3. Choose the midpoint of each subinterval. In this case,

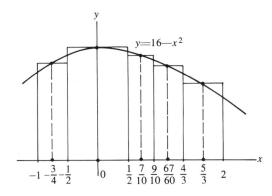

Figure 5.20

$$S_5(m) = f(-\tfrac{3}{4})[-\tfrac{1}{2} - (-1)] + f(0)[\tfrac{1}{2} - (-\tfrac{1}{2})] + f(\tfrac{7}{10})[\tfrac{9}{10} - \tfrac{1}{2}]$$
$$+ f(\tfrac{67}{60})[\tfrac{4}{3} - \tfrac{9}{10}] + f(\tfrac{5}{3})[2 - \tfrac{4}{3}]$$
$$= (16 - \tfrac{9}{16})\tfrac{1}{2} + (16 - 0)1 + (16 - \tfrac{49}{100})\tfrac{2}{5}$$
$$+ (16 - \tfrac{4489}{900})\tfrac{13}{30} + (16 - \tfrac{25}{9})\tfrac{2}{3}$$
$$= 43.51.$$

Since each of the five addends in $S_5(m)$ equals the area of one of the five rectangles shown in Figure 5.20, then $S_5(m)$ approximates A'.

4. Choose a random point in each interval; let us choose $-\tfrac{4}{7}, \tfrac{1}{10}, \tfrac{3}{4}, 1, \tfrac{3}{2}$.

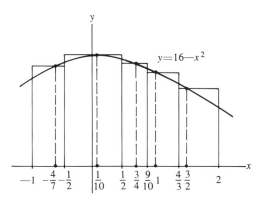

Figure 5.21

In this case,

$$S_5(ran) = f(-\tfrac{4}{7})[-\tfrac{1}{2} - (-1)] + f(\tfrac{1}{10})[\tfrac{1}{2} - (-\tfrac{1}{2})]$$
$$+ f(\tfrac{3}{4})[\tfrac{9}{10} - \tfrac{1}{2}] + f(1)[\tfrac{4}{3} - \tfrac{9}{10}] + f(\tfrac{3}{2})[2 - \tfrac{4}{3}]$$
$$= (16 - \tfrac{16}{49})\tfrac{1}{2} + (16 - \tfrac{1}{100})1 + (16 - \tfrac{9}{16})\tfrac{2}{5}$$
$$+ (16 - 1)\tfrac{13}{30} + (16 - \tfrac{9}{4})\tfrac{2}{3}$$
$$= 45.67.$$

Since each of the five addends in $S_5(ran)$ equals the area of one of the five rectangles shown in Figure 5.21, then $S_5(ran)$ approximates A'.

Again, one can continue this procedure: Given a positive integer n, choose a partition of $[-1, 2]$ having n subintervals; then compute the sums $S_n(l)$, $S_n(r)$, $S_n(m)$, $S_n(ran)$. This has been done for various values of n, and the results are tabulated below. Since it serves no purpose, the particular partition points in each case are not listed.

(5.2.1)   Values of $S_n(l)$, $S_n(r)$, $S_n(m)$, $S_n(ran)$, where $[-1, 2]$ has been divided into n subintervals of varying lengths:

n	$S_n(l)$	$S_n(r)$	$S_n(m)$	$S_n(ran)$
5	45.61	43.82	43.51	45.67
10	46.91	42.10	45.23	46.70
25	45.47	44.43	45.02	44.90
50	45.13	44.85	45.005	45.002
100	45.095	44.897	45.001	44.988
200	45.042	44.957	45.0003	45.011
300	45.034	44.965	45.0002	45.0036
400	45.0194	44.9801	45.0001	44.9980
500	45.0173	44.9825	45.00006	45.00004
600	45.0146	44.9892	45.00003	44.9989
700	45.0123	44.9910	45.00001	44.9985
800	45.0112	44.9921	45.000009	44.9993
1000	45.0101	44.9928	45.000006	45.00002

Since each of these columns appears to be closing in on 45, we conclude that $A' = 45$.

Before we summarize our work in Example 3 and proceed to a more general case, it will be helpful to introduce some convenient notation. The **sigma notation** is used to represent a finite sum of numbers; namely, if n is a positive integer, then

$$a_1 + a_2 + \cdots + a_n$$

will be written as

$$\sum_{i=1}^{n} a_i.$$

Here i is called the **index** and a_i the **general term**. In expanding any summation written in sigma notation, all one has to do is successively substitute into the general term each positive integer in the given range and add the resulting terms. For example,

$$\sum_{i=1}^{5} i = 1 + 2 + 3 + 4 + 5 = 15$$

$$\sum_{i=1}^{4} i^2 = 1 + 4 + 9 + 16 = 30$$

$$\sum_{i=3}^{6} (i + 2) = 5 + 6 + 7 + 8 = 26.$$

It is easy to verify the following facts about the sigma notation:

$$\sum_{i=1}^{n} ka_i = k \sum_{i=1}^{n} a_i$$

$$\sum_{i=1}^{n} (a_i + b_i) = \sum_{i=1}^{n} a_i + \sum_{i=1}^{n} b_i$$

$$\sum_{i=1}^{n} a_i = \sum_{i=1}^{m} a_i + \sum_{i=m+1}^{n} a_i \qquad (1 \le m \le n).$$

Also, given a partition

$$a = x_0 < x_1 < x_2 < \cdots < x_n = b$$

of the interval $[a, b]$, we shall use the **delta notation**

$$\Delta x_i$$

to represent the length

$$x_i - x_{i-1}$$

of the ith subinterval $[x_{i-1}, x_i]$, where $i = 1, 2, \ldots, n$.

With this notation in hand, let us summarize the work done in Example 3. Let $f(x) = 16 - x^2$; let points

$$-1 = x_0 < x_1 < x_2 < \cdots < x_{n-1} < x_n = 2$$

be a partition of $[-1, 2]$, where n is a positive integer, and for each $i = 1, 2, \ldots, n$ let t_i be any point in subinterval $[x_{i-1}, x_i]$ (see Figure 5.22). Then

$$f(t_1)(x_1 - x_0) + f(t_2)(x_2 - x_1) + \cdots + f(t_n)(x_n - x_{n-1})$$

$$= f(t_1) \Delta x_1 + f(t_2) \Delta x_2 + \cdots + f(t_n) \Delta x_n$$

$$= \sum_{i=1}^{n} f(t_i) \Delta x_i$$

is an approximation of the area of region R', and this approximation gets

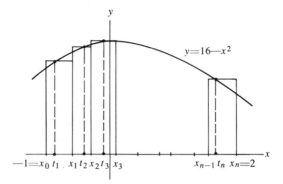

Figure 5.22

better as the norm of the partition becomes smaller. However, Table 5.2.1 indicates that

$$\sum_{i=1}^{n} f(t_i)\,\Delta x_i$$

can be made as close to 45 as we like by choosing any partition of $[-1, 2]$ whose norm is sufficiently small. Therefore we conclude that the area of R' equals 45.

Now we shall use this approach to treat the general case mentioned at the beginning of this section.

Let R be the plane region bounded by the curves

$$x = a, \qquad x = b, \qquad y = 0, \qquad y = f(x),$$

where $a < b$ and f is a continuous, nonnegative function on $[a, b]$. Let the points

$$a = x_0 < x_1 < x_2 < \cdots < x_{n-1} < x_n = b$$

be a partition of $[a, b]$, where n is a positive integer, and for each $i = 1, 2, \ldots, n$ let t_i be any point in the subinterval $[x_{i-1}, x_i]$ (see Figure 5.23).

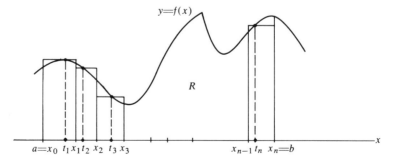

Figure 5.23

Then

$$f(t_1)(x_1 - x_0) + f(t_2)(x_2 - x_1) + \cdots + f(t_n)(x_n - x_{n-1}) = \sum_{i=1}^{n} f(t_i)\,\Delta x_i$$

is an approximation of the area of the region R, and this approximation gets better as the norm of the partition becomes smaller.

The renowned Greek thinker Archimedes (287–212 B.C.) used inscribed rectangles and circumscribed rectangles—although not the coordinatized plane—to determine the areas of certain regions. In particular, from his work we know that the area of the region S given by

$$0 \leq y \leq x^2$$

$$0 \leq x \leq b$$

is

$$\frac{b^3}{3} \quad \text{(see Figure 5.24)}.$$

In Section 5.6 we shall verify this result.

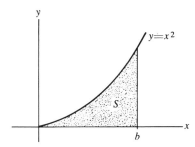

Figure 5.24

EXERCISES

1. Approximate the area of the region bounded by the curves $y = x + 1$, $y = 0$, $x = 1$, $x = 3$ by computing $S_4(l)$ and $S_4(r)$ using the partition $1, \frac{3}{2}, 2, \frac{5}{2}, 3$ of $[1, 3]$. How do your approximations compare with the exact answer? [Recall that the area of a trapezoid with bases b_1, b_2 and height h is $\frac{1}{2}h(b_1 + b_2)$.]

2. Approximate the area of the region bounded by the curves $y = \sqrt{4 - x^2}$ and $y = 0$ by the following.

    (a) Compute $S_4(m)$ using the partition $-2, -1, 0, 1, 2$ of the interval $[-2, 2]$.
    (b) Compute $S_8(r)$ using the partition $-2, -\frac{3}{2}, -1, -\frac{1}{2}, 0, \frac{1}{2}, 1, \frac{3}{2}, 2$ of the interval $[-2, 2]$.
    How do your approximations compare with the exact answer? [Recall that the area of a circle of radius r is πr^2.]

3. Approximate the area of the region bounded by the curves $y = 2x^2$, $y = 0$, and $x = 5$ by the following.

   (a) Compute $S_5(l)$ and $S_5(r)$ using the partition 0, 1, 2, 3, 4, 5 of the interval $[0, 5]$.

   (b) Compute $S_6(m)$ using the partition 0, $\frac{1}{2}$, $\frac{3}{2}$, 2, $\frac{8}{3}$, 4, 5 of the interval $[0, 5]$.

4. Approximate the area of the region bounded by the curves $y = 4/(1 + x^2)$, $x = 0$, and $x = 1$ by computing $S_4(l)$ and $S_4(r)$ using the partition 0, $\frac{1}{4}$, $\frac{1}{2}$, $\frac{3}{4}$, 1 of the interval $[0, 1]$.

5. The method of approximating area given in the text can be modified by using areas of trapezoids instead of rectangles. For example, let $y = f(x) > 0$ and take the regular partition $a_0 = x_0, x_1, x_2, \ldots, x_n = b$ of $[a, b]$ and let $y_k = f(x_k)$. The area of the shaded trapezoid in Figure 5.25 is $\frac{1}{2}(y_k - y_{k+1})h$. Therefore the area A is approximately $\frac{1}{2}(y_0 + y_1)h + \frac{1}{2}(y_1 + y_2)h + \cdots + \frac{1}{2}(y_{n-1} + y_n)h = (\frac{1}{2}y_0 + y_1 + y_2 + \cdots + y_{n-1} + \frac{1}{2}y_n)h$. This method of approximation is called the **trapezoidal rule**. Use this method in Exercises 1 and 3.

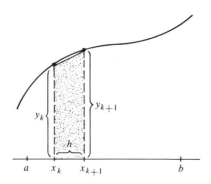

Figure 5.25

6. Express each of the following sums using the sigma notation.

   (a) $1 + 2 + 3 + 4$. (b) $2 + 4 + 6 + 8 + 10$.
   (c) $3 + 4 + 5 + 6 + 7$. (d) $9 + 16 + 25 + 36$.
   (e) $16 + 36 + 64 + 100 + 144 + 196$.

7. Let $f(x) = 2x^2$. Compute each of the following sums.

   (a) $\sum_{i=1}^{4} f(i)$. (b) $\sum_{i=1}^{3} f(2i)$.
   (c) $\sum_{i=1}^{5} f(i)\,\Delta x_i$, where $\Delta x_i = 2$. (d) $\sum_{i=1}^{5} f(i + 1)\,\Delta x_i$, where $\Delta x_i = 2$.
   (e) $\sum_{i=1}^{5} f(i)\,\Delta x_i$, where $\Delta x_i = 2i$.

8. Verify that for any positive integer n, the following hold.

   (a) $\sum_{i=1}^{n} ka_i = k \sum_{i=1}^{n} a_i$.
   (b) $\sum_{i=1}^{n} (a_i + b_i) = \sum_{i=1}^{n} a_i + \sum_{i=1}^{n} b_i$.
   (c) $\sum_{i=1}^{n} a_i = \sum_{i=1}^{m} a_i + \sum_{i=m+1}^{n} a_i$ $(1 \leq m \leq n)$.

9. Show that for any positive integer n, $\sum_{i=1}^{n} i = [n(n + 1)]/2$.

5.3 THE DEFINITE INTEGRAL

The method of vertical rectangles discussed in the preceding section can be generalized to define another of the key concepts of the calculus—the definite integral. This we now proceed to do, followed by an important fact concerning the definite integral (5.3.3) and a few of its important properties.

5.3.1

Definition   Let f be a function defined on $[a, b]$ and let L be a number with the following property:

$$\sum_{i=1}^{n} f(t_i)\, \Delta x_i$$

can be made as close to L as we like by choosing any partition

$$a = x_0 < x_1 < x_2 < \cdots < x_n = b$$

of $[a, b]$ whose norm is sufficiently small, where t_i is any point in $[x_{i-1}, x_i]$, $i = 1, 2, \ldots, n$.

Then the number L is called **the definite integral of f on $[a, b]$** and is denoted by the symbol

$$\int_a^b f(x)\, dx.$$

In this case we say that f is **integrable on $[a, b]$**.

Remarks

1. The above definition applies when $a < b$. For the sake of convenience we define

$$\int_a^a f(x)\, dx = 0$$

and

$$\int_a^b f(x)\, dx = -\int_b^a f(x)\, dx$$

when $a > b$.

2. The sum

$$\sum_{i=1}^{n} f(t_i)\, \Delta x_i$$

is called a **Riemann sum** after the German mathematician George B. Riemann (1826–1866). In the case that f is integrable on $[a, b]$, if we think of the Riemann sums as closing in on L as the norms of the partitions shrink toward zero,

$$\sum_{i=1}^{n} f(t_i)\, \Delta x_i \to L,$$

then it is quite natural to denote this number L by the symbol

$$\int_a^b f(x)\, dx.$$

As we shall see, this notation, due to Leibniz, is often helpful in evaluating the integral. In the expression

$$\int_a^b f(x)\, dx$$

the term $f(x)$ is sometimes called the **integrand**.

3. Definition 5.3.1 is not as precise a definition of the definite integral as is possible. Those who desire a more precise statement are referred to the optional supplement following the Exercises to this section.

EXAMPLE 1   Let f be the function defined on $[a, b]$ by

$$f(x) = k,$$

where k is a constant. Then for any partition

$$a = x_0 < x_1 < x_2 < \cdots < x_n = b$$

of $[a, b]$ and any choice of

$$t_i \text{ in } [x_{i-1}, x_1] \qquad (i = 1, 2, \ldots, n)$$

we have

$$\sum_{i=1}^{n} f(t_i)\, \Delta x_i = \sum_{i=1}^{n} k\, \Delta x_i$$

$$= k \sum_{i=1}^{n} \Delta x_i$$

$$= k(x_1 - x_0 + x_2 - x_1 + \cdots + x_{n-2} - x_{n-1}$$
$$ + x_n - x_{n-1})$$

$$= k(x_n - x_0)$$

$$= k(b - a).$$

Since each Riemann sum is equal to the constant

$$k(b - a),$$

it follows from Definition 5.3.1 that

(5.3.2) $$\int_a^b k \, dx = k(b - a)$$

for any constant k.

The following example shows that the number L in 5.3.1 need not exist.

EXAMPLE 2   Consider the function f defined on $[0, 1]$ by the following: For any x in $[0, 1]$,

$$f(x) = \begin{cases} 1, & x \text{ rational,} \\ 0, & x \text{ irrational.} \end{cases}$$

Suppose that f is integrable on $[0, 1]$ and that

$$\int_0^1 f(x) \, dx = L.$$

Then, by Definition 5.3.1,

$$\sum_{i=1}^n f(t_i) \, \Delta x_i$$

can be made as close to L as we wish by choosing any partition of $[0, 1]$ whose norm is sufficiently small. In particular, we can get the Riemann sum within $\frac{1}{2}$ of L,

$$\left| \sum_{i=1}^n f(t_i) \, \Delta x_i - L \right| < \frac{1}{2},$$

by choosing any partition of $[0, 1]$ whose norm is small enough.

Let P be any partition of $[0, 1]$. Regardless of the smallness of the norm of P, we can choose each t_i to be rational, in which case

$$\sum_{i=1}^n f(t_i) \, \Delta x_i = \sum_{i=1}^n \Delta x_i = 1.$$

Hence

$$\left| \sum_{i=1}^n f(t_i) \, \Delta x_i - L \right| < \frac{1}{2}$$

$$|1 - L| < \frac{1}{2}$$

$$-\frac{1}{2} < L - 1 < \frac{1}{2}$$

$$\frac{1}{2} < L < \frac{3}{2}.$$

On the other hand, again regardless of the smallness of the norm of P, we can choose each t_i to be irrational, in which case

$$\sum_{i=1}^{n} f(t_i) \, \Delta x_i = 0.$$

Thus

$$\left| \sum_{i=1}^{n} f(t_i) \, \Delta x_i - L \right| < \frac{1}{2}$$

$$|0 - L| < \frac{1}{2}$$

$$-\frac{1}{2} < L < \frac{1}{2}.$$

However, it is a contradiction to say that

$$L > \tfrac{1}{2} \quad \text{and} \quad L < \tfrac{1}{2}.$$

Therefore there exists no such number L, and f is not integrable on $[0, 1]$.

The question of which functions are integrable is a difficult one. The following important result is a partial answer.

5.3.3

Fact

If f is continuous on $[a, b]$, then f is integrable on $[a, b]$.

We say this is a partial answer to the stated question because there are noncontinuous functions which are integrable. An example is the greatest integer function

$$f(x) = [x]$$

on any interval $[a, b]$. (See Exercises 9 and 10 at the end of this section.)

The proof of 5.3.3 is difficult and will be omitted here. However, the result is certainly reasonable in the case of f being continuous and nonnegative on $[a, b]$ (see Figure 5.26), for then, if A is the area of R,

$$\sum_{i=1}^{n} f(t_i) \, \Delta x_i$$

can be made as close to A as we like by choosing any partition of $[a, b]$ whose norm is small enough. Thus

$$\int_a^b f(x) \, dx = A.$$

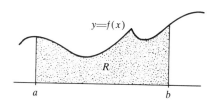

Figure 5.26

Because of 5.3.3 we know that the number

$$L = \int_a^b f(x)\, dx$$

exists whenever f is continuous on $[a, b]$. In this case we may compute L by using any particular sequence of partitions of $[a, b]$ having shrinking norms and by choosing the points t_i to be particular points in the subintervals.

EXAMPLE 3   Consider the function f defined on $[a, b]$ by

$$f(x) = x.$$

Since f is continuous on $[a, b]$, by 5.3.3, f is integrable on $[a, b]$. We can determine

$$L = \int_a^b x\, dx$$

as follows.

For partition P given by

$$a = x_0 < x_1 < x_2 < \cdots < x_n = b,$$

choose t_i to be the midpoint of $[x_{i-1}, x_i]$; that is,

$$t_i = \frac{x_i + x_{i-1}}{2} \qquad (i = 1, 2, \ldots, n).$$

Then we have

$$\sum_{i=1}^n f(t_i)\, \Delta x_i = \sum_{i=1}^n \left(\frac{x_i + x_{i-1}}{2} \right)(x_i - x_{i-1})$$

$$= \frac{1}{2} \sum_{i=1}^n (x_i^2 - x_{i-1}^2)$$

$$= \frac{1}{2}(x_1^2 - x_0^2 + x_2^2 - x_1^2 + \cdots + x_{n-1}^2 - x_{n-2}^2$$

$$+ x_n^2 - x_{n-1}^2)$$

$$= \frac{1}{2}(x_n^2 - x_0^2)$$

$$= \frac{1}{2}(b^2 - a^2).$$

Thus by Definition 5.3.1,

(5.3.4) $$\int_a^b x \, dx = \frac{b^2 - a^2}{2}.$$

EXAMPLE 4   Consider the function f defined on $[-1, 3]$ by
$$f(x) = x^2 - 1.$$

Since f is continuous on $[-1, 3]$, by Fact 5.3.3 f is integrable on $[-1, 3]$. One method of finding the number
$$L = \int_{-1}^3 (x^2 - 1) \, dx$$

is as follows.

For a positive integer n, consider the partition

$$-1 < -1 + 4\left(\frac{1}{n}\right) < -1 + 4\left(\frac{2}{n}\right) < -1 + 4\left(\frac{3}{n}\right) < \cdots$$

$$\cdots < -1 + 4\left(\frac{n}{n}\right) = 3$$

of $[-1, 3]$, so that each subinterval has length $4/n$. For each $i = 1, 2, 3, \ldots, n$ choose

$$t_i = -1 + 4\left(\frac{i}{n}\right);$$

that is, t_i is the right end point of the ith subinterval (see Figure 5.27). Then

$$\sum_{i=1}^n f(t_i) \, \Delta x_i = \sum_{i=1}^n (t_i^2 - 1) \, \Delta x_i$$

$$= \sum_{i=1}^n \left[\left(-1 + 4\left(\frac{i}{n}\right)\right)^2 - 1\right]\left(\frac{4}{n}\right)$$

$$= \sum_{i=1}^n \left(\frac{16}{n^2} i^2 - \frac{8}{n} i\right)\left(\frac{4}{n}\right)$$

$$= \frac{64}{n^3} \sum_{i=1}^n i^2 - \frac{32}{n^2} \sum_{i=1}^n i\;*$$

$$= \frac{64}{n^3}\left[\frac{n(n+1)(2n+1)}{6}\right] - \frac{32}{n^2}\left[\frac{n(n+1)}{2}\right]$$

$$= \frac{16}{3} + \frac{16}{n} + \frac{32}{3n^2}.$$

	t_1	t_2	t_3			t_n

-1 $-1+4(\frac{1}{n})$ $-1+4(\frac{2}{n})$ $-1+4(\frac{n-1}{n})$ $-1+4(\frac{n}{n})=3$

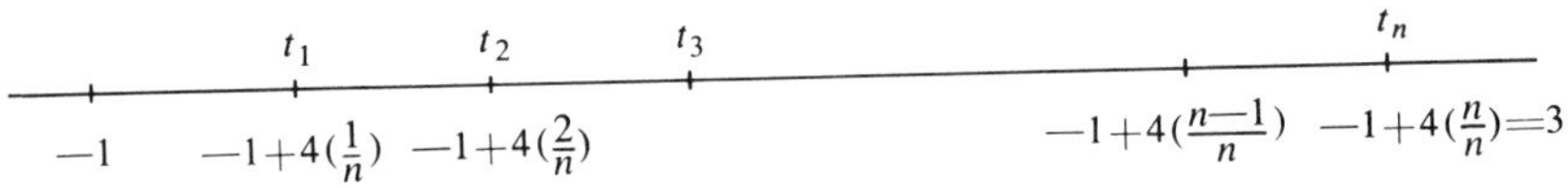

Figure 5.27

$\quad * \sum_{i=1}^n i^2 = [n(n+1)(2n+1)]/6$ and $\sum_{i=1}^n i = [n(n+1)]/2$.

Since $16/n$ and $32/3n^2$ can each be made as close to zero as we like by choosing any partition of $[-1, 3]$ whose norm is sufficiently small (this forces n to be large),

$$\sum_{i=1}^{n} f(t_i) \, \Delta x_i = \frac{16}{3} + \frac{16}{n} + \frac{32}{3n^2}$$

can be made as close to $\frac{16}{3}$ as we like by choosing any partition of $[-1, 3]$ whose norm is sufficiently small. Therefore

$$\int_{-1}^{3} (x^2 - 1) \, dx = \frac{16}{3}.$$

In general it is extremely difficult and tedious to evaluate

$$\int_{a}^{b} f(x) \, dx$$

using Definition 5.3.1. Fortunately, in many cases there is an easier way to determine this number, as we shall see in Section 5.4.

We conclude this section by stating several properties of the definite integral and illustrating how they can be used in computations.

5.3.5

Properties of the Definite Integral

1. If f is integrable on $[a, b]$, then for any constant k, the function kf is integrable on $[a, b]$ and

$$\int_{a}^{b} kf(x) \, dx = k \int_{a}^{b} f(x) \, dx.$$

2. If f, g are integrable on $[a, b]$, then the function $f + g$ is integrable on $[a, b]$ and

$$\int_{a}^{b} [f(x) + g(x)] \, dx = \int_{a}^{b} f(x) \, dx + \int_{a}^{b} g(x) \, dx.$$

3. If f is integrable on $[a, c]$ and $[c, b]$, then f is integrable on $[a, b]$ and

$$\int_{a}^{b} f(x) \, dx = \int_{a}^{c} f(x) \, dx + \int_{c}^{b} f(x) \, dx.$$

4. If f is integrable on $[a, b]$ and m, M are numbers such that

$$m \leq f(x) \leq M$$

for each x in $[a, b]$, then

$$m(b - a) \leq \int_{a}^{b} f(x) \, dx \leq M(b - a).$$

Properties 1, 2, and 3 follow from Definition 5.3.1 and the three facts about sigma notation listed in Section 5.2. For instance, to verify 2, let

$$\int_a^b f(x)\, dx = L \qquad \text{and} \qquad \int_a^b g(x)\, dx = M.$$

Then

$$\sum_{i=1}^n f(t_i)\, \Delta x_i$$

can be made as close to L as we like by choosing any partition of $[a, b]$ whose norm is sufficiently small, and

$$\sum_{i=1}^n g(t_i)\, \Delta x_i$$

can be made as close to M as we like by choosing any partition of $[a, b]$ whose norm is sufficiently small. Hence

$$\sum_{i=1}^n [f(t_i) + g(t_i)]\, \Delta x_i = \sum_{i=1}^n f(t_i)\, \Delta x_i + \sum_{i=1}^n g(t_i)\, \Delta x_i$$

can be made as close to $L + M$ as we like by choosing any partition of $[a, b]$ whose norm is sufficiently small. That is,

$$\int_a^b [f(x) + g(x)]\, dx = L + M.$$

Complete proofs of Properties 2 and 4 are given in the optional supplement to this section.

EXAMPLE 5
$$\int_1^2 5x\, dx = 5 \int_1^2 x\, dx \qquad [5.3.5(1)]$$

$$= 5 \left(\frac{2^2 - 1^2}{2} \right) \qquad (5.3.4)$$

$$= \tfrac{15}{2}.$$

EXAMPLE 6
$$\int_2^3 (3x - 1)\, dx = \int_2^3 3x\, dx + \int_2^3 (-1)\, dx \qquad [5.3.5(2)]$$

$$= 3 \int_2^3 x\, dx + \int_2^3 (-1)\, dx \qquad [5.3.5(1)]$$

$$= 3 \left(\frac{3^2 - 2^2}{2} \right) + (-1)(3 - 2) \qquad (5.3.4, 5.3.3)$$

$$= \tfrac{13}{2}.$$

EXAMPLE 7

$$\int_{-3}^{1} |x| \, dx = \int_{-3}^{0} |x| \, dx + \int_{0}^{1} |x| \, dx \qquad [5.3.5(3)]$$

$$= \int_{-3}^{0} -x \, dx + \int_{0}^{1} x \, dx$$

$$= -\int_{-3}^{0} x \, dx + \int_{0}^{1} x \, dx \qquad [5.3.5(1)]$$

$$= -\left(\frac{0^2 - (-3)^2}{2}\right) + \left(\frac{1^2 - 0^2}{2}\right)$$

$$= 5.$$

EXAMPLE 8   Find numbers p and q such that

$$p \le \int_{-1}^{2} (9 - x^2) \, dx \le q.$$

Consider the graph of $y = 9 - x^2$ on $[-1, 2]$ as given in Figure 5.28. On $[-1, 2]$, $9 - x^2$ has maximum value of 9 at $x = 0$ and minimum value of 5 at $x = 2$. Hence, by 5.3.5(4),

$$5(3) \le \int_{-1}^{2} (9 - x^2) \, dx \le 9(3);$$

that is,

$$15 \le \int_{-1}^{2} (9 - x^2) \, dx \le 27.$$

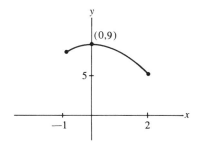

Figure 5.28

In the next section we shall establish a technique which will enable us to quickly show that

$$\int_{-1}^{2} (9 - x^2) \, dx = 24.$$

EXERCISES

1. Use 5.3.2, 5.3.4, or 5.3.5 to evaluate each of the following.

   (a) $\int_0^1 2 \, dx.$

   (b) $\int_1^3 -x/5 \, dx.$

   (c) $\int_{-2}^{-1} (x + 3) \, dx.$

   (d) $\int_{-3}^1 (6 - x/4) \, dx.$

2. Evaluate the following.

   (a) $\int_0^2 |x - 1| \, dx.$

   (b) $\int_{-1}^1 f(x) \, dx,$ where $f(x) = \begin{cases} -1, & -1 \le x < 0, \\ x - 1, & 0 \le x \le 1. \end{cases}$

3. Evaluate $\int_{-1}^2 (x - 1) \, dx$, $\int_2^4 (x - 1) \, dx$, and $\int_{-1}^4 (x - 1) \, dx$, and use these results to illustrate 5.3.5(3).

4. (a) Find numbers p and q such that $p \le \int_1^3 \sqrt{1 + x^4} \, dx \le q.$

   (b) Find numbers p and q such that $p \le \int_1^2 (1/x) \, dx \le q.$

   (c) Find numbers p and q such that $p \le \int_{-3}^3 [1/(1 + x^2)] \, dx \le q.$

5. Evaluate $\int_0^2 [|x| + |1 - x|] \, dx.$

6. Consider the function f defined on $[a, b]$ by $f(x) = x$ and the partition $a < a + [(b - a)/n] < a + 2[(b - a)/n] < \cdots < a + n[(b - a)/n] = b$ of $[a, b]$.

   (a) Determine the Riemann sum if t_i is chosen as the right end point of the subinterval $[x_{i-1}, x_i]$; that is, $t_i = a + i[(b - a)/n]$, where $i = 1, 2, \ldots, n$. [Hint: $\sum_{i=1}^n i = [n(n + 1)]/2.$]

   (b) Show that the Riemann sum in part (a) can be made as close to $(b^2 - a^2)/2$ as we like by choosing n sufficiently large.

   (c) Conclude from part (b) that $\int_a^b x \, dx = (b^2 - a^2)/2.$

7. Consider the function defined on $[0, b]$, $b > 0$, by $f(x) = x^2$ and the partition $0 < b/n < 2b/n < 3b/n < \cdots < nb/n = b$, of $[0, b]$.

   (a) Determine the Riemann sum if t_i is chosen as the left end point of the subinterval $[x_{i-1}, x_i]$; that is, $t_i = [(i - 1)b]/n$, where $i = 1, 2, 3, \ldots, n$.

   (b) Show that the Riemann sum in part (a) can be made as close to $b^3/3$ as we like by choosing n sufficiently large.

   (c) Conclude from part (b) that $\int_0^b x^2 \, dx = b^3/3.$

8. Consider the function defined on $[0, b]$, $b > 0$, by $f(x) = 2x + 3$.

   (a) For a partition given by $0 = x_0 < x_1 < x_2 < \cdots < x_n = b$, choose t_i to be the midpoint of $[x_{i-1}, x_i]$; that is, $t_i = (x_i + x_{i-1})/2$. Simplify the Riemann sum $\sum_{i=1}^n f(t_i) \Delta x_i.$

   (b) Conclude from part (a) that $\int_0^b (2x + 3) \, dx = b^3 + 3b.$

9. Give reasons for the steps indicated.

(a) Let $f(x) = k$ for x in $[a, b]$, where k is a constant (see Figure 5.29). Then $\int_a^b f(x)\, dx = k(b - a)$. Why?

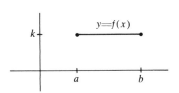

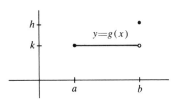

Figure 5.29   Figure 5.30

(b) Let

$$g(x) = \begin{cases} k, & a \le x < b, \\ h, & x = b, \end{cases}$$

where k, h are constants (see Figure 5.30). Let $a = x_0 < x_1 < x_2 < \cdots < x_n = b$ be a partition of $[a, b]$, and for each $i = 1, 2, \ldots, n$, choose t_i in $[x_{i-1}, x_i]$. Let

$$S = \sum_{i=1}^{n} g(t_i)\, \Delta x_i$$

and

$$S' = \sum_{i=1}^{n-1} g(t_i)\, \Delta x_i + g(x_{n-1})\, \Delta x_n$$

$$= \sum_{i=1}^{n-1} k\, \Delta x_i + k\, \Delta x_n$$

$$= \sum_{i=1}^{n} k\, \Delta x_i$$

$$= k(b - a).$$

Then

$$\left| \sum_{i=1}^{n} g(t_i)\, \Delta x_i - k(b - a) \right| = |S - S'| \quad \text{(why?)}$$

$$= |g(t_n) - g(x_{n-1})|\, \Delta x_n \quad \text{(why?)}$$

$$= |g(t_n) - k|\, \Delta x_n \quad \text{(why?)}$$

$$\le |h - k|\, \Delta x_n \quad \text{(why?)}.$$

Therefore $\sum_{i=1}^{n} g(t_i)\, \Delta x_i$ can be made as close to $k(b - a)$ as we like by choosing any partition of $[a, b]$ whose norm is small enough. Why? Hence, g is integrable on $[a, b]$ and $\int_a^b g(x)\, dx = k(b - a)$. Why?

10. Consider the greatest integer function $h(x) = [x]$ defined for all x.

    (a) For any integer n, show that h is integrable on $[n, n + 1]$ and that $\int_n^{n+1} h(x)\, dx = n$. [*Hint:* Use Exercise 9.]

    (b) For integers m, n (where $m < n$), show that h is integrable on $[m, n]$ and that $\int_m^n h(x)\, dx = m + (m + 1) + (m + 2) + \cdots + (n - 1)$. [*Hint:* Use part (a) and 3.5.3(3).]

    (c) For any numbers a, b (where $a < b$), show that h is integrable on $[a, b]$. In particular, $\int_a^b h(x)\, dx = (m - 1)(m - a) + m + (m + 1) + \cdots + (n - 1) + n(b - n)$, where $m - 1 < a \le m \le n \le b < n + 1$, m and n integers.

11. Let function f be defined by

$$f(x) = \begin{cases} x, & x \text{ rational,} \\ 0, & x \text{ irrational.} \end{cases}$$

Is f integrable on $[0, 1]$?

Supplement (optional)

Here we first give a more precise definition of the definite integral and then use it to prove Properties 2 and 4 in 5.3.5.

5.3.6

Definition (Alternative Form)

Let f be a function defined on $[a, b]$ and let L be a number with the following property: For each $\varepsilon > 0$ there exists a $\delta > 0$ such that

$$\left| \sum_{i=1}^{n} f(t_i)\, \Delta x_i - L \right| < \varepsilon$$

whenever

$$a = x_0 < x_1 < x_2 < \cdots < x_n = b$$

is a partition P of $[a, b]$ with $|P| < \delta$, and t_i is any point in $[x_{i-1}, x_i]$, $i = 1, 2, \ldots, n$.*

    Then L is called the **definite integral of f on $[a, b]$** and is denoted by the symbol

$$\int_a^b f(x)\, dx.$$

Proof of 2: Let f, g be integrable on $[a, b]$, say

$$\int_a^b f(x)\, dx = L \quad \text{and} \quad \int_a^b g(x) = M.$$

* $|P|$ denotes the norm of partition P.

To see that $f + g$ is integrable on $[a, b]$ and that

$$\int_a^b [f(x) + g(x)] = L + M,$$

let $\varepsilon > 0$ be given. Then there exists a $\delta_1 > 0$ such that

$$\left| \sum_{i=1}^n f(t_i) \, \Delta x_i - L \right| < \frac{\varepsilon}{2}$$

whenever

$$a = x_0 < x_1 < \cdots < x_n = b$$

is a partition P of $[a, b]$ with $|P| < \delta_1$ and t_i is any point in $[x_{i-1}, x_i]$, $i = 1, 2, \ldots, n$. Also, there exists a $\delta_2 > 0$ such that

$$\left| \sum_{i=1}^n g(t_i) \, \Delta x_i - M \right| < \frac{\varepsilon}{2}$$

whenever

$$a = x_0 < x_1 < \cdots < x_n = b$$

is a partition P of $[a, b]$ with $|P| < \delta_2$ and t_i is any point in $[x_{i-1}, x_i]$, $i = 1, 2, \ldots, n$.

Choose $\delta = \min\{\delta_1, \delta_2\}$. Then whenever

$$a = x_0 < x_1 < \cdots < x_n = b$$

is a partition P of $[a, b]$ with $|P| < \delta$ and t_i is any point in $[x_{i-1}, x_i]$, we have

$$\left| \sum_{i=1}^n [f(t_i) + g(t_i)] \, \Delta x_i - (L + M) \right|$$

$$= \left| \left(\sum_{i=1}^n f(t_i) \, \Delta x_i - L \right) + \left(\sum_{i=1}^n g(t_i) \, \Delta x_i - M \right) \right|$$

$$\leq \left| \sum_{i=1}^n f(t_i) \, \Delta x_i - L \right| + \left| \sum_{i=1}^n g(t_i) \, \Delta x_i - M \right|$$

$$< \frac{\varepsilon}{2} + \frac{\varepsilon}{2} = \varepsilon.$$

Proof of 4: We shall prove that

$$\int_a^b f(x) \, dx \leq M(b - a);$$

the proof that

$$m(b - a) \leq \int_a^b f(x) \, dx$$

is similar.

Suppose the contrary—that

$$\int_a^b f(x) \, dx > M(b - a).$$

Then

$$\int_a^b f(x) \, dx - M(b - a) > 0$$

$$\int_a^b f(x) \, dx - \int_a^b M \, dx > 0$$

$$\int_a^b [f(x) - M] \, dx > 0;$$

say that

$$\int_a^b [f(x) - M] \, dx = L > 0.$$

By definition, for $\varepsilon = L > 0$ there exists a $\delta > 0$ such that

(5.3.7) $$\left| \sum_{i=1}^n [f(t_i) - M] \, \Delta x_i - L \right| < L$$

whenever

$$a = x_0 < x_1 < x_2 < \cdots < x_n = b$$

is a partition P of $[a, b]$ with $|P| < \delta$ and t_i is any point in $[x_{i-1}, x_i]$, $i = 1, 2, \ldots, n$. However, for any such partition and any such points t_i, by hypothesis

$$f(t_i) \le M$$

and so

$$f(t_i) - M \le 0$$

$$[f(t_i) - M] \, \Delta x_i \le 0$$

$$\sum_{i=1}^n [f(t_i) - M] \, \Delta x_i \le 0$$

$$\sum_{i=1}^n [f(t_i) - M] \, \Delta x_i - L \le -L < 0$$

$$\left| \sum_{i=1}^n [f(t_i) - M] \, \Delta x_i - L \right| \ge L.$$

However, this contradicts 5.3.7. Therefore

$$\int_a^b f(x) \, dx \le M(b - a).$$

5.4 THE FUNDAMENTAL THEOREM OF CALCULUS

As we have indicated in the preceding section, computing definite integrals by means of the definition is difficult even for relatively simple functions. In this section we shall develop a powerful result which will enable us to quickly evaluate a given definite integral whenever we can determine an antiderivative of the integrand. Motivation for this result is obtained by considering the area function A defined on $[a, b]$ by

$$A(x) = \int_a^x f(t)\, dt \qquad \text{(see Figure 5.31)}.$$

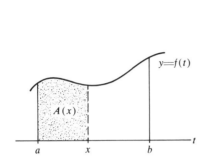

Figure 5.31

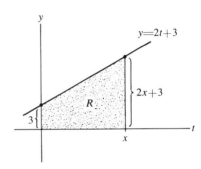

Figure 5.32

EXAMPLE 1   Let x be a fixed positive number, and consider the region R given by

$$0 \le y \le 2t + 3$$

$$0 \le t \le x \qquad \text{(see Figure 5.32)}.$$

The area of a trapezoid of altitude a and bases b, b' (see Figure 5.33) is

$$ab + \frac{1}{2} a(b' - b) = \frac{a}{2}(b + b').$$

Thus the area of region R above is

$$\frac{x}{2}(3 + 2x + 3) = x^2 + 3x.$$

Figure 5.33

Also, from our work in the preceding section, we know that the area of R is

$$\int_0^x (2t + 3) \, dt.$$

Therefore

$$\int_0^x (2t + 3) \, dt = x^2 + 3x.$$

It is interesting to pause and examine the last equation in Example 1, for observe that

$$\int_0^x (2t + 3) \, dt = x^2 + 3x$$

is an antiderivative of

$$2x + 3.$$

The last expression, $2x + 3$, is just the above integrand with t replaced by x. It was the discovery of Newton and Leibniz in the seventeenth century that *this astounding result is true in a more general situation.* Namely, if f is a continuous function, then

$$\int_a^x f(t) \, dt$$

is an antiderivative of

$$f(x).$$

This result formed the cornerstone in their development of modern-day calculus. It indicates the intimate relationship that exists between the differential calculus and the integral calculus, a relationship formally stated in the Fundamental Theorem of Calculus.

5.4.1
Fundamental Theorem of Calculus (first form)

If f is continuous on $[a, b]$, then the function G defined on $[a, b]$ by

$$G(x) = \int_a^x f(t) \, dt$$

is an antiderivative of f; that is,

$$G'(x) = f(x).$$

More briefly, *this theorem says that continuous functions have antiderivatives.* Its proof is given in the optional supplement to this section.

Now let f be continuous on $[a, b]$ and let F be any antiderivative of f. By 5.4.1,

$$\int_a^x f(t)\, dt$$

is also an antiderivative of f, and by 4.7.4 there exists a constant C such that

$$\int_a^x f(t)\, dt = F(x) + C$$

for all x in $[a, b]$. In particular,

1. When $x = a$,

$$0 = \int_a^a f(t)\, dt = F(a) + C$$

and so

$$C = -F(a).$$

2. When $x = b$,

$$\int_a^b f(t)\, dt = F(b) + C$$

$$= F(b) - F(a).$$

5.4.2
Fundamental Theorem of Calculus (second form)

If f is continuous on $[a, b]$ and F is any antiderivative of f, then

$$\int_a^b f(x)\, dx = F(b) - F(a).$$

The second form of the Fundamental Theorem is a powerful computational tool provided that one can find an antiderivative of the integrand $f(x)$. Whenever it is convenient, we shall use the notation

$$F(x)\Big|_a^b$$

to stand for

$$F(b) - F(a).$$

EXAMPLE 2   Determine $\int_1^3 x^2\, dx$. Since $F(x) = x^3/3$ is an antiderivative of x^2, we have

$$\int_1^3 x^2\, dx = F(3) - F(1) = 9 - \frac{1}{3} = \frac{26}{3}.$$

EXAMPLE 3   Determine $\int_{-2}^{-1} x^3\, dx$. Since $F(x) = x^4/4$ is an antiderivative of x^3,

$$\int_{-2}^{-1} x^3\, dx = F(-1) - F(-2) = \frac{1}{4} - 4 = -\frac{15}{4}.$$

EXAMPLE 4   Find $\int_0^1 \sqrt{x}\, dx$. Since $F(x) = \frac{2}{3}x^{3/2}$ is an antiderivative of $\sqrt{x}$, then

$$\int_0^1 \sqrt{x}\, dx = F(1) - F(0) = \frac{2}{3} - 0 = \frac{2}{3}.$$

EXAMPLE 5   Evaluate $\int_{-1}^0 (x^3 + 4x + 2)\, dx$. Since

$$\frac{x^4}{4} + 2x^2 + 2x$$

is an antiderivative of

$$x^3 + 4x + 2,$$

it follows that

$$\int_{-1}^0 (x^3 + 4x + 2)\, dx = \left(\frac{x^4}{4} + 2x^2 + 2x \right)\Bigg|_{-1}^0$$

$$= 0 - \left(\frac{1}{4} + 2 - 2 \right)$$

$$= -\frac{1}{4}.$$

EXAMPLE 6   Determine $\int_{-2}^{-1} (2x + 1)^2\, dx$. We seek an antiderivative of

$$(2x + 1)^2.$$

By 4.12.4 we know that an antiderivative of

$$(2x + 1)^2 \cdot 2$$

is

$$\frac{(2x + 1)^3}{3},$$

and so, by 4.12.5 an antiderivative of

$$(2x + 1)^2$$

is

$$\frac{(2x + 1)^3}{6}.$$

Hence

$$\int_{-2}^{-1} (2x + 1)^2 \, dx = \left. \frac{(2x + 1)^3}{6} \right|_{-2}^{-1}$$

$$= \frac{1}{6}(-1)^3 - \frac{1}{6}(-3)^3$$

$$= \frac{13}{3}.$$

EXAMPLE 7 Evaluate $\int_{-3}^{2} f(x) \, dx$, where

$$f(x) = \begin{cases} x^2, & -3 \le x \le 1, \\ 2 - x, & 1 < x \le 2. \end{cases}$$

An antiderivative of

$$x^2$$

is

$$\frac{x^3}{3},$$

and an antiderivative of

$$2 - x$$

is

$$2x - \frac{x^2}{2}.$$

Hence, using 5.3.5(3),

$$\int_{-3}^{2} f(x) \, dx = \int_{-3}^{1} f(x) \, dx + \int_{1}^{2} f(x) \, dx$$

$$= \int_{-3}^{1} x^2 \, dx + \int_{1}^{2} (2 - x) \, dx$$

$$= \left. \frac{x^3}{3} \right|_{-3}^{1} + \left(2x - \frac{x^2}{2} \right) \Big|_{1}^{2}$$

$$= \left[\frac{1}{3} - \left(\frac{-27}{3} \right) \right] + \left[2 - \frac{3}{2} \right]$$

$$= \frac{59}{6}.$$

EXERCISES

1. Evaluate the following integrals.

(a) $\int_{-2}^{-1} (3x + 2)\, dx.$

(b) $\int_{-1}^{0} 3x^2\, dx.$

(c) $\int_{1}^{3} (4x^2 + 2x)\, dx.$

(d) $\int_{1}^{2} (x + 1)^2\, dx.$

(e) $\int_{-1}^{1} (2x - 3)^3\, dx.$

(f) $\int_{0}^{1} (3x^3 - 2x + 5)\, dx.$

(g) $\int_{1}^{4} (2x^{3/2} - x^{1/3} + x - 4)\, dx.$

(h) $\int_{4}^{9} (2/\sqrt{x})\, dx.$

(i) $\int_{-5}^{-3} (1/x^2)\, dx.$

(j) $\int_{1}^{3} \sqrt{2x + 7}\, dx.$

(k) $\int_{0}^{1} [3x^2/(1 + x^3)^2]\, dx.$

(l) $\int_{0}^{1} x\sqrt{1 + 2x^2}\, dx.$

2. Evaluate each of the following integrals.

(a) $\int_{-1}^{2} f(x)\, dx$, where $f(x) = \begin{cases} -1, & -1 \le x < 0, \\ x^2 - 1, & 0 \le x \le 2. \end{cases}$

(b) $\int_{-1}^{1} f(x)\, dx$, where $f(x) = \begin{cases} x^{2/3}, & -1 \le x < 0, \\ \sqrt{x}, & 0 \le x \le 1. \end{cases}$

(c) $\int_{-2}^{3} f(x)\, dx$, where $f(x) = \begin{cases} x - 1, & -2 \le x \le 1, \\ (x - 1)^3, & 1 < x \le 3. \end{cases}$

(d) $\int_{0}^{1} |3x - 2|\, dx.$

(e) $\int_{-3}^{3} |(x + 2)(x - 1)|\, dx.$

3. Find $F'(x)$ if the following hold.

(a) $F(x) = \int_{0}^{x} (t^2/\sqrt{1 + t^2})\, dt.$

(b) $F(x) = x^2 \int_{0}^{x} \sqrt{1 + t^2}\, dt.$

(c) $F(x) = (1 + x^2) \int_{0}^{x} [1/(1 + t^2)]\, dt.$

(d) $F(x) = \int_{0}^{1} \sqrt[3]{7x^2 + 6x + 1}\, dx.$

4. Let $f(x) = \int_{-1}^{x} \sqrt{1 - t}\, dt$. Determine $f'(x)$ and $\int_{-1}^{1} f(x)\, dx$.

5. Consider the region given by $0 \le y \le 2x + 4$ and $-2 \le x \le 0$.

(a) For any x in $[-2, 0]$, find a formula $H(x)$ which gives the area of the region R bounded by the vertical line through x and the curves $y = 0$ and $y = 2x + 4$ (see Figure 5.34).

(b) Verify that $H(x) = \int_{-2}^{x} (2x + 4)\, dx$ and that $H'(x) = 2x + 4$.

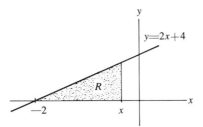

Figure 5.34

6. (a) Consider the function $f(x) = x^2 - 4$ and the partition $0, \frac{1}{3}, \frac{1}{4}, 1, \frac{3}{2}, \frac{5}{3}, \frac{9}{4}, \frac{14}{5}, 3$ of the interval $[0, 3]$. Using the left end point of each subinterval, compute the associated Riemann sum.

(b) Evaluate $\int_{0}^{3} (x^2 - 4)\, dx$ and compare the answer with the answer to part (a).

7. (a) Evaluate $\int_{1}^{2} x\, dx$, $\int_{1}^{2} (x^2 + 2)\, dx$, and $\int_{1}^{2} (x^3 + 2x)\, dx$.

(b) Is it true in general that $\int_{a}^{b} f(x)g(x)\, dx = [\int_{a}^{b} f(x)\, dx][\int_{a}^{b} g(x)\, dx]$?

8. Is it true that, given f' continuous, $\dfrac{d}{dx} \displaystyle\int_{a}^{x} f(t)\, dt = \displaystyle\int_{a}^{x} f'(t)\, dt$? Explain.

9. Let f' and g' be continuous on $[a, b]$. Evaluate the following.

(a) $\displaystyle\int_{a}^{b} [f'(x) + g'(x)]\, dx.$

(b) $\displaystyle\int_{a}^{b} [f'(x)g(x) + f(x)g'(x)]\, dx.$

(c) $\displaystyle\int_{a}^{b} \{[g(x)f'(x) - f(x)g'(x)]/[g(x)]^2\}\, dx.$

10. Let f' and g' be continuous on $[a, b]$. Show that $\int_{a}^{b} f(x)g'(x)\, dx = f(x)g(x)\big|_{a}^{b} - \int_{a}^{b} f'(x)g(x)\, dx$. This is known as the **integration by parts formula**.

11. If f is continuous on $[a, b]$, show that $\left| \int_{a}^{b} f(x)\, dx \right| \le \int_{a}^{b} |f(x)|\, dx$. [*Hint:* $-|f(x)| \le f(x) \le |f(x)|$.]

12. Show that if f is integrable on $[a, b]$ and $f(x) \ge 0$ for each x in $[a, b]$, then $\int_{a}^{b} f(x)\, dx \ge 0$. [*Hint:* Use 5.3.5(4).]

13. Show that if f and g are integrable on $[a, b]$ and $f(x) \le g(x)$ for each x in $[a, b]$, then $\int_{a}^{b} f(x)\, dx \le \int_{a}^{b} g(x)\, dx$. [*Hint:* Apply Exercise 12 to the function $g - f$.]

14. (a) Let $B(x) = \int_a^{\phi(x)} f(t)\, dt$, where f is continuous and ϕ is differentiable. Show that $B'(x) = f(\phi(x))\phi'(x)$. [*Hint:* $B(x) = G(\phi(x))$, where $G(x) = \int_a^x f(t)\, dt$.]

  (b) Use part (a) to determine $B'(x)$ when

  (1) $B(x) = \int_3^{4x^2} (9t - 1)\, dt$.

  (2) $B(x) = \int_a^{2x} f(t)\, dt$.

15. (a) Let $D(x) = \int_{\psi(x)}^a f(t)\, dt$, where f is continuous and ψ is differentiable. Show that $D'(x) = -f(\psi(x))\psi'(x)$. [*Hint:* $\int_{\psi(x)}^a f(t)\, dt = -\int_a^{\psi(x)} f(t)\, dt$.]

  (b) Use part (a) to determine $D'(x)$ when

  (1) $D(x) = \int_{3-x^3}^1 (7t - t^2)\, dt$.

  (2) $D(x) = \int_{3x^2}^a f(t)\, dt$.

16. (a) Let $E(x) = \int_{\psi(x)}^{\phi(x)} f(t)\, dt$, where f is continuous and ϕ, ψ are differentiable. Show that $E'(x) = f(\phi(x))\phi'(x) - f(\psi(x))\psi'(x)$. [*Hint:* $\int_{\psi(x)}^{\phi(x)} f(t)\, dt = \int_a^{\phi(x)} f(t)\, dt - \int_a^{\psi(x)} f(t)\, dt$.]

  (b) Use part (a) to determine $E'(x)$ when

  (1) $E(x) = \int_x^{x^2} \sqrt{t - 1}\, dt$.

  (2) $E(x) = \int_{x^4}^{1-x} f(t)\, dt$.

Supplement (optional)

In this supplement we shall prove the Fundamental Theorem of Calculus (first form), but first we need a lemma.

5.4.3

Mean Value Theorem for Integrals

If f is continuous on $[a, b]$, then there exists a point c in $[a, b]$ such that

$$\int_a^b f(x)\, dx = (b - a)f(c).$$

Proof: Since f is continuous on $[a, b]$, then by the Maximum-Minimum Theorem [3.3.4(1)] there exist points p, q in $[a, b]$ at which f attains minimum and maximum values, respectively. That is, for each x in $[a, b]$,

$$f(p) \le f(x) \quad \text{and} \quad f(q) \ge f(x).$$

Letting $m = f(p)$ and $M = f(q)$, then for each x in $[a, b]$ we have

$$m \le f(x) \le M.$$

Therefore by 5.3.5(4),

$$m(b - a) \le \int_a^b f(x) \, dx \le M(b - a)$$

or

$$m \le \frac{1}{b - a} \int_a^b f(x) \, dx \le M.$$

Recall that $m = f(p)$, $M = f(q)$, where points p, q are in $[a, b]$; f is continuous on the closed interval determined by p and q. Hence by the Intermediate Value Theorem [3.3.4(2)] there exists a point c between p and q such that

$$\frac{1}{b - a} \int_a^b f(x) \, dx = f(c).$$

Thus there exists a point c in $[a, b]$ such that

$$\int_a^b f(x) \, dx = (b - a)f(c).$$

Proof of 5.4.1: Let x be a fixed point in $[a, b)$. Choose $h > 0$ and small enough so that

$$x + h \le b.$$

Then $[x, x + h]$ is inside $[a, b]$ and f is continuous on $[x, x + h]$. By the Mean Value Theorem for Integrals, there exists a point c_h in $[x, x + h]$ such that

(5.4.4) $$\int_x^{x+h} f(t) \, dt = [(x + h) - x]f(c_h)$$

$$= h(f(c_h)).$$

By 5.3.5(3),

$$\int_a^{x+h} f(t) \, dt = \int_a^x f(t) \, dt + \int_x^{x+h} f(t) \, dt,$$

and so

$$\int_x^{x+h} f(t) \, dt = \int_a^{x+h} f(t) \, dt - \int_a^x f(t) \, dt$$

$$= G(x + h) - G(x).$$

Thus Equation 5.4.4 can be written as

$$G(x + h) - G(x) = hf(c_h)$$

or

$$\frac{G(x + h) - G(x)}{h} = f(c_h).$$

We recognize the left-hand side of the last equation as the difference quotient for the function G. Therefore let us take the limit of both sides of this equation as

$$h \downarrow 0$$

observing that this forces

$$c_h \downarrow x \qquad \text{(see Figure 5.35)}.$$

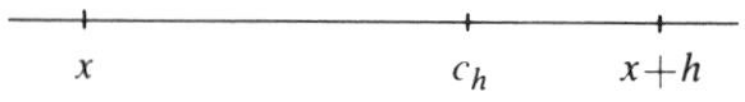

Figure 5.35

Hence

$$\lim_{h \downarrow 0} \frac{G(x + h) - G(x)}{h} = \lim_{h \downarrow 0} f(c_h)$$

$$= \lim_{c_h \downarrow x} f(c_h)$$

$$= f(x),$$

where the last equality holds because f is continuous at x.

In a similar fashion, by taking $h < 0$ and $[x + h, x]$ inside $[a, b]$, one can show that

$$\lim_{h \uparrow 0} \frac{G(x + h) - G(x)}{h} = f(x).$$

Therefore

$$G'(x) = \lim_{h \to 0} \frac{G(x + h) - G(x)}{h} = f(x).$$

5.5   CHANGE OF VARIABLE: A TECHNIQUE FOR EVALUATING THE INTEGRAL

Before proceeding with applications of the definite integral, we pause here to develop the change of variable method for evaluating the integral. This method often simplifies the evaluation of an integral when an antiderivative of the integrand is not immediately obvious. The procedure involves changing the variable in the given integrand in such a way that the resulting integrand has an antiderivative which is immediately obvious.

We begin by considering a function g with a continuous derivative on the interval $[a, b]$. Its graph might appear as in Figure 5.36. Notice that $g(a) = c$ and $g(b) = d$. Suppose that f is a continuous function on $[c, d']$; then $f(g(x))$ is continuous on $[a, b]$. If $F'(x) = f(x)$ on $[c, d]$, then

$$\int_c^d f(u) \, du = F(d) - F(c).$$

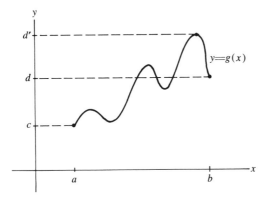

Figure 5.36

Moreover, if $\phi(x) = F(g(x))$, then

$$\phi'(x) = F'(g(x))g'(x) = f(g(x))g'(x).$$

Therefore

$$\int_a^b f(g(x))g'(x) \, dx = \phi(x)\Big|_a^b = F(g(x))\Big|_a^b$$

$$= F(g(b)) - F(g(a))$$

$$= F(d) - F(c).$$

Thus we have

(5.5.1) $$\int_a^b f(g(x))g'(x) \, dx = \int_c^d f(u) \, du$$

subject to the conditions on f and g previously mentioned.

The integral on the right-hand side of Equation 5.5.1 can be obtained by changing the variable in the integral on the left-hand side of 5.5.1 as follows. Let

$$u = g(x)$$

be the equation representing the change of variable. Note that

$$g(a) = c \quad \text{and} \quad g(b) = d,$$

which indicates that a is changed to c and b is changed to d in the limits of integration. Second, we observe that

$$g'(x) = \frac{du}{dx} \, .$$

Therefore we have

$$\int_a^b f(g(x))g'(x) \, dx = \int_c^d f(u) \frac{du}{dx} \, dx.$$

If we make the convention that

$$\frac{du}{dx} dx = du,$$

it follows that

$$\int_a^b f(g(x))g'(x)\, dx = \int_c^d f(u)\, du,$$

which is precisely Equation 5.5.1. This is one illustration of the convenience of Leibniz's notation for the integral.

It is this procedure just mentioned which is followed, rather than memorization of Equation 5.5.1. We shall illustrate this procedure by means of several examples.

EXAMPLE 1 Evaluate $\int_0^1 (2x + 1)^3\, dx$.

Let $u = 2x + 1$; then $du = \dfrac{du}{dx} dx = 2\, dx$ or $dx = \frac{1}{2}\, du$. Furthermore, if $x = 0$, then $u = 1$, and if $x = 1$, then $u = 3$. Therefore

$$\int_0^1 (2x + 1)^3\, dx = \int_1^3 u^3\, \tfrac{1}{2}\, du$$

$$= \frac{1}{2} \left(\frac{u^4}{4}\right)\Bigg|_1^3$$

$$= \frac{1}{2}\left[\frac{81}{4} - \frac{1}{4}\right] = 10.$$

EXAMPLE 2 Evaluate $\int_1^2 (x^2/\sqrt{9 - x^3})\, dx$.

Let $u = 9 - x^3$; then $du = \dfrac{du}{dx} dx = (-3x^2)\, dx$ and $x^2\, dx = -\frac{1}{3}\, du$. Moreover, if $x = 1$, then $u = 8$, and if $x = 2$, then $u = 1$. Hence

$$\int_1^2 \frac{x^2\, dx}{\sqrt{9 - x^3}} = \int_8^1 \frac{-\frac{1}{3}\, du}{\sqrt{u}} = -\frac{1}{3}\int_8^1 \frac{du}{\sqrt{u}}$$

$$= \frac{1}{3}\int_1^8 \frac{du}{\sqrt{u}}$$

$$= \frac{1}{3}\, 2\sqrt{u}\,\Bigg|_1^8$$

$$= \frac{2}{3}(\sqrt{8} - 1).$$

EXAMPLE 3   Evaluate $\int_0^1 (x^3/\sqrt{4 + 5x^2})\, dx$.

Let $u = 4 + 5x^2$; then $du = 10x\, dx$ or $x\, dx = \frac{1}{10}\, du$. If $x = 0$, then $u = 4$, and if $x = 1$, then $u = 9$. Therefore

$$
\int_0^1 \frac{x^3\, dx}{\sqrt{4 + 5x^2}} = \int_0^1 \frac{x^2 x\, dx}{\sqrt{4 + 5x^2}}
$$

$$
= \int_4^9 \frac{[(u - 4)/5]\frac{1}{10}\, du}{\sqrt{u}}
$$

$$
= \frac{1}{50} \int_4^9 \frac{u - 4}{\sqrt{u}}\, du
$$

$$
= \frac{1}{50} \int_4^9 \left(\sqrt{u} - \frac{4}{\sqrt{u}}\right) du
$$

$$
= \frac{1}{50} \left(\frac{2}{3} u^{3/2} - 8u^{1/2}\right)\Big|_4^9
$$

$$
= \frac{1}{50} \left[\frac{2}{3}(27) - 8(3) - \frac{2}{3}(8) + 8(2)\right]
$$

$$
= \frac{7}{75}.
$$

EXERCISES

1. Evaluate each of the following definite integrals.

(a) $\int_1^2 2x\sqrt{x^2 + 3}\, dx$.

(b) $\int_0^1 [(5x^4 + 6x + 1)/(x^5 + 2x^3 + x + 17)^{10}]\, dx$.

(c) $\int_0^2 (x - 1)^3\, dx$.

(d) $\int_0^4 \sqrt{1 + 2x}\, dx$.

(e) $\int_{-3}^{-2} [1/(2 - 3x)^2]\, dx$.

(f) $\int_0^1 [x/(8x^2 + 1)^{1/3}]\, dx$.

(g) $\int_4^9 (x/\sqrt{1 + x})\, dx$.

(h) $\int_{-1}^0 x^2\sqrt{2 + x}\, dx$.

(i) $\int_0^{1/2} (x^3/\sqrt{1 - x})\, dx$.

2. Let f be an even function on $[-a, a]$; that is, $f(-x) = f(x)$ for all x in $[-a, a]$. Show that $\int_{-a}^a f(x)\, dx = 2\int_0^a f(x)\, dx$. [*Hint:* Let $u = -x$ in $\int_{-a}^0 f(x)\, dx$.]

3. Let f be an odd function on $[-a, a]$; that is, $f(-x) = -f(x)$ for all x in $[-a, a]$. Show that $\int_{-a}^{a} f(x)\, dx = 0$. [*Hint:* Let $u = -x$ in $\int_{-a}^{0} f(x)\, dx$.]

4. Let f be continuous on $[a, b]$. Show that for any number c, $\int_{a}^{b} f(x)\, dx = \int_{a+c}^{b+c} f(x - c)\, dx$. Interpret this result geometrically.

5. Suppose that function f is defined on **R** and that there exists a number p with the property that $f(x) = f(x + p)$ for each x in **R**. (The number p is called a **period for** f.)

   (a) Show that $f(x) = f(x - p)$ for each x in **R**.

   (b) If f is integrable on every closed interval $[a, b]$, show that $\int_{a}^{b} f(x)\, dx = \int_{a+p}^{b+p} f(x)\, dx$. [*Hint:* Use Exercise 4.]

6. Show that $\int_{a}^{b} f(cx)\, dx = 1/c \int_{ac}^{bc} f(u)\, du$.

5.6 AREAS

This section and the following three sections illustrate various uses of the definite integral. Since we have already seen that there is a natural relationship between the integral and area, we devote the first of these application sections to solving several types of area problems by means of the integral.

EXAMPLE 1   For $b > 0$,

$$\int_{0}^{b} x^2\, dx = \frac{x^3}{3}\bigg|_{0}^{b} = \frac{b^3}{3}.$$

The area of the shaded region in Figure 5.37 is $b^3/3$. This verifies the result of Archimedes that was mentioned at the end of Section 5.2.

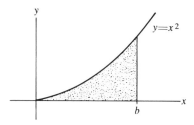

Figure 5.37

EXAMPLE 2

$$\int_{-4}^{-1} x^3\, dx = \frac{x^4}{4}\bigg|_{-4}^{-1} = -\frac{255}{4}.$$

The area of the shaded region in Figure 5.38 is $\frac{255}{4}$.

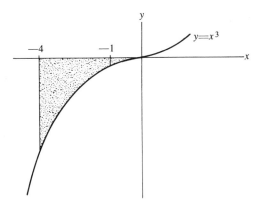

Figure 5.38

EXAMPLE 3

$$\int_1^2 (4 - x^2)\, dx = \left(4x - \frac{x^3}{3}\right)\Bigg|_1^2 = \left(8 - \frac{8}{3}\right) - \left(4 - \frac{1}{3}\right) = \frac{5}{3}$$

$$\int_2^5 (4 - x^2)\, dx = \left(4x - \frac{x^3}{3}\right)\Bigg|_2^5 = \left(20 - \frac{125}{3}\right) - \left(8 - \frac{8}{3}\right) = -\frac{81}{3}$$

$$\int_1^5 (4 - x^2)\, dx = \left(4x - \frac{x^3}{3}\right)\Bigg|_1^5 = \left(20 - \frac{125}{3}\right) - \left(4 - \frac{1}{3}\right) = -\frac{76}{3}.$$

In Figure 5.39 the area of region I is $\frac{5}{3}$; the area of region II is $\frac{81}{3}$; and the area of region I minus the area of region II is $-\frac{76}{3}$.

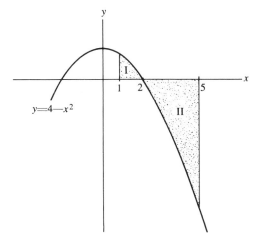

Figure 5.39

The integral can also be used to find the area of a region whose boundary does not include a portion of the x-axis.

EXAMPLE 4   Find the area A of the region R bounded by the curves

$$x = 1, \qquad x = 2, \qquad y = x^3, \qquad y = -x \qquad \text{(see Figure 5.40)}.$$

$x^4/4$ is an antiderivative of x^3 and $-x^2/2$ is an antiderivative of $-x$, and so

$$\int_1^2 x^3 \, dx = \left.\frac{x^4}{4}\right|_1^2 = 4 - \frac{1}{4} = \frac{15}{4}$$

and

$$\int_1^2 -x \, dx = \left.-\frac{x^2}{2}\right|_1^2 = -2 + \frac{1}{2} = -\frac{3}{2}.$$

Hence, in Figure 5.40, $\frac{15}{4}$ is the area of region I, and $-(-\frac{3}{2}) = \frac{3}{2}$ is the area of region II. Therefore $A = \frac{15}{4} + \frac{3}{2} = \frac{21}{4}$.

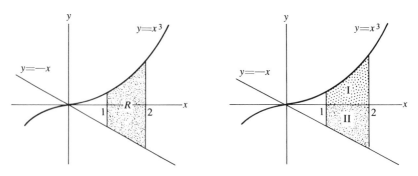

Figure 5.40

EXAMPLE 5   Determine the area A of a region R bounded by the curves

$$x = 1, \qquad y = 8x, \qquad y = \frac{1}{x^2}.$$

Note that

1. The point of intersection $(\frac{1}{2}, 4)$ may be found by solving simultaneously the equations $y = 8x$ and $y = 1/x^2$.
2. In Figure 5.41, region III is bounded by the given curves plus the curve $y = 0$ (x-axis) and hence is not part of the desired region.

Since $4x^2$ and $-1/x^2$ are antiderivatives of $8x$ and $1/x^2$, respectively, then

$$\int_{1/2}^1 8x \, dx = \left.4x^2\right|_{1/2}^1 = 4 - 1 = 3$$

and

$$\int_{1/2}^1 \frac{1}{x^2} \, dx = \left.-\frac{1}{x}\right|_{1/2}^1 = -1 + 2 = 1.$$

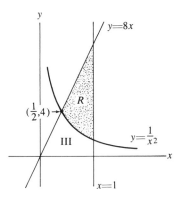
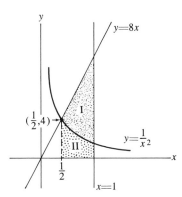

Figure 5.41

In Figure 5.41, 3 is the area of region I-II, and 1 is the area of the region II. Therefore

$$A = 3 - 1 = 2.$$

Consider Examples 4 and 5 in retrospect; it is interesting to note and easy to check that

$$\int_1^2 (x^3 - (-x))\, dx = \frac{21}{4}$$

and that

$$\int_{1/2}^1 \left(8x - \frac{1}{x^2}\right) dx = 2.$$

In fact, the following more general result is true.

5.6.1
Fact

If f, g are continuous functions on $[a, b]$ and for each x in $[a, b]$,

$$f(x) \geq g(x),$$

then the area of the region bounded by the curves

$$y = f(x), \qquad x = a,$$
$$y = g(x), \qquad x = b,$$

is

$$\int_a^b [f(x) - g(x)]\, dx \qquad \text{(see Figure 5.42)}.$$

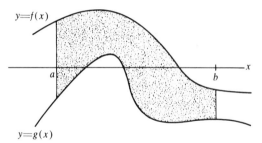

Figure 5.42

EXAMPLE 6   Find the area A of the region R bounded by the curves

$$y = 2x + 1, \qquad y = x^2 - 2 \qquad \text{(see Figure 5.43)}.$$

Again the points of intersection $(-1, -1)$ and $(3, 7)$ may be found by solving simultaneously the equations $y = x^2 - 2$ and $y = 2x + 1$. Since R is bounded by the curves

$$y = 2x + 1, \qquad x = -1,$$
$$y = x^2 - 2, \qquad x = 3,$$

where $2x + 1 \geq x^2 - 2$ for each x in $[-1, 3]$, then by Fact 5.6.1

$$A = \int_{-1}^{3} [(2x + 1) - (x^2 - 2)] \, dx$$

$$= \int_{-1}^{3} (3 + 2x - x^2) \, dx.$$

Now $3x + x^2 - x^3/3$ is an antiderivative of $3 + 2x - x^2$, and so by the Fundamental Theorem

$$A = \int_{-1}^{3} (3 + 2x - x^2) \, dx$$

$$= 3x + x^2 - \frac{x^3}{3}\bigg|_{-1}^{3}$$

$$= (9 + 9 - 9) - (-3 + 1 + \tfrac{1}{3})$$

$$= \tfrac{32}{3}.$$

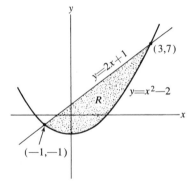

Figure 5.43

EXAMPLE 7   Determine the area A of the region R bounded by the curves

$$y = 5x, \qquad y = x^3 - 12x, \qquad x = 1.$$

Solving simultaneously the equations

$$y = 5x \qquad \text{and} \qquad y = x^3 - 12x$$

gives the points of intersection

$$(-\sqrt{17}, -5\sqrt{17}), \qquad (0, 0), \qquad (\sqrt{17}, 5\sqrt{17}).$$

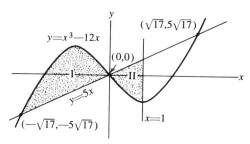

Figure 5.44

In Figure 5.44, region I is bounded by the curves

$$y = x^3 - 12x, \qquad x = -\sqrt{17},$$
$$y = 5x, \qquad\qquad x = 0,$$

where $x^3 - 12x \geq 5x$ for all x in $[-\sqrt{17}, 0]$, and so by Fact 5.6.1,

$$\text{Area region I} = \int_{-\sqrt{17}}^{0} [(x^3 - 12x) - 5x]\, dx$$

$$= \int_{-\sqrt{17}}^{0} (x^3 - 17x)\, dx.$$

Since $x^4/4 - \frac{17}{2}x^2$ is an antiderivative of $x^3 - 17x$, by the Fundamental Theorem

$$\text{Area region I} = \int_{-\sqrt{17}}^{0} (x^3 - 17x)\, dx$$

$$= \left(\frac{x^4}{4} - \frac{17}{2}x^2 \right)\Bigg|_{-\sqrt{17}}^{0}$$

$$= 0 - \left(\frac{17^2}{4} - \frac{17^2}{2} \right)$$

$$= \frac{17^2}{4}$$

$$= \frac{289}{4}.$$

Similarly, region II is bounded by the curves

$$y = 5x, \qquad x = 0,$$

$$y = x^3 - 12x, \qquad x = 1,$$

where $5x \geq x^3 - 12x$ for each x in $[0, 1]$, and so

$$\text{Area region II} = \int_0^1 [5x - (x^3 - 12x)] \, dx$$

$$= \int_0^1 (17x - x^3) \, dx.$$

Since $\frac{17}{2}x^2 - x^4/4$ is an antiderivative of $17x - x^3$, we have

$$\text{Area region II} = \int_0^1 (17x - x^3) \, dx$$

$$= \frac{17}{2} x^2 - \frac{x^4}{4} \Big|_0^1$$

$$= \left(\frac{17}{2} - \frac{1}{4} \right) - 0$$

$$= \frac{33}{4}.$$

Finally, then

$$A = \text{Area region I} + \text{Area region II}$$

$$= \frac{289}{4} + \frac{33}{4}$$

$$= \frac{161}{2}.$$

EXAMPLE 8   Compute the area of the region R bounded by the curves

$$x = 0, \qquad y = \sqrt{x}, \qquad y = 2.$$

Method A (see Figure 5.45): Integrating over an interval of the x-axis and using 5.6.1, we have

$$A = \int_0^4 (2 - \sqrt{x}) \, dx.$$

Since $2x - \frac{2}{3}x^{3/2}$ is an antiderivative of $2 - \sqrt{x}$,

$$A = \int_0^4 (2 - \sqrt{x})\, dx$$

$$= \left(2x - \frac{2}{3}x^{3/2}\right)\bigg|_0^4$$

$$= \left(8 - \frac{16}{3}\right) - (0)$$

$$= \frac{8}{3}.$$

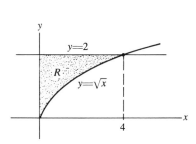

Figure 5.45   Figure 5.46

Method B (see Figure 5.46): Integrating over an interval on the y-axis,

$$A = \int_0^2 y^2\, dy.$$

Since $y^3/3$ is an antiderivative of y^2, we have

$$A = \int_0^2 y^2\, dy$$

$$= \frac{y^3}{3}\bigg|_0^2$$

$$= \frac{8}{3} - 0$$

$$= \frac{8}{3}.$$

EXAMPLE 9   Determine the area A_m of the region R_m bounded by the curves

$$x = 1, \qquad x = m, \qquad y = 0, \qquad y = \frac{1}{x^{2/3}},$$

where m is fixed and $m > 1$ (see Figure 5.47).

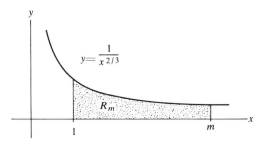

Figure 5.47

Since $3x^{1/3}$ is an antiderivative of $1/x^{2/3}$,

$$\begin{aligned}
A_m &= \int_1^m \frac{1}{x^{2/3}} \, dx \\
&= 3x^{1/3} \Big|_1^m \\
&= 3m^{1/3} - 3.
\end{aligned}$$

EXAMPLE 10   Determine the area B_m of the region S_m bounded by the curves

$$x = 1, \qquad x = m, \qquad y = 0, \qquad y = \frac{1}{x^{3/2}},$$

where m is fixed and $m > 1$ (see Figure 5.48).
Since $-2/\sqrt{x}$ is an antiderivative of $1/x^{3/2}$,

$$\begin{aligned}
B_m &= \int_1^m \frac{1}{x^{3/2}} \, dx \\
&= -\frac{2}{\sqrt{x}} \Big|_1^m \\
&= -\frac{2}{\sqrt{m}} + 2.
\end{aligned}$$

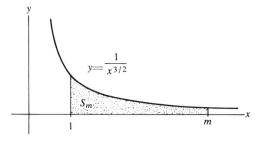

Figure 5.48

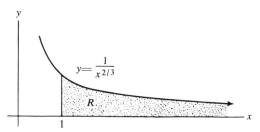

Figure 5.49

Remark

It is interesting to consider the results of Examples 9 and 10 as m grows larger. Since

$$\lim_{m \to +\infty} A_m = \lim_{m \to +\infty} (3m^{1/3} - 3) = +\infty,$$

it is reasonable to say that in Figure 5.49 the region R of infinite extent does not have finite area, whereas, because

$$\lim_{m \to +\infty} B_m = \lim_{m \to +\infty} \left(2 - \frac{2}{\sqrt{m}}\right) = 2,$$

it is reasonable to say that in Figure 5.50 the region S of infinite extent does have a finite area, namely 2. Thus *it is possible to assign a finite area to a region of infinite extent.*

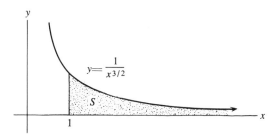

Figure 5.50

EXAMPLE 11   Find the area A_ε of the region R_ε bounded by the curves

$$x = \varepsilon, \qquad x = 1, \qquad y = 0, \qquad y = \frac{1}{x^{4/5}},$$

where ε is fixed and $0 < \varepsilon < 1$ (see Figure 5.51). Again, since $5x^{1/5}$ is an antiderivative of $1/x^{4/5}$,

$$A_\varepsilon = \int_\varepsilon^1 \frac{1}{x^{4/5}} \, dx$$

$$= 5x^{1/5} \big|_\varepsilon^1$$

$$= 5 - 5\varepsilon^{1/5}.$$

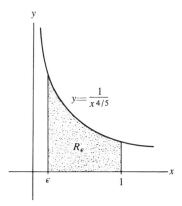

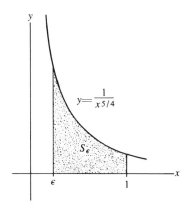

Figure 5.51   Figure 5.52

EXAMPLE 12   Find the area B_ε of the region S_ε bounded by the curves

$$x = \varepsilon, \qquad x = 1, \qquad y = 0, \qquad y = \frac{1}{x^{5/4}},$$

where ε is fixed and $0 < \varepsilon < 1$ (see Figure 5.52). Again, since $-4/x^{1/4}$ is an antiderivative of $1/x^{5/4}$, we have

$$B_\varepsilon = \int_\varepsilon^1 \frac{1}{x^{5/4}} \, dx$$

$$= -\frac{4}{x^{1/4}} \Big|_\varepsilon^1$$

$$= -4 + \frac{4}{\varepsilon^{1/4}}.$$

Remark

This time it is interesting to consider the results of Examples 11 and 12 as ε grows smaller. Since

$$\lim_{\varepsilon \downarrow 0} A_\varepsilon = \lim_{\varepsilon \downarrow 0}(5 - 5\varepsilon^{1/5}) = 5,$$

it is reasonable to say that in Figure 5.53 the region R of infinite extent has finite area, namely 5. Therefore *again we see that it is possible to assign a finite area to a region of infinite extent.*

Since

$$\lim_{\varepsilon \downarrow 0} B_\varepsilon = \lim_{\varepsilon \downarrow 0}\left(\frac{4}{\varepsilon^{1/4}} - 4\right) = +\infty,$$

it is reasonable to say that in Figure 5.54 the region S of infinite extent does not have finite area.

EXERCISES

1. In each case, graph the given equations and find the area of the region bounded by those curves.

(a) $x + y = 9$, $x = 0$, $y = 0$.

(b) $y = 3 + x^2$, $y = 0$, $x = -3$, $x = -2$.

(c) $y = x^3 - 8$, $y = 0$, $x = 0$, $x = 2$.

(d) $y = 1/x^2$, $x = \frac{1}{2}$, $x = 3$, $y = 0$.

(e) $y = \frac{1}{2}\sqrt{x}$, $y = 0$, $x = 4$, $x = 9$.

(f) $y = -x^{2/3}$, $x = -4$, $y = 0$.

(g) $y = (3x + 2)^2$, $y = 0$, $x = -1$, $x = 2$.

(h) $y = x^3$, $y = 0$, $x = -1$, $x = 2$.

2. Find the area of the region bounded by the following curve(s).

(a) $y = x^2$ and $y = 4$.   (b) $y = x^4 - 16$, $y = 0$.

(c) $y = x^2 - 1$ and $y = 1 - x^2$.   (d) $y = x^2$ and $x = y^2$.

(e) $|x| + |y| = 1$.   (f) $y = |x|$ and $y = 2$.

(g) $y = x^3$ and $x = y^3$.   (h) $y = [x]$, $x = \frac{3}{2}$, $x = \frac{5}{4}$.

(i) $y = \sqrt{x}$, $x = 9$, $y = 0$.   (j) $y^2 - 2y = -x$, $x = 0$.

(k) $y = (x + 1)^2 - 3$, $y = x$.   (l) $y = x^{2/3}$, $y = 9$.

(m) $y = x^{1/3}$, $y = x$.   (n) $y = (2x - 1)^4$, $x = 0$, $y = 0$.

(o) $x + y^2 = 0$, $x - y = 0$.

3. Use integration to show that the triangle drawn in Figure 5.55 with base b height h has area $\frac{1}{2}bh$.

4. Interpret the integral $\int_0^a \sqrt{a^2 - x^2}\, dx$ as the area of a well-known region in order to obtain its value.

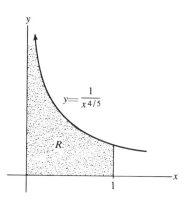

Figure 5.53

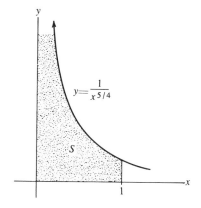

Figure 5.54

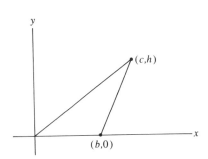

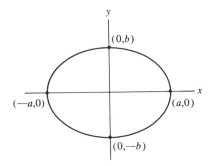

Figure 5.55   Figure 5.56

5. Find the area of the region inside the ellipse drawn in Figure 5.56:
   $x^2/a^2 + y^2/b^2 = 1$. [*Hint:* The top half of the ellipse has the equation
   $y = (b/a)\sqrt{a^2 - x^2}$; see Exercise 4.]

6. Let $v(t)$ be the velocity of a point moving on the x-axis. Suppose that the
   graph of v is as drawn in Figure 5.57. Show that the area of the shaded region
   is the total distance traveled on the time interval $[a, b]$ and that $\int_a^b |v(t)|\, dt$ is
   this distance. What does $\int_a^b v(t)\, dt$ represent?

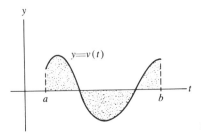

Figure 5.57

7. Interpret geometrically the results in Exercises 2 and 3 at the end of Section 5.5.

8. (a) Determine the area A_m bounded by the curves $x = 1$, $x = m$, $y = 0$,
       $y = 1/x^{3/4}$ ($m > 1$).
   (b) Evaluate $\lim\limits_{m \to +\infty} A_m$.
   (c) Is it possible to assign a finite area to the infinite strip bounded by the
       curves $x = 1$, $y = 0$, $y = 1/x^{3/4}$? If so, what?

9. (a) Determine the area B_m bounded by the curves $x = 1$, $x = m$, $y = 0$,
       $y = 1/x^{4/3}$ ($m > 1$).
   (b) Evaluate $\lim\limits_{m \to +\infty} B_m$.
   (c) Is it possible to assign a finite area to the infinite strip bounded by the
       curves $x = 1$, $y = 0$, $y = 1/x^{4/3}$? If so, what?

10. (a) Determine the area A_ε bounded by the curves $x = \varepsilon$, $x = 1$, $y = 0$, $y = 1/x^{5/6}$ $(0 < \varepsilon < 1)$.

(b) Evaluate $\lim_{\varepsilon \downarrow 0} A_\varepsilon$.

(c) Is it possible to assign a finite area to the infinite strip bounded by the curves $x = 0$, $x = 1$, $y = 0$, $y = 1/x^{5/6}$? If so, what?

11. (a) Determine the area B_ε bounded by the curves $x = \varepsilon$, $x = 1$, $y = 0$, $y = 1/x^{6/5}$ $(0 < \varepsilon < 1)$.

(b) Evaluate $\lim_{\varepsilon \downarrow 0} B_\varepsilon$.

(c) Is it possible to assign a finite area to the infinite strip bounded by the curves $x = 0$, $x = 1$, $y = 0$, $y = 1/x^{6/5}$? If so, what?

12. Can one assign a finite area to the infinite strip bounded by the curves $x = 2$, $y = 0$, $y = x/(x^2 + 1)^2$? If so, what?

5.7 VOLUMES

Let S be a solid (that is, a three-dimensional region) and l a coordinatized line passing through S, as in Figure 5.58. For each x in $[a, b]$ define a function A by

$$A(x) = \text{Area of the cross section of } S$$
$$\text{formed by passing a plane through } x$$
$$\text{and perpendicular to the line } l \qquad (\text{see Figure 5.59}).$$

Figure 5.58

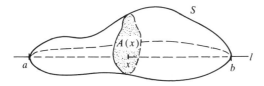

Figure 5.59

If A is a continuous function on $[a, b]$, then we can determine the volume V of S as follows.

Let points

$$a = x_0 < x_1 < \cdots < x_{n-1} < x_n = b$$

be a partition of $[a, b]$, where n is a positive integer, and for each $i = 1, 2, 3, \ldots, n$ let t_i be any point in the subinterval $[x_{i-1}, x_i]$. Let V_i be the

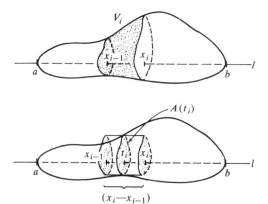

Figure 5.60

volume of that slice of S formed by passing planes perpendicular to l through x_{i-1} and x_i. Then, taking note of Figure 5.60,

$$V_i \approx A(t_i)(x_i - x_{i-1})$$

and

$$
\begin{aligned}
V &= V_1 + V_2 + \cdots + V_n \\
&\approx A(t_1)(x_1 - x_0) + A(t_2)(x_2 - x_1) + \cdots + A(t_n)(x_n - x_{n-1}) \\
&= \sum_{i=1}^{n} A(t_i)\, \Delta x_i.
\end{aligned}
$$

Since A is continuous on $[a, b]$, this approximation gets better as the norm of the partition gets smaller. Thus

$$\sum_{i=1}^{n} A(t_i)\, \Delta x_i$$

can be made as close to V as we wish by choosing any partition of $[a, b]$ whose norm is sufficiently small. In light of Definition 5.3.1,

$$V = \int_a^b A(x)\, dx.$$

To summarize, note the following.

5.7.1

Volume

If for each x in $[a, b]$, $A(x)$ is the cross-sectional area formed by passing a plane through x and perpendicular to l and if A is continuous on $[a, b]$, then the volume of S is

$$\int_a^b A(x)\, dx \qquad \text{(see Figure 5.61).}$$

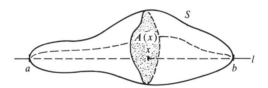

Figure 5.61

EXAMPLE 1   Use 5.7.1 to justify the familiar fact that the volume of a rectangular box is the product of its three dimensions.

Situate the box S as in Figure 5.62, where line l is perpendicular to those two faces which it intersects. Then for any x in $[a, b]$

$$A(x) = hw,$$

which is a continuous function on $[0, b]$. Therefore by 5.7.1 the volume V of the box is $\int_0^b hw\ dx$. Finally, since hwx is an antiderivative of hw, we have

$$V = \int_0^b hw\ dx = hwx\Big|_0^b = hwb.$$

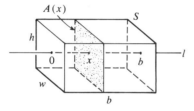

Figure 5.62   Figure 5.63

EXAMPLE 2   A pyramid is situated as in Figure 5.63, where for each x in $[0, h]$ the perpendicular cross section is an isosceles right triangle with legs of length x. Find the volume V of P.

In this case, for each x in $[0, h]$

$$A(x) = \tfrac{1}{2}x^2,$$

which is a continuous function on $[0, h]$. Also,

$$\tfrac{1}{6}x^3$$

is an antiderivative of $\tfrac{1}{2}x^2$, and so

$$V = \int_0^h A(x)\ dx = \int_0^h \frac{1}{2}x^2\ dx = \frac{1}{6}x^3\Big|_0^h = \frac{1}{6}h^3.$$

EXAMPLE 3   Let R be a plane region bounded by the curves

$$x = a, \qquad x = b,$$
$$y = 0, \qquad y = f(x),$$

where $a < b$ and f is a continuous, nonnegative function on $[a, b]$ (see Figure 5.64). If R is rotated about the x-axis, find the volume V of the resulting solid S.

The solid S is shown in Figure 5.65, where a perpendicular cross section through any x in $[a, b]$ is a circle of radius $f(x)$; hence its area is

$$A(x) = \pi[f(x)]^2.$$

This function is a continuous function on $[a, b]$, and so by 5.7.1

$$V = \int_a^b \pi[f(x)]^2 \, dx.$$

The fact that a thin slice of S formed by two planes perpendicular to the x-axis appears to be a disk yields a name for this formula.

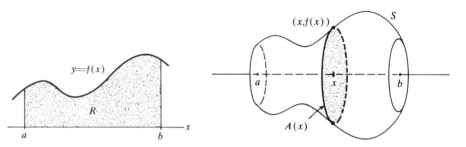

Figure 5.64   Figure 5.65

5.7.2

Disk Method

Let the region R be bounded by the curves

$$x = a, \qquad x = b,$$
$$y = 0, \qquad y = f(x),$$

where $a < b$ and f is a continuous, nonnegative function on $[a, b]$ (see Figure 5.66).

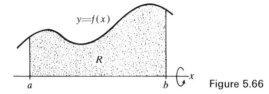

Figure 5.66

If R is rotated about the x-axis, then the volume of the resulting solid is

$$\pi \int_a^b [f(x)]^2 \, dx.$$

Let R be the type of region just described in 5.7.2, and in addition assume that $0 \leq a < b$. We have a formula for the volume of the solid of revolution formed by rotating R about the x-axis (namely, the disk method); now we seek a formula for the volume V of the solid formed by rotating R about the y-axis.

Again, let points

$$a = x_0 < x_1 < x_2 < \cdots < x_{n-1} < x_n = b$$

be a partition of $[a, b]$, where n is a positive integer, and for each $i = 1, 2, \ldots, n$ let t_i be the midpoint of $[x_{i-1}, x_i]$; thus, $t_i = \frac{1}{2}(x_{i-1} + x_i)$. Let R_i be that portion of R above $[x_{i-1}, x_i]$ and let V_i be the volume of that shell formed by rotating R_i about the y-axis (see Figure 5.67). The volume of that shell formed by rotating rectangle r_i about the y-axis is

$$\pi(x_i^2)f(t_i) - \pi(x_{i-1}^2)f(t_i) = \pi f(t_i)(x_i^2 - x_{i-1}^2)$$

$$= 2f(t_i)\left(\frac{x_{i-1} + x_i}{2}\right)(x_i - x_{i-1})$$

$$= 2\pi f(t_i)(t_i)(x_i - x_{i-1}).$$

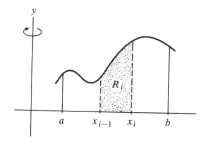 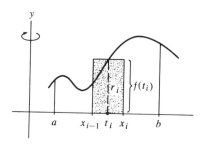

Figure 5.67

Hence

$$V_i \approx 2t_i f(t_i)(x_i - x_{i-1})$$

and

$$V = V_1 + V_2 + \cdots + V_n$$

$$\approx 2\pi t_1 f(t_1)(x_1 - x_0) + 2\pi t_2 f(t_2)(x_2 - x_1) + \cdots$$

$$+ 2\pi t_n f(t_n)(x_n - x_{n-1})$$

$$= \sum_{i=1}^{n} 2\pi t_i f(t_i)\, \Delta x_i.$$

Since the function $2\pi x f(x)$ is continuous on $[a, b]$, this approximation gets better as the norm of the partition gets smaller. Thus

$$\sum_{i=1}^{n} 2\pi t_i f(t_i)\, \Delta x_i$$

can be made as close to V as we wish by choosing any partition of $[a, b]$ whose norm is sufficiently small. Because of Definition 5.3.1

$$V = 2\pi \int_a^b xf(x) \, dx.$$

5.7.3

Shell Method   Let the region R be bounded by the curves

$$x = a, \qquad x = b,$$

$$y = 0, \qquad y = f(x),$$

where $0 \leq a < b$ and f is a continuous, nonnegative function on $[a, b]$ (see Figure 5.68).

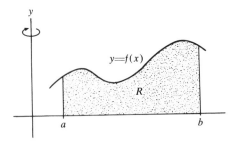

Figure 5.68

If R is rotated about the y-axis, then the volume of the resulting solid is

$$2\pi \int_a^b xf(x) \, dx.$$

EXAMPLE 4   Some familiar solids are obtained when the regions in Figure 5.69 are rotated about the x-axis.

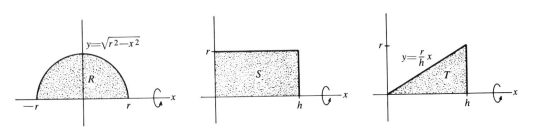

Figure 5.69

A sphere of radius r results when R is rotated, and its volume is

$$\pi \int_{-r}^{r} (r^2 - x^2) \, dx.$$

Since $r^2x - x^3/3$ is an antiderivative of $r^2 - x^2$, we have

$$\pi \int_{-r}^{r} (r^2 - x^2) \, dx = \left(r^2x - \frac{x^3}{3} \right)\Big|_{-r}^{r}$$

$$= \pi[\tfrac{2}{3}r^3 + \tfrac{2}{3}r^3]$$

$$= \tfrac{4}{3}\pi r^3.$$

A cylinder of height h and radius of base r results when S is rotated, and its volume is

$$\pi \int_{0}^{h} r^2 \, dx.$$

Since r^2x is an antiderivative of r^2, we have

$$\pi \int_{0}^{h} r^2 \, dx = \pi[r^2x|_0^h]$$

$$= \pi[r^2h - 0]$$

$$= \pi r^2 h.$$

A cone of height h and radius of base r results when T is rotated, and its volume is

$$\pi \int_{0}^{h} \left(\frac{r}{h} x \right)^2 \, dx.$$

Since $(r^2/3h^2)x^3$ is an antiderivative of $[(r/h)x]^2$, we have

$$\pi \int_{0}^{h} \left(\frac{r}{h} x \right)^2 \, dx = \pi \left[\frac{r^2}{3h^2} x^3 \Big|_0^h \right]$$

$$= \pi[\tfrac{1}{3}r^2h - 0]$$

$$= \tfrac{1}{3}\pi r^2 h.$$

EXAMPLE 5   Let R be bounded by the curves

$$x = 1, \quad x = 2, \quad y = 0, \quad y = \frac{1}{x} \quad \text{(see Figure 5.70)}.$$

Let V_x, V_y be the volumes of those solids formed when R is rotated about the x- and y-axes, respectively. Then

$$V_x = \pi \int_{1}^{2} \frac{1}{x^2} \, dx \quad \text{and} \quad V_y = 2\pi \int_{1}^{2} 1 \, dx.$$

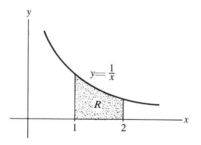

Figure 5.70

Also, $-1/x$ and x are antiderivatives of $1/x^2$ and 1, respectively, and

$$V_x = \pi \int_1^2 \frac{1}{x^2} \, dx = \pi \left[-\frac{1}{x} \Big|_1^2 \right]$$

$$= \pi[-\tfrac{1}{2} + 1]$$

$$= \frac{\pi}{2}$$

and

$$V_y = 2\pi \int_1^2 1 \, dx = 2\pi[x \big|_1^2]$$

$$= 2\pi[2 - 1]$$

$$= 2\pi.$$

EXAMPLE 6   Consider the region S bounded by the curves

$$x = 0, \qquad y = -x + 2, \qquad y = x^3.$$

Let V_x, V_y be the volumes of those solids formed by rotating S about the x- and y-axes, respectively.

In Figure 5.71, when region I-II is rotated about the x-axis, its volume is

$$\pi \int_0^1 (-x + 2)^2 \, dx = \pi \int_0^1 (x^2 - 4x + 4) \, dx$$

$$= \pi \left[\left(\frac{x^3}{3} - 2x^2 + 4x \right) \Big|_0^1 \right]$$

$$= \pi[\tfrac{7}{3} - 0]$$

$$= \tfrac{7}{3}\pi.$$

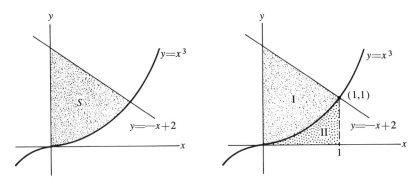

Figure 5.71

When region II is rotated about the x-axis, its volume is

$$\pi \int_0^1 (x^3)^2 \, dx = \pi \int_0^1 x^6 \, dx$$

$$= \pi \left[\frac{x^7}{7} \bigg|_0^1 \right]$$

$$= \pi(\tfrac{1}{7} - 0)$$

$$= \tfrac{1}{7}\pi.$$

Therefore

$$V_x = \tfrac{7}{3}\pi - \tfrac{1}{7}\pi$$

$$= \tfrac{46}{21}\pi.$$

Similarly, one computes V_y as

$$2\pi \int_0^1 x(-x + 2) \, dx = 2\pi \int_0^1 (-x^2 + 2x) \, dx$$

$$= 2\pi \left[\left(-\frac{x^3}{3} + x^2 \right) \bigg|_0^1 \right]$$

$$= 2\pi[\tfrac{2}{3} - 0]$$

$$= \tfrac{4}{3}\pi.$$

Also,

$$2\pi \int_0^1 x(x^3) \, dx = 2\pi \int_0^1 x^4 \, dx$$

$$= 2\pi \left[\frac{x^5}{5} \bigg|_0^1 \right]$$

$$= 2\pi[\tfrac{1}{5} - 0]$$

$$= \tfrac{2}{5}\pi.$$

Therefore

$$V_y = \tfrac{4}{3}\pi - \tfrac{2}{5}\pi = \tfrac{14}{15}\pi.$$

EXAMPLE 7   Let R_m be the region of Example 9 in the preceding section; it is drawn in Figure 5.72. If R_m is rotated about the x-axis and V_m is the volume of the resulting solid, then

$$V_m = \pi \int_1^m \left(\frac{1}{x^{2/3}}\right)^2 dx = \pi \int_1^m \frac{1}{x^{4/3}} \, dx.$$

Now $-3/x^{1/3}$ is an antiderivative of $1/x^{4/3}$, and so

$$V_m = \pi \int_1^m \frac{1}{x^{4/3}} \, dx = \pi \left[-\frac{3}{x^{1/3}} \bigg|_1^m \right]$$

$$= \pi \left[-\frac{3}{m^{1/3}} + 3 \right]$$

$$= 3\pi \left[1 - \frac{1}{m^{1/3}} \right].$$

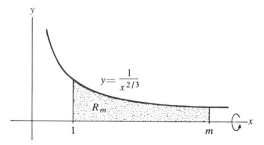

Figure 5.72

Remark

Let us carry this example one step further. Since

$$\lim_{m \to +\infty} V_m = \lim_{m \to +\infty} \left[3\pi \left(1 - \frac{1}{m^{1/3}} \right) \right] = 3\pi,$$

it is reasonable to say that in Figure 5.73 the region R of infinite extent when rotated about the x-axis yields a solid with a finite volume, namely 3π.

EXAMPLE 8   Let S_ε be the region of Example 12 in the preceding section. It is drawn in Figure 5.74. If S is rotated about the y-axis and V_ε is the volume of the resulting solid, then

$$V_\varepsilon = 2\pi \int_\varepsilon^1 x \left(\frac{1}{x^{5/4}}\right) dx = 2\pi \int_\varepsilon^1 \frac{1}{x^{1/4}} \, dx.$$

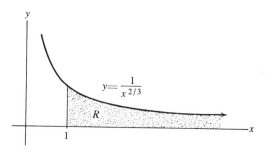

Figure 5.73

Now $\frac{4}{3}x^{3/4}$ is an antiderivative of $1/x^{1/4}$, and so

$$V_\varepsilon = 2\pi \int_\varepsilon^1 \frac{1}{x^{1/4}}\, dx = 2\pi\left[\tfrac{4}{3}x^{3/4}\Big|_\varepsilon^1\right]$$

$$= 2\pi\left[\tfrac{4}{3} - \tfrac{4}{3}\varepsilon^{3/4}\right]$$

$$= \frac{8\pi}{3}\left[1 - \varepsilon^{3/4}\right].$$

Remark

Once again, let us carry this one step further. Since

$$\lim_{\varepsilon\downarrow 0} V_\varepsilon = \lim_{\varepsilon\downarrow 0}\left[\frac{8\pi}{3}(1 - \varepsilon^{3/4})\right] = \frac{8\pi}{3},$$

then it is reasonable to say that in Figure 5.75 the region S of infinite extent when rotated about the y-axis yields a solid with a finite volume, namely $8\pi/3$.

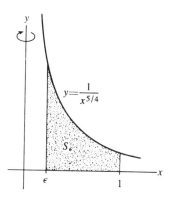

Figure 5.74

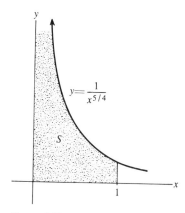

Figure 5.75

Observe that in the last two remarks, regions R and S each had infinite area, yet when rotated about certain axes they produced solids having finite volumes.

EXERCISES

1. Find the volumes V_x and V_y of the solids generated when the following regions are rotated around the x- and y-axes, respectively.

(a) $y = x^2$, $y = 0$, $x = 2$.   (b) $y = 1 - |x|$, $y = 0$, $x = 0$.

(c) $y = \sqrt{4 - x^2}$, $y = 0$, $x = 0$.   (d) $y = \sqrt{x}$, $x = 4$, $y = 0$.

(e) $y = x^2$, $x = y^2$.   (f) $y = x^{2/3}$, $y = 16$, $x = 0$.

(g) $y = x^2$, $y = 9$.   (h) $y = |x|$, $y = 2$.

(i) $y = x^2 + 2x + 1$, $y = 1$.   (j) $y^2 - 2y = -x$, $x = 0$.

2. Find the volumes of the solids generated when the shaded region in Figure 5.76 is rotated around the x-axis and the y-axis.

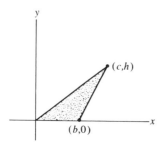

Figure 5.76

3. Find the volume of the solid generated when the region bounded by $x + y = 3$, $x = 0$, $y = 0$ is rotated as follows.

(a) About the x-axis.   (b) About the y-axis.

(c) About the line $y = 3$.   (d) About the line $x = 3$.

(e) About the line $x = 4$.

4. Let R be the region bounded by $x^2 + (y - b)^2 = a^2$, where $a < b$. Find the volume of the solid (called a **torus**) formed by rotating R about the x-axis. [*Hint:* $\int_{-a}^{a} \sqrt{a^2 - x^2}\, dx = \pi a^2/2$.]

5. A drain pipe of length 10 has circular cross sections with radius r varying according to the formula $r(x) = (x + 1)/2$ (see Figure 5.77). Find the volume of the drain pipe.

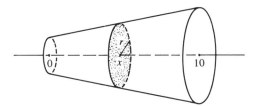

Figure 5.77

6. A pyramid is situated as in Figure 5.78, where for each y in $[0, h]$ the perpendicular cross section is an equilateral triangle with sides of length y. Find the volume of the pyramid.

7. A container of height 5 has rectangular horizontal cross sections; it is drawn in Figure 5.79. For any y in $[0, 5]$ the cross section is a rectangle of dimensions $3(y + 1)$ and $y + 2$. Find the volume of the container.

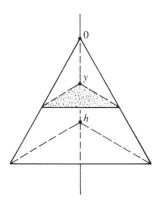

Figure 5.78

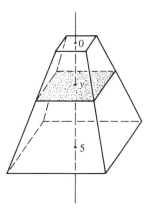

Figure 5.79

8. Let R_m be the region having area A_m in Exercise 8 of Section 5.6.

   (a) Determine the volume V_m of the solid formed by rotating R_m about the x-axis.

   (b) Evaluate $\lim\limits_{m \to +\infty} V_m$.

   (c) Is it possible to assign a finite volume to the solid formed by rotating about the x-axis the infinite strip bounded by the curves $x = 1$, $y = 0$, $y = 1/x^{3/4}$? If so, what?

9. Let S_ε be the region having area B_ε in Exercise 11 of Section 5.6.

   (a) Determine the volume V_ε of the solid formed by rotating S_ε about the y-axis.

   (b) Evaluate $\lim\limits_{\varepsilon \downarrow 0} V_\varepsilon$.

(c) Is it possible to assign a finite volume to the solid formed by rotating about the *y*-axis the infinite strip bounded by the curves $x = 0$, $x = 1$, $y = 0$, $y = 1/x^{6/5}$? If so, what?

10. Consider the ellipsoid in Figure 5.80. For any *t* in the interval $[-1, 1]$ the cross section perpendicular to the *t*-axis is an ellipse with the equation $[x^2/4(1 - t^2)] + [y^2/9(1 - t^2)] = 1$; this ellipse is drawn in Figure 5.81. Determine the volume of the ellipsoid. [*Hint:* Recall (Exercise 5 of Section 5.6) that the area inside the ellipse $(x^2/a^2) + (y^2/b^2) = 1$ is πab. Use 5.7.1.]

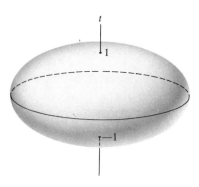

Figure 5.80   Figure 5.81

5.8 WORK

Let *l* be a coordinatized line. If a constant force *k* is applied to some object at every point in the interval $[a, b]$ of *l*, then the product

$$\text{Force} \times \text{Distance} = k(b - a)$$

is defined by the physicist to be the **work** performed over the given interval $[a, b]$.

EXAMPLE 1   If one pushes an object a distance of 2 feet while applying a constant force of 5 pounds, then

$$5 \times 2 = 10 \text{ foot-pounds}$$

is the work performed (see Figure 5.82).

Figure 5.82

Now we generalize this situation. Let *l* be a coordinatized line, $[a, b]$ an interval of *l*, and *f* a function defined for each *x* in $[a, b]$ by

$$f(x) = \text{Force applied at the point } x \qquad (\text{see Figure 5.83}).$$

Figure 5.83

This function f, sometimes referred to as a **force function**, need not be constant on $[a, b]$. The question is how to extend the primitive notion of work described in the opening paragraph (where a constant force is applied) so as to enable us to find the work W performed over $[a, b]$ by a possibly variable force. If f is continuous on $[a, b]$, then we can determine the work W as follows.

Let points

$$a = x_0 < x_1 < x_2 < \cdots < x_{n-1} < x_n = b$$

be a partition P of $[a, b]$, where n is a positive integer, and for each $i = 1, 2, \ldots, n$ let t_i be any point in the subinterval $[x_{i-1}, x_i]$ (see Figure 5.84).

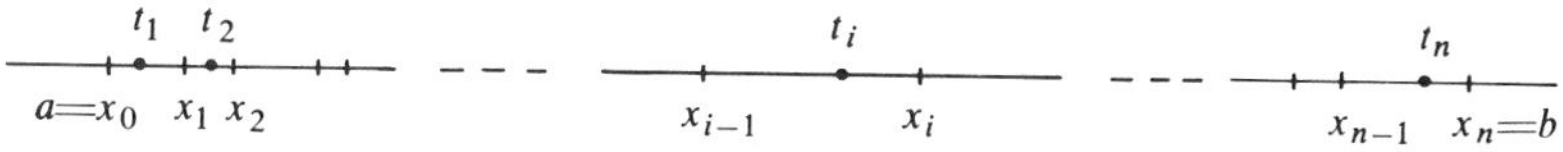

Figure 5.84

Let W_i be the work performed over $[x_{i-1}, x_i]$. Now the force applied at the point t_i is $f(t_i)$, which may indeed differ from values of f at other points in $[x_{i-1}, x_i]$, but nevertheless

$$W_i \approx f(t_i)(x_i - x_{i-1})$$

and

$$W = W_1 + W_2 + \cdots + W_n$$
$$\approx f(t_1)(x_1 - x_0) + f(t_2)(x_2 - x_1) + \cdots + f(t_n)(x_n - x_{n-1})$$
$$= \sum_{i=1}^{n} f(t_i)\, \Delta x_i.$$

Since f is continuous on $[a, b]$, then the approximation gets better as the norm of the partition gets smaller. Thus

$$\sum_{i=1}^{n} f(t_i)\, \Delta x_i$$

can be made as close to W as we wish by choosing any partition of $[a, b]$ whose norm is sufficiently small. In light of Definition 5.3.1

$$W = \int_a^b f(x)\, dx.$$

Thus the work done over $[a, b]$ is just the integral of the force function over $[a, b]$.

5.8.1

Work

Let l be a coordinatized line, $[a, b]$ an interval of l, and for each x in $[a, b]$ let $f(x)$ be the force applied to an object at the point x. If f is a continuous function on $[a, b]$, then the work performed over $[a, b]$ is

$$\int_a^b f(x) \, dx.$$

EXAMPLE 2

In the case that the force function is constant, statement 5.8.1 gives the expected result, for if a constant force k is applied everywhere on $[a, b]$, then the work W done over $[a, b]$ is

$$\int_a^b k \, dx.$$

However, kx is an antiderivative of k, and so

$$W = \int_a^b k \, dx = kx \Big|_a^b$$

$$= kb - ka$$

$$= k(b - a).$$

EXAMPLE 3

If the force $F(x) = 2x^2 - 1$ is applied at the point x in $[2, 4]$, then the work W performed over this interval is

$$\int_2^4 (2x^2 - 1) \, dx.$$

Since $\frac{2}{3}x^3 - x$ is an antiderivative of $2x^2 - 1$, we have

$$W = \int_2^4 (2x^2 - 1) \, dx = \left(\frac{2}{3}x^3 - x\right)\Big|_2^4$$

$$= \left(\frac{128}{3} - 4\right) - \left(\frac{16}{3} - 2\right)$$

$$= \frac{106}{3}.$$

EXAMPLE 4

A steel spring stretched x units beyond its natural length will exert a force given by

$$f(x) = kx,$$

where the constant k depends on the particular spring and units used. This is known as **Hooke's law**.

A certain spring exerts a force of 1 pound when stretched $\frac{1}{2}$ foot beyond its natural length. What is the work W done in stretching the spring 1 foot beyond its natural length? (See Figure 5.85.)

Since $1 = f(\frac{1}{2}) = k(\frac{1}{2})$, we have $k = 2$ and

$$f(x) = 2x \quad .$$

Hence

$$W = \int_0^1 2x \, dx$$

$$= x^2 \big|_0^1$$

$$= 1 \text{ foot-pound.}$$

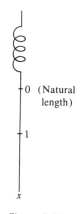

0 (Natural length)

1

x

Figure 5.85

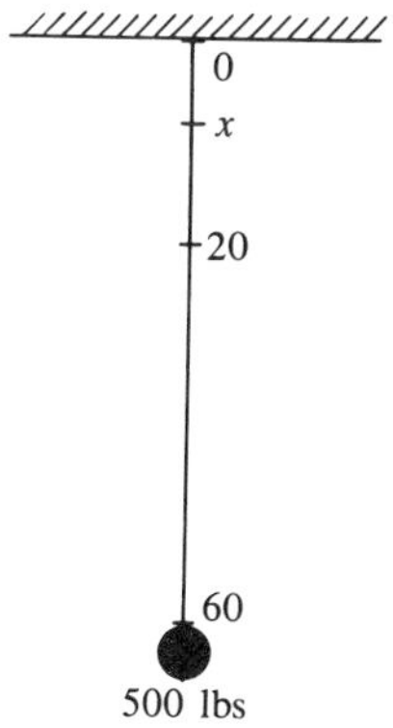

0

x

20

60

500 lbs

Figure 5.86

EXAMPLE 5   An iron ball weighing 500 pounds hangs from a 60-foot cable which weighs 2 pounds per foot; determine the work W done in winding up 20 feet of the cable.

Referring to Figure 5.86, at any x in $[0, 20]$ the force exerted by the ball and cable below that point is

$$500 + 2(60 - x).$$

Hence

$$W = \int_0^{20} [500 + 2(60 - x)] \, dx$$

$$= \int_0^{20} [620 - 2x] \, dx$$

$$= (620x - x^2) \big|_0^{20}$$

$$= 620(20) - 400$$

$$= 12{,}000 \text{ foot-pounds.}$$

While the integral given in 5.8.1 may not be directly applicable to many work problems, the technique used in its derivation is often useful, as the following example indicates.

EXAMPLE 6   Find the work W required to pump out a conical tank of height 10 feet and radius 5 feet if it is filled with a liquid of density k pounds per cubic foot (see Figure 5.87).

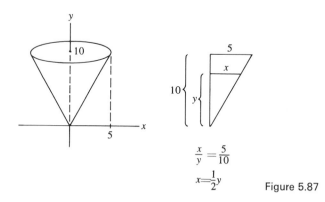

$$\frac{x}{y} = \frac{5}{10}$$

$$x = \frac{1}{2}y$$

Figure 5.87

Let

$$0 = y_0 < y_1 < \cdots < y_{n-1} < y_n = 10$$

be a partition of the vertical axis $[0, 10]$ of the cone, where n is a positive integer, and for each $i = 1, 2, \ldots, n$ let t_i be any point in $[y_{i-1}, y_i]$. Let W_i be the work done in pumping out the slice S_i of the cone formed by the horizontal planes through y_{i-1} and y_i. Now, referring to Figure 5.88, the volume of the cylinder s_i is

$$\pi(\tfrac{1}{2}t_i)^2(y_i - y_{i-1}),$$

which approximates the volume of the slice S_i, and so

$$k\pi(\tfrac{1}{2}t_i)^2(y_i - y_{i-1})$$

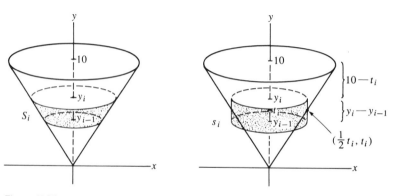

Figure 5.88

approximates the weight of the slice S_i. Also,

$$10 - t_i$$

approximates the distance that the slice S_i must be raised (that is, pumped).
Thus

$$W_i \approx k\pi \left(\frac{t_i}{2}\right)^2 (10 - t_i)(y_i - y_{i-1})$$

and

$$W = W_1 + W_2 + \cdots + W_n$$

$$\approx k\pi \left(\frac{t_1}{2}\right)^2 (10 - t_1)(y_1 - y_0) + \cdots + k\pi \left(\frac{t_n}{2}\right)^2 (10 - t_n)(y_n - y_{n-1})$$

$$= \sum_{i=1}^{n} k\pi \left(\frac{t_i}{2}\right)^2 (10 - t_i) \, \Delta y_i.$$

The above approximation to the weight of S_i and the distance it must be moved
get better as the norm of the partition gets smaller. Hence

$$\sum_{i=1}^{n} k\pi \left(\frac{t_i}{2}\right)^2 (10 - t_i) \, \Delta y_i$$

can be made as close to W as we wish by choosing any partition of $[0, 10]$
whose norm is sufficiently small. Because of Definition 5.3.1,

$$W = \int_0^{10} k\pi \left(\frac{y}{2}\right)^2 (10 - y) \, dy.$$

Finally,

$$W = \int_0^{10} k\pi \left(\frac{y}{2}\right)^2 (10 - y) \, dy$$

$$= \frac{k\pi}{4} \int_0^{10} (10y^2 - y^3) \, dy$$

$$= \frac{k\pi}{4} \left[\left(\frac{10}{3} y^3 - \frac{y^4}{4}\right)\Big|_0^{10} \right]$$

$$= \frac{k\pi}{4} \left(\frac{10,000}{3} - \frac{10,000}{4}\right)$$

$$= \frac{625}{3} k\pi \text{ foot-pounds.}$$

EXERCISES

1. A particle moves along the x-axis and is subjected to a force $f(x) = k/x^2$.
   If $f(x) = 2$ when $x = 3$, find the work done on $[1, 4]$.

2. Two men are pushing a car up an incline 5 yards long. For each x in $[0, 5]$,
   they exert a combined force of $2x^2 + x$ pounds. Find the work done in
   pushing the car.

3. A given spring exerts a force of 1 pound when stretched one foot beyond its natural length. Find the work done in stretching the spring as follows.

(a) One-half foot beyond its natural length.

(b) One foot beyond its natural length.

(c) Two feet beyond its natural length.

4. In stretching a given spring it takes twice as much work to stretch it from a length of 3 feet to 4 feet as it does to stretch it from 2 feet to 3 feet. Find the natural length of the spring.

5. A 10-foot cable weighs 1 pound per foot. Find the work done in lifting the cable so that it hangs from a beam of the following heights.

(a) Ten feet.   (b) Twenty feet.

6. A 20-foot cable weighing 1 pound per foot has a 25-pound weight attached to one end. Find the work done in lifting the cable, with the weight at the bottom, from a beam 30 feet high.

7. A particle of mass m moves in one direction along the x-axis under the influence of a force $f(x) = mv(dv/dx)$ (given by **Newton's second law**), where v is the velocity (as a function of position x) of the particle. Let $K(x) = \frac{1}{2}mv(x)^2$ be the kinetic energy of the particle. Show that the work done on $[a, b]$ is the change in kinetic energy, that is, $K(b) - K(a)$.

8. Find the work W required to pump out a cylindrical tank of height 8 feet and radius 4 feet if it is filled with a liquid of density k pounds per cubic foot.

9. Find the work W required to pump out a cylindrical tank of height 8 feet and radius 4 feet if it is half filled with a liquid of density k pounds per cubic foot.

10. Find the work W required to pump out a conical tank of height 10 feet and radius 5 feet if it is filled up to a depth of 6 feet with a liquid of density k pounds per cubic foot.

5.9 TOTAL VALUE, EXPECTED VALUE

If a businessman earns a fixed profit of P dollars on each of k days, then the product

$$Pk$$

gives the **total profit** over the span of k days. However, what if the profit is not fixed and varies continuously from one instant to the next over a period of time? How can one find the total profit over such a period?

Let us generalize on such a situation. Let $a \geq 0$ and let f be a function defined for each t in $[a, b]$ by

$$f(t) = \text{Amount of a certain commodity}$$
$$\text{earned (or spent, borrowed,}$$
$$\text{traded, etc.) at time } t.$$

Observe that f need not be constant on $[a, b]$. We seek the total amount A of the commodity earned (or spent, etc.) between times a and b.

As we have done several times before, let

$$a = t_0 < t_1 < \cdots < t_{n-1} < t_n = b$$

be a partition P of $[a, b]$, where n is a positive integer, and for each $i = 1, 2, \ldots, n$ let z_i be any point in the subinterval $[t_{i-1}, t_i]$. Let A_i be the amount of the commodity earned (or spent, etc.) on the time interval $[t_{i-1}, t_i]$. Now the amount earned (or spent, etc.) at time z_i is $f(z_i)$, which may indeed differ from values of f at other times in $[t_{i-1}, t_i]$, but nevertheless

$$A_i \approx f(z_i)(t_i - t_{i-1})$$

and

$$\begin{aligned} A &= A_1 + A_2 + \cdots + A_n \\ &\approx f(z_1)(t_1 - t_0) + f(z_2)(t_2 - t_1) + \cdots + f(z_n)(t_n - t_{n-1}) \\ &= \sum_{i=1}^{n} f(z_i)\, \Delta t_i. \end{aligned}$$

If f is continuous on $[a, b]$, then this approximation gets better as the norm of the partition gets smaller. In that case

$$\sum_{i=1}^{n} f(z_i)\, \Delta t_i$$

can be made as close to A as we wish by choosing any partition of $[a, b]$ whose norm is sufficiently small. Thus by Definition 5.3.1

$$A = \int_a^b f(t)\, dt.$$

5.9.1
Total Value

Let $a \geq 0$ and for each t in $[a, b]$ let $f(t)$ be the amount of a certain commodity earned (or spent, borrowed, traded, etc.) at time t. If f is a continuous function on $[a, b]$, then the total amount of the commodity earned (or spent, etc.) between times a and b is

$$\int_a^b f(t)\, dt.$$

EXAMPLE 1

An analysis of its many investments indicates what a particular bank will be earning over the next 6 months. For any t in $[0, 6]$

$$f(t) = \tfrac{1}{10}t + 4$$

is the number of dollars (in thousands) earned at time t.

By 5.9.1 the total earnings E over the next 6 months is

$$\int_0^6 (\tfrac{1}{10}t + 4)\, dt.$$

Since $(\tfrac{1}{20}t^2 + 4t)$ is an antiderivative of $\tfrac{1}{10}t + 4$,

$$
\begin{aligned}
E &= \int_0^6 \left(\frac{1}{10}t + 4\right) dt \\
&= \left(\frac{1}{20}t^2 + 4t\right)\Bigg|_0^6 \\
&= \frac{36}{20} + 24 \\
&= \frac{129}{5}.
\end{aligned}
$$

EXAMPLE 2 A savings and loan association has estimated its savings account withdrawals over the coming year. For any t in $[0, 12]$

$$w(t) = -\tfrac{175}{6}t^2 + 325t + 900$$

is the number of dollars (in thousands) withdrawn at time t.

If this estimate is accurate, then the total amount W of money withdrawn over the next 12 months will be, by 5.9.1,

$$\int_0^{12} \left(-\frac{175}{6}t^2 + 325t + 900\right) dt.$$

Since $(-\tfrac{175}{18}t^3 + \tfrac{325}{2}t^2 + 900t)$ is an antiderivative of $(-\tfrac{175}{6}t^2 + 325t + 900)$,

$$
\begin{aligned}
W &= \int_0^{12} \left(-\frac{175}{6}t^2 + 325t + 900\right) dt \\
&= \left(-\frac{175}{18}t^3 + \frac{325}{2}t^2 + 900t\right)\Bigg|_0^{12} \\
&= -\frac{175}{18}(1728) + \left(\frac{325}{2}\right)(144) + (900)(12) \\
&= 17{,}400.
\end{aligned}
$$

EXAMPLE 3 On a certain stock exchange it is estimated that the shares traded over the next year will be given by the formula

$$(t - 1)^3 - t + 3.$$

That is, for any t in $[0, 1]$

$$s(t) = (t - 1)^3 - t + 3$$

is the number of shares (in billions) traded at time t.

By 5.9.1 the total number S of shares traded during that year is

$$\int_0^1 [(t-1)^3 - t + 3]\, dt.$$

Finally, $[\frac{1}{4}(t-1)^4 - \frac{1}{2}t^2 + 3t]$ is an antiderivative of $[(t-1)^3 - t + 3]$, and so

$$
\begin{aligned}
S &= \int_0^1 [(t-1)^3 - t + 3]\, dt \\
&= \left(\frac{1}{4}(t-1)^4 - \frac{1}{2}t^2 + 3t \right)\Big|_0^1 \\
&= \left(-\frac{1}{2} + 3 \right) - \left(\frac{1}{4} \right) \\
&= \frac{9}{4}.
\end{aligned}
$$

In the first three examples we have examined problems involving total value, whereas now we shall consider the concept of expected value, in which the idea of probability plays an important role. The **probability** that an event will occur is a number p, where

$$0 \le p \le 1$$

and p indicates the likelihood that the event will occur—the closer p is to 1, the greater the likelihood that the event will occur.

EXAMPLE 4   In a deck of 52 playing cards there are 4 aces and 4 jacks. Thus the probability of drawing either an ace or a jack in one draw is

$$\frac{\text{Number of favorable outcomes}}{\text{Total number of outcomes}} = \frac{8}{52} = \frac{2}{13}.$$

If p is the probability of receiving value d, then the product

$$pd$$

is called the **expected value**.

EXAMPLE 5   A grocer has six watermelons selling for \$2 each. He estimates that the probability of selling a first one is $\frac{9}{10}$, but the probability of selling successive ones decreases by $\frac{1}{10}$ for each watermelon sold (that is, the probability of selling a second watermelon is $\frac{9}{10} - \frac{1}{10} = \frac{8}{10}$, the probability of selling a third watermelon is $\frac{8}{10} - \frac{1}{10} = \frac{7}{10}$, etc.).

Thus his expected receipts from selling six watermelons is

$$\tfrac{9}{10}(\$2) + \tfrac{8}{10}(\$2) + \tfrac{7}{10}(\$2) + \tfrac{6}{10}(\$2) + \tfrac{5}{10}(\$2) + \tfrac{4}{10}(\$2) = \$7.80.$$

Now we generalize on this discrete situation. Let $b > 0$ and let x in $[0, b]$ represent the xth item of a certain commodity which is being marketed. Define functions f, P on $[0, b]$ by

$$f(x) = \text{Money earned (or lost, spent, etc.) in selling the } x^{\text{th}} \text{ item}$$

$$P(x) = \text{Probability of selling the } x^{\text{th}} \text{ item.}$$

What profit E might be expected from marketing amount b of this commodity? Again, let

$$0 = x_0 < x_1 < x_2 < \cdots < x_{n-1} < x_n = b$$

be a partition P of $[0, b]$, where n is a positive integer, and for each $i = 1, 2, \ldots, n$ let t_i be any point in the subinterval $[x_{i-1}, x_i]$. Now

$$P(t_i)f(t_i)$$

is the expected value of selling item t_i, and this expected value may indeed vary over $[x_{i-1}, x_i]$, but nevertheless

$$E_i \approx P(t_i)f(t_i)(x_i - x_{i-1})$$

and

$$E = E_1 + E_2 + \cdots + E_n$$

$$\approx P(t_1)f(t_1)(x_1 - x_0) + P(t_2)f(t_2)(x_2 - x_1) + \cdots$$

$$+ P(t_n)f(t_n)(x_n - x_{n-1})$$

$$= \sum_{i=1}^{n} P(t_i)f(t_i)\, \Delta x_i.$$

If P and f are continuous functions on $[0, b]$, then so is the product Pf, and the above approximation gets better as the norm of the partition gets smaller. In this case

$$\sum_{i=1}^{n} P(t_i)f(t_i)\, \Delta x_i$$

can be made as close to E as we wish by choosing any partition of $[0, b]$ whose norm is sufficiently small. Therefore by Definition 5.3.1

$$E = \int_0^b P(x)f(x)\, dx.$$

5.9.2
Expected Value

Let $b > 0$. For each x in $[0, b]$ let $f(x)$ be the money earned (or lost, spent, etc.) in selling the xth item of a certain commodity which is being marketed, and let $P(x)$ be the probability of selling that x^{th} item. If f and P are continuous on $[0, b]$, then the expected value from marketing amount b of the commodity is

$$\int_0^b P(x)f(x)\, dx.$$

EXAMPLE 6

A farmer is marketing his potato crop of 50 tons at a time when demand is great and he is assured of selling his entire crop. If he prices the potatoes so that his profit from selling each ton is fixed at $160, what is the expected profit E from marketing this entire crop?

Since the farmer receives a fixed profit of $160 for each ton, then for each x in $[0, 50]$ the profit is given by

$$f(x) = 160,$$

and since he is assured of selling each ton, then for each x in $[0, 50]$ the probability is given by
$$P(x) = 1.$$

Hence by 5.9.2

$$E = \int_0^{50} (1)(160)\, dx = \int_0^{50} 160\, dx.$$

Since $160x$ is an antiderivative of 160, then

$$E = \int_0^{50} 160\, dx$$

$$= 160x \Big|_0^{50}$$

$$= 160(50)$$

$$= 8000.$$

That is, the expected profit is $8000.

EXAMPLE 7

A farmer is marketing a peach crop of 2000 pounds and prices it so as to receive a fixed profit of 7 cents per pound. However, he realizes that his product is perishable and expects that for each x in $[0, 2000]$

$$P(x) = 1 - \left(\frac{x}{3000}\right)^2$$

is the probability of selling the xth pound. (That is, as x increases, the more difficult it becomes to sell the xth pound.) What is the expected profit E from marketing this crop?

Since the profit is fixed at 7 cents per pound, for each x in $[0, 2000]$ the profit is given by

$$f(x) = 0.07,$$

and we are given the probability of selling the xth pound as

$$P(x) = 1 - \left(\frac{x}{3000}\right)^2,$$

where x is in $[0, 2000]$. Thus by 5.9.2

$$E = \int_0^{2000} \left[1 - \left(\frac{x}{3000}\right)^2\right](0.07)\, dx$$

$$= 0.07 \int_0^{2000} \left[1 - \left(\frac{x}{3000}\right)^2\right] dx.$$

Since $x - x^3/3(3000)^2$ is an antiderivative of $1 - x^2/(3000)^2$,

$$E = 0.07 \int_0^{2000} \left[1 - \frac{x^2}{(3000)^2}\right] dx$$

$$= 0.07 \left[\left(x - \frac{x^3}{3(3000)^2}\right)\Big|_0^{2000}\right]$$

$$= 0.07 \left[2000 - \frac{(2000)^3}{3(3000)^2}\right]$$

$$= 119.$$

Therefore the expected profit is $119.

EXAMPLE 8   Another farmer is also marketing a peach crop of 2000 pounds. Realizing that his product is perishable, he plans to stabilize at $\frac{9}{10}$ the probability of selling each pound by lowering the price (and hence his profit) as more pounds are sold. Specifically, for each x in $[0, 2000]$ the profit is given by

$$f(x) = 0.07 - \left(\frac{x}{1000}\right)(0.01)$$

$$= \frac{7000 - x}{100,000}.$$

Determine the expected profit E from marketing this crop.

Here, for each x in $[0, 2000]$ we are told that the profit is

$$f(x) = \frac{7000 - x}{100,000},$$

and the probability of selling the xth item is given by

$$P(x) = \tfrac{9}{10}.$$

Hence by 5.9.2

$$E = \int_0^{2000} \left(\frac{9}{10}\right)\left(\frac{7000 - x}{100,000}\right) dx$$

$$= \frac{9}{1,000,000} \int_0^{2000} (7000 - x)\, dx.$$

Since $7000x - x^2/2$ is an antiderivative of $7000 - x$,

$$E = \frac{9}{1,000,000} \int_0^{2000} (7000 - x)\, dx$$

$$= \frac{9}{1,000,000} \left[\left(7000x - \frac{x^2}{2}\right)\Big|_0^{2000}\right]$$

$$= \frac{9}{1,000,000} \left[(7000)(2000) - \frac{(2000)^2}{2}\right]$$

$$= 108.$$

Therefore the expected profit is $108.

EXERCISES

1. A study has been made of traffic on a particular expressway between the hours of 4 P.M. and 6 P.M. It is determined that at time t, where $4 \leq t \leq 6$, $6t^2$ additional vehicles use the expressway. Find the total number of vehicles using the expressway between 4 P.M. and 6 P.M.

2. In a certain city the telephone company determined that, between 1 P.M. and 4 P.M. on a weekday, the number (in hundreds) of calls made at time t is $4 + 4t - t^2$, where $1 \leq t \leq 4$. Find the total number of calls in the given time period.

3. In another city the telephone company determined that, between 1 P.M. and 4 P.M. on a weekday, the number (in hundreds) of calls made at time t is $4t - t^2$, $1 \leq t \leq 2$, and $5 - \frac{1}{2}t$, $2 < t \leq 4$. Find the total number of calls made in the given time period.

4. Over a certain 10-day period it is determined that the cash outflow (in thousands of dollars) of a corporation at time t is $100 - \frac{1}{5}t$, where $0 \leq t \leq 10$. Find the corporation's total cash outflow over this period.

5. An auto salesman has seven new autos, each selling at $3000, in his showroom. He is certain of selling one of these, but the probability of selling successive ones decreases by $\frac{1}{10}$ for each auto sold. Determine his expected receipts for selling seven autos.

6. A service station owner has just received a shipment of 2000 gallons of gasoline from his supplier. If he is certain of selling the entire shipment on that day and his profit is 3 cents per gallon, find his expected profits from the sale of the shipment on that day.

7. Another service station owner has received a shipment of 2000 gallons of gasoline; he also receives 3 cents on each gallon sold. However, he is not certain of selling the entire 2000 gallons on that day, and he has found that the probability of selling the xth gallon is $1 - (x/2500)^2$. Find his expected profit from the sale of the shipment on that day.

8. In marketing an overripe grape crop of 1000 pounds, a farmer will lose 2 cents for each pound sold. If the probability of selling the xth pound is $1 - x/1100$, determine the expected loss from the sale of his crop.

9. In marketing 1500 pounds of bananas the seller adjusts the price periodically so that the profit from selling the xth pound is $3[(1600 - x)/1500]$ cents, and the probability of selling the xth pound is $\frac{9}{10}[(1600 - x)/1500]$. Find the expected profit from selling the bananas.

10. Justify the following definition: Let $b > 0$. For each x in $[0, b]$ let $f(x)$ be the money earned in selling the xth item of a certain commodity. If f is continuous on $[0, b]$, then we say that the total money earned in selling amount b of the commodity is $\int_0^b f(x)\, dx$.

11. Let $b > 0$. For each x in $[0, b]$ let $F(x)$ be the total money earned in selling the first x items of a certain commodity. Show that for any x in $[0, b]$, $F'(x)$ is the money earned in selling the xth item. [*Hint:* Use Exercise 10 and the Fundamental Theorem of Calculus.]

12. Let $b > 0$. For each x in $[0, b]$ let $F(x)$ be the total money earned in selling the first x items of a certain commodity, and let $p(x)$ be the probability of selling the xth item. Show that the expected earnings from selling amount b of the commodity is $\int_0^b p(x)F'(x)\, dx$. [*Hint:* Use Exercise 11 and 5.9.2.]

13. For each x in $[0, 100]$ let $F(x) = 100 + x^2/4$ be the total money earned in selling the first x items of a certain commodity, and let $p(x) = 1 - x/150$ be the probability of selling the xth item. Find the expected earnings from selling the given commodity.

5.10   THE INDEFINITE INTEGRAL

As we have seen in the Fundamental Theorem of Calculus, when a function f has an antiderivative there is a relationship between the definite integral of f and *any* antiderivative of f. This being the case, it is customary to denote the general form of an antiderivative by using notation related to the integral.

Specifically, recall that for a continuous function f with an antiderivative F we have

$$\int_a^b f(x)\, dx = F(x)\Big|_a^b.$$

Since $F(x) + C$ is also an antiderivative of f for any constant C, we have

$$\int_a^b f(x)\, dx = F(x) + C\Big|_a^b.$$

Elimination of any reference to limits a and b in the above equation gives

(5.10.1) $$\int f(x)\, dx = F(x) + C.$$

The symbol on the left in Equation 5.10.1 is called the **indefinite integral of** f; it represents *all* antiderivatives of f.

The variable x appearing in Equation 5.10.1 is of no significance; any variable would be proper to use. Therefore we may also write

$$\int f(u)\, du = F(u) + C.$$

The following three formulas are helpful in evaluating indefinite integrals:

(5.10.2) $$\int kf(x)\, dx = k \int f(x)\, dx;$$

that is, *the integral of a constant k times a function is the constant k times the integral of the function.*

(5.10.3) $$\int [f(x) + g(x)]\, dx = \int f(x)\, dx + \int g(x)\, dx;$$

that is, *the integral of a sum is the sum of the integrals.*

(5.10.4) $$\int f(g(x))g'(x)\, dx = \int f(u)\, du,$$

where $u = g(x)$.

The first two of these formulas are restatements of 4.12.5 and 4.12.6, respectively. The third, a *change of variable formula for indefinite integrals*, is obtained by omitting the symbols a, b, c, d in Equation 5.5.1.

EXAMPLE 1

$$\int (x^3 - 2x + 7)\, dx = \int x^3\, dx - 2 \int x\, dx + 7 \int dx$$

$$= \frac{x^4}{4} - 2 \left(\frac{x^2}{2} \right) + 7(x) + C$$

$$= \frac{x^4}{4} - x^2 + 7x + C.$$

EXAMPLE 2

$$\int \left(x^{4/5} - \frac{1}{x^2} \right) dx = \int x^{4/5}\, dx - \int x^{-2}\, dx$$

$$= \frac{5}{9} x^{9/5} - (-1)x^{-1} + C$$

$$= \frac{5}{9} x^{9/5} + \frac{1}{x} + C.$$

EXAMPLE 3   Evaluate $\int (x^2/\sqrt{1 + x^3})\, dx$.

Let $u = 1 + x^3$; then $du = \dfrac{du}{dx}\, dx = 3x^2\, dx$, or $x^2\, dx = \tfrac{1}{3}\, du$. Then

$$\int \frac{x^2\, dx}{\sqrt{1 + x^3}} = \int \frac{\tfrac{1}{3}\, du}{\sqrt{u}} = \frac{1}{3} \int \frac{du}{\sqrt{u}} = \frac{1}{3}\, [2\sqrt{u}] + C.$$

Since $u = 1 + x^3$, we have

$$\int \frac{x^2\, dx}{\sqrt{1 + x^3}} = \frac{2}{3} \sqrt{1 + x^3} + C.$$

EXAMPLE 4   Evaluate $\int x(1 + x)^8\, dx$.

Let $u = 1 + x$; then $du = dx$. Therefore

$$\int x(1 + x)^8\, dx = \int (u - 1)u^8\, du$$

$$= \int (u^9 - u^8)\, du$$

$$= \frac{u^{10}}{10} - \frac{u^9}{9} + C$$

$$= \frac{1}{10}\, (1 + x)^{10} - \frac{1}{9}\, (1 + x)^9 + C.$$

EXAMPLE 5   Evaluate $\int (x^3/\sqrt{5 + x^2})\, dx$.

Let $u = 5 + x^2$; then $du = 2x\, dx$, $x^2 = u - 5$, and $x\, dx = \tfrac{1}{2}\, du$. Hence

$$\int \frac{x^3}{\sqrt{5 + x^2}}\, dx = \int \frac{(u - 5)\tfrac{1}{2}}{\sqrt{u}}\, du$$

$$= \frac{1}{2} \int \left[\sqrt{u} - \frac{5}{\sqrt{u}} \right] du$$

$$= \frac{1}{2} \int \sqrt{u}\, du - \frac{5}{2} \int \frac{1}{\sqrt{u}}\, du$$

$$= \frac{1}{2} \left(\frac{2}{3}\, u^{3/2} \right) - \frac{5}{2}\, (2u^{1/2}) + C$$

$$= \frac{1}{3}\, u^{3/2} - 5u^{1/2} + C.$$

Finally, since $u = 5 + x^2$, we have

$$\int \frac{x^3}{\sqrt{5 + x^2}}\, dx = \frac{1}{3}\, (5 + x^2)^{3/2} - 5(5 + x^2)^{1/2} + C.$$

EXERCISES

In Exercises 1–12, evaluate the given integral.

1. $\displaystyle\int (3 - 2x + x^2)\, dx.$ **2.** $\displaystyle\int (4x^3 - 1/x^2)\, dx.$

3. $\displaystyle\int (\sqrt{x} + 2x^{1/3} + 7)\, dx.$ **4.** $\displaystyle\int (3x + 7)^2\, dx.$

5. $\displaystyle\int (1/\sqrt[3]{9x - 1})\, dx.$ **6.** $\displaystyle\int (x + 1)(4x^2 + 8x + 3)\, dx.$

7. $\displaystyle\int (x^2/\sqrt{1 - x^3})\, dx.$ **8.** $\displaystyle\int [x/(3 + 5x)^{1/2}]\, dx.$

9. $\displaystyle\int x\sqrt{8x - 1}\, dx.$ **10.** $\displaystyle\int (x^5/\sqrt{1 + x^3})\, dx.$

11. $\displaystyle\int (x^2\sqrt{2 + x})\, dx.$ **12.** $\displaystyle\int (x^3/\sqrt{1 - x})\, dx.$

13. Solve each of the following *differential equations*.

(a) $dy/dx = 5.$
(b) $dy/dx = \frac{1}{3}x^2.$
(c) $dy/dx = \sqrt{x} - 1/\sqrt{x}.$

14. Let f, g, F, G be functions such that $F'(x) = f(x)$ and $G'(y) = g(y)$. Consider the differential equation $dy/dx = f(x)/g(y)$, $g(y) \neq 0$. If we regard the symbol dy/dx as dy divided by dx, then we can manipulate the given differential equation as follows: $dy/dx = f(x)/g(y)$, $g(y)\, dy = f(x)\, dx$, $\int g(y)\, dy = \int f(x)\, dx$, and $G(y) = F(x) + C$. Assuming that the last equation defines y implicitly as a function of x, verify that $dy/dx = f(x)/g(y)$. [Thus we call $G(y) = F(x) + C$ the **solution** to the **separable differential equation** $dy/dx = f(x)/g(y)$.]

15. Use the procedure described in Exercise 14 to solve the following separable differential equations.

(a) $dy/dx = 1/y.$ (b) $dy/dx = x/y.$
(c) $dy/dx = y^2/x^2.$ (d) $dy/dx = x\sqrt{y}.$

Chapter 6

LOGARITHM AND
EXPONENTIAL FUNCTIONS

In the preceding chapters all our examples and exercises have dealt exclusively with **algebraic functions**, that is, functions that can be built in a finite number of steps using only the standard algebraic operations of addition, subtraction, multiplication, division, raising to powers, and taking roots. It happens, however, that functions not of this type appear naturally in the mathematical analysis of many important problems. Such nonalgebraic functions are called **transcendental functions**. Two such functions—the *natural logarithm function* and the *natural exponential function*—are the object of study in this chapter.

6.1 EXPONENTS AND LOGARITHMS

By this time we are familiar with the operation of *raising a number to a power*; for example,

$$2^5 = 2 \cdot 2 \cdot 2 \cdot 2 \cdot 2 = 32$$

$$3^{-2} = \frac{1}{3^2} = \frac{1}{3 \cdot 3} = \frac{1}{9}$$

$$5^{1/3} = \sqrt[3]{5}$$

$$6^{2/5} = \sqrt[5]{6^2} = \sqrt[5]{6 \cdot 6} = \sqrt[5]{36}.$$

The number in the power position is called an **exponent**. More generally, for a given number $a > 0$ we use the following notation:

$$a^m = \underbrace{a \cdot a \cdot a \cdots a}_{m \text{ factors}} \qquad \text{if } m \text{ is a positive integer}$$

$$a^0 = 1$$

$$a^{-m} = \frac{1}{a^m} \qquad \text{if } m \text{ is a positive integer}$$

$$a^{1/m} = \sqrt[m]{a} \qquad \text{if } m \text{ is a positive integer}$$

$$a^{m/n} = \sqrt[n]{a^m} \qquad \text{if } m, n \text{ are integers and } n > 0.$$

The French mathematician Nicole Oresme (circa 1323) was the first to use fractional exponents, although not in the "modern" notation we have introduced above. One of Newton's mentors, John Wallis, contributed significantly to the development and use of exponents in mathematics.

With this notation at our disposal we can list some properties of exponents which hold for any rational exponents r and s. (Recall that a rational number is a number that can be written as a ratio of integers m/n.)

6.1.1
Rules for
Exponents   Given $a > 0$, then

$$a^r a^s = a^{r+s}$$

$$\frac{a^r}{a^s} = a^{r-s}$$

$$(a^r)^s = a^{rs}.$$

EXAMPLE 1

$$2^{1/2} \cdot 2^{5/2} = 2^{1/2 + 5/2} = 2^{6/2} = 2^3 = 8$$

$$\frac{(100)^{3/2}}{100} = (100)^{3/2 - 1} = (100)^{1/2} = 10$$

$$(8^{1/2})^{4/3} = 8^{4/6} = 8^{2/3} = 4.$$

Since we are familiar with numbers which are rational powers of a positive number, let us now turn our attention to those numbers which are represented by a positive number raised to an irrational power. We begin by asking the question: What number do we call $3^{\sqrt{2}}$? Recall from our discussion in the Introduction that the sequence of rational numbers

$$1, \quad 1.4, \quad 1.41, \quad 1.414, \quad 1.4142, \ldots$$

closes in on a number, which is called the $\sqrt{2}$. We now consider the sequence of rational numbers

$$3^1, \quad 3^{1.4}, \quad 3^{1.41}, \quad 3^{1.414}, \quad 3^{1.4142}, \ldots \quad .$$

Does this sequence of numbers close in on a number? Since we already know how to raise a positive number to a rational power, let us compute 3 raised to these rational powers.

r	1	1.4	1.41	1.414	1.4142	1.41421	1.414213	1.4142135	1.41421356	1.414213562
3^r	3	4.7	4.71	4.727	4.7287	4.72879	4.728801	4.7288041	4.72880437	4.728804386

We note that as r increases, 3^r increases but does not increase without bound. In fact, 3^r seems to be bounded above by 5. Indeed, 3^r closes in on a number

$$4.728804\ldots,$$

which is called

$$3^{\sqrt{2}}.$$

EXAMPLE 2   Approximate 2^π to six decimal places. Since the sequence

$$3, \quad 3.1, \quad 3.14, \quad 3.141, \quad 3.1415, \quad 3.14159, \ldots$$

closes in on π, we consider the following table.

r	3	3.1	3.14	3.141	3.1415	3.14159	3.141592	3.1415926	3.14159265	3.141592653
2^r	8	8.6	8.82	8.821	8.8244	8.82496	8.824974	8.8249775	8.82497781	8.824977823

Therefore, $2^\pi \approx 8.824977$.

In light of these examples, given a positive number a we are able to speak of a^c, where c is any number, rational or irrational.

We shall accept the fact that *the rules for exponents listed in 6.1.1 are valid for any numbers r and s.* To see that

$$a^{r+s} = a^r a^s$$

for any r, s, consider the sequence of rationals

$$r_1, r_2, r_3, \ldots$$

which closes in on r, and the sequence of rationals

$$s_1, \; s_2, \; s_3, \ldots$$

which closes in on s. Then the sequence

$$a^{r_1}a^{s_1}, \qquad a^{r_2}a^{s_2}, \qquad a^{r_3}a^{s_3}, \ldots$$

closes in on $a^r a^s$, and the sequence

$$a^{r_1+s_1}, \qquad a^{r_2+s_2}, \qquad a^{r_3+s_3}, \ldots$$

closes in on a^{r+s}.

Since these last two sequences represent the same sequence, it follows that they close in on the same number. Therefore

$$a^{r+s} = a^r a^s.$$

The idea of logarithms is closely related to exponents. Consider the following sequence of rational powers of 2:

$$2^3, \qquad 2^{35/10}, \qquad 2^{358/100}, \qquad 2^{3584/1000}, \qquad 2^{35849/10000}, \ldots$$

which closes in on 12. Observe that there exists an increasing sequence of rationals

$$3, \quad \frac{35}{10}, \quad \frac{358}{100}, \quad \frac{3584}{1000}, \quad \frac{35{,}849}{10{,}000}, \ldots$$

which is bounded above by 4 and which closes in on a number, say u, such that

$$2^u = 12.$$

More generally, let a be a fixed positive number with $a \neq 1$. If x is any positive number, then there exists an increasing bounded sequence of rationals

$$r_1, \qquad r_2, \qquad r_3, \ldots$$

such that

$$a^{r_1}, \qquad a^{r_2}, \qquad a^{r_3}, \ldots$$

closes in on x. If

$$r_1, \qquad r_2, \ldots$$

closes in on u, then

$$a^u = x.$$

In summary, given $x > 0$, there exists a u such that

$$a^u = x;$$

that is, u *is the exponent to which* a *must be raised in order to obtain* x. The number u is called **logarithm of x to the base a** and is denoted by

$$u = \log_a x.$$

Therefore

$$(6.1.2) \qquad u = \log_a x \qquad \Leftrightarrow \qquad a^u = x.$$

EXAMPLE 3

$$\log_2(32) = 5 \qquad \text{since } 2^5 = 32$$
$$\log_3(\tfrac{1}{9}) = -2 \qquad \text{since } 3^{-2} = \tfrac{1}{9}$$
$$\log_6(1) = 0 \qquad \text{since } 6^0 = 1$$
$$\log_{1/4}(\tfrac{1}{2}) = \tfrac{1}{2} \qquad \text{since } (\tfrac{1}{4})^{1/2} = \tfrac{1}{2}.$$

If, in the equation $a^u = x$ of 6.1.2, we replace u by $\log_a x$, then we obtain

$$a^{\log_a x} = x.$$

Likewise, if in $\log_a x = u$, we replace x by a^u, we obtain

$$\log_a(a^u) = u.$$

EXAMPLE 4 Simplify $2^{(\log_2 6)+4}$.

$$2^{(\log_2 6)+4} = (2^{\log_2 6})(2^4)$$
$$= (6)(16)$$
$$= 96.$$

The rules of exponents listed in 6.1.1 give rise to similar ones for logarithms.

6.1.3

Rules for Logarithms Given $a > 0$, then for any positive numbers x, y and any number p,

1. $\log_a(xy) = \log_a x + \log_a y$.
2. $\log_a(x/y) = \log_a x - \log_a y$.
3. $\log_a x^p = p \log_a x$.

To show that 1 is true, we let $u = \log_a(xy)$, $u_1 = \log_a x$, and $u_2 = \log_a y$; then $xy = a^u$, $x = a^{u_1}$, and $y = a^{u_2}$. Therefore

$$a^u = xy = a^{u_1}a^{u_2} = a^{u_1 + u_2};$$

that is,

$$a^u = a^{u_1 + u_2}.$$

Taking the $\log_a$ of both sides, we obtain

$$\log_a(a^u) = \log_a(a^{u_1 + u_2})$$

$$u = u_1 + u_2$$

$$\log_a(xy) = \log_a x + \log_a y.$$

The others are shown in a similar fashion.

EXAMPLE 5   Simplify $\log_{10} 3000$.

$$\log_{10} 3000 = \log_{10}(3)(10^3)$$

$$= \log_{10} 3 + \log_{10}(10^3)$$

$$= \log_{10} 3 + 3.$$

Logarithms are the invention of the Scottish mathematician John Napier (1550–1617); he first published his work on logarithms, or *ratio numbers*, in a 1614 paper titled "A Description of the Wonderful Law of Logarithms." It was immediately acclaimed and adopted throughout Europe because of its assistance in shortening computations required in astronomy, engineering, navigation, etc. The rules of logarithms reduce difficult problems involving multiplication and division into simpler ones involving addition and subtraction. The next example, while not difficult in itself, illustrates the method.

EXAMPLE 6   Find $N = [(81)(62)]/(53)$.
First we find $\log_{10} N$:

$$\log_{10} N = \log_{10} \frac{(81)(62)}{(53)}$$

$$= \log_{10}(81)(62) - \log_{10}(53)$$

$$= \log_{10} 81 + \log_{10} 62 - \log_{10} 53$$

$$= 1.9085 + 1.7924 - 1.7243.$$

The last step is accomplished with the use of logarithm tables. Therefore

$$\log_{10} N = 1.9766.$$

Returning to the tables, we see that

$$\log_{10}(94.8) = 1.9766;$$

hence

$$N = 94.8.$$

As we have seen in 6.1.2, it makes sense to talk of a logarithm to base a, where a is any fixed positive number. Nevertheless, in high school the student works primarily with the so-called **common logarithms**, or logarithms to base 10. This is done to facilitate computations of the type illustrated in Example 6. Common logarithms, not included in Napier's original paper, were introduced at the suggestion of the English mathematics professor Henry Briggs (1561–1631).

In our study of the calculus, however, another base number appears to be natural in order to facilitate the differentiation and integration of exponential and logarithm functions. These ideas will be pursued in the next three sections.

EXERCISES

1. Evaluate each of the following.

(a) 2^3.

(b) 3^4.

(c) 3^{-4}.

(d) 3^{-2}.

(e) 10^4.

(f) 10^{-3}.

(g) $16^{-1/4}$.

(h) $(\frac{1}{2})^1$.

(i) $(\frac{1}{2})^0$.

(j) $27^{2/3}$.

(k) $(-2)^3$.

(l) $(\frac{1}{2})^4$.

(m) $(b^2)^2$.

(n) $3^{1/2}(27)^{1/6}$.

(o) $2^3/(32)^{2/5}$.

2. Given $a^2 = A$ and $a^3 = B$, where $a > 0$, find the following in terms of A and B:
$a^4, a^5, a^7, a^9, a^{15}, a^{20}, a^{1/2}, a^{2/3}, a^{-2}, a^{-7/3}$.

3. In each of the following, determine x, if any, satisfying the given equation.

(a) $2^x = 4$.

(b) $3^{-x} = 9$.

(c) $2^3 \cdot 2^x = 16$.

(d) $(4^x)^2 = 4$.

(e) $2^3/2^x = 8$.

(f) $2^x = -1$.

4. Approximate each of the following to four decimal places.

(a) $3^{\sqrt{2}+1}$.

(b) $3^{2\sqrt{2}}$.

(c) $2^{2-\pi}$.

(d) $\sqrt{2^\pi}$.

5. Find the values of the following.

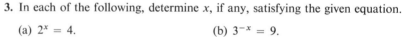

(a) $\log_2 8$.

(b) $\log_4 16$.

(c) $\log_3 \frac{1}{9}$.

(d) $\log_{10} 10^4$.

(e) $\log_{10} (0.001)$.

(f) $\log_{16}(4)$.

(g) $\log_{16}(\frac{1}{2})$.

(h) $\log_{27}(9)$.

(i) $\log_{1/2}(\frac{1}{16})$.

(j) $\log_\pi(\pi^2)$.

(k) $\log_{\sqrt{3}}(\frac{1}{9})$.

6. Given $\log_a 2 = A, \log_a 3 = B, \log_a 5 = C$, find the following in terms of A, B, C.

(a) $\log_a 4$.

(b) $\log_a 6$.

(c) $\log_a 15$.

(d) $\log_a 30$.

(e) $\log_a 45$.

(f) $\log_a \frac{4}{5}$.

(g) $\log_a 100$.

(h) $\log_a \frac{6}{5}$.

(i) $\log_a 180$.

7. In each of the following, determine x, if any, which satisfy the given equation.

(a) $\log_2(x + 1) = 1$.   (b) $\log_3 x/3 = 2$.

(c) $\log_{10} x + \log_{10} 5 = 1$.   (d) $\log_4 x^2 - \log_4 100 = \frac{1}{2}$.

(e) $\log_2(x^2 - 2) = 1$.

8. Verify rules (2) and (3) of 6.1.3.

9. Let f be defined by $f(x) = 10^x$. Show that f is increasing on **R**; that is, $x_1 < x_2 \Rightarrow 10^{x_1} < 10^{x_2}$.

10. Let g be defined by $g(x) = \log_{10} x$. Show that g is increasing on $\mathbf{R}^+$.

6.2 THE NUMBER e

Initially, we shall investigate continuity and differentiability of the function f defined by

$$f(x) = \log_{10} x,$$

and thereby discover that $f'(x)$ depends on a new number denoted by e—which is the primary object of our investigation in this section.

We now proceed to examine the continuity of f defined by $f(x) = \log_{10} x$. To begin, let us examine continuity at 1:

$$\lim_{x \to 1} \log_{10} x.$$

x	$\log_{10} x$	x	$\log_{10} x$
0.1	-1	2	0.3
0.5	-0.3	1.3	0.11
0.9	-0.046	1.1	0.041
0.92	-0.036	1.09	0.037
0.942	-0.0259	1.05	0.021
0.993	-0.0031	1.004	0.0017
0.998	-0.00087	1.0007	0.0003
0.9999	-0.000043	1.0002	0.00009
0.99996	-0.000017	1.00003	0.00001
0.999992	-0.0000035	1.000001	0.0000004

Hence

$$\lim_{x \to 1} \log_{10} x = 0 = \log_{10} 1;$$

that is, $f(x) = \log_{10} x$ is continuous at 1.

Moreover,

$$\lim_{x \to c} [\log_{10} x - \log_{10} c] = \lim_{x \to c} \log_{10} \left(\frac{x}{c}\right).$$

Let $u = x/c$; then

$$x \to c \quad \Leftrightarrow \quad u \to 1.$$

Hence

$$\lim_{x \to c} \log_{10} \left(\frac{x}{c} \right) = \lim_{u \to 1} \log_{10} u = 0.$$

Therefore

$$\lim_{x \to c} [\log_{10} x - \log_{10} c] = 0.$$

6.2.1
Important Fact

For any positive number c,

$$\lim_{x \to c} \log_{10} x = \log_{10} c.$$

In other words, $\log_{10} x$ *is continuous at each point* c.

Shortly, we shall have to deal with the limit of a composite function, namely

$$\lim_{x \to +\infty} \log_{10}(f(x)), \qquad \text{where } f(x) > 0.$$

If $\lim_{x \to +\infty} f(x) = c > 0$, then $f(x)$ can be made as close to c as we wish by choosing x sufficiently large. Since $\log_{10} x$ is continuous at c, it is reasonable to assert that $\log_{10} f(x)$ can be made as close to $\log_{10} c$ as we wish by choosing x sufficiently large.

6.2.2
Fact

Suppose that $f(x) > 0$ for each x and that

$$\lim_{x \to +\infty} f(x)$$

exists and is positive. Then

$$\lim_{x \to +\infty} \log_{10} f(x) = \log_{10} (\lim_{x \to +\infty} f(x)).$$

Also, this result holds if $+\infty$ is replaced by $-\infty$.

EXAMPLE 1   What is $\lim_{x \to +\infty} \log_{10}[(2x^2 + 5x + 1)/(20x^2 - 1)]$?

Since

$$\lim_{x \to +\infty} \frac{2x^2 + 5x + 1}{20x^2 - 1} = \lim_{x \to +\infty} \frac{2 + 5/x + 1/x^2}{20 - 1/x^2} = \frac{1}{10},$$

we have

$$\lim_{x \to +\infty} \log_{10}\left(\frac{2x^2 + 5x + 1}{20x^2 - 1}\right) = \log_{10}\left(\frac{1}{10}\right)$$

$$= \log_{10} 1 - \log_{10} 10$$

$$= -1.$$

For $f(x) = \log_{10} x$, let us examine

$$\frac{f(x + h) - f(x)}{h}.$$

We have

$$\frac{\log_{10}(x + h) - \log_{10} x}{h} = \frac{\log_{10}[(x + h)/x]}{h}$$

$$= \frac{1}{h} \log_{10}\left(1 + \frac{h}{x}\right)$$

$$= \log_{10}\left(1 + \frac{h}{x}\right)^{1/h}.$$

Now let $m = x/h$, so that

$$\log_{10}\left(1 + \frac{h}{x}\right)^{1/h} = \log_{10}\left(1 + \frac{1}{m}\right)^{m/x}$$

$$= \log_{10}\left[\left(1 + \frac{1}{m}\right)^m\right]^{1/x}$$

$$= \frac{1}{x} \log_{10}\left(1 + \frac{1}{m}\right)^m.$$

Therefore, to summarize,

$$\frac{\log_{10}(x + h) - \log_{10} x}{h} = \frac{1}{x} \log_{10}\left(1 + \frac{1}{m}\right)^m.$$

Since $m = x/h$, where $x > 0$,

$$h \downarrow 0 \Leftrightarrow m \to +\infty$$

and

$$h \uparrow 0 \Leftrightarrow m \to -\infty.$$

If one were to make a hasty guess at the value of $(1 + 1/m)^m$ as $m \to +\infty$, he might say 1 (because $1 + 1/m$ closes in on 1) or he might say $+\infty$ (because the exponent is getting larger). In fact, neither of these guesses is correct since both $1 + 1/m$ and the exponent m are changing simultaneously. Some values of $(1 + 1/m)^m$ are given in the following tables.

m	$(1 + 1/m)^m$	m	$(1 + 1/m)^m$
1	2	-2	4
5	2.48832	-5	3.05175781
10	2.59374246	-10	2.86797199
20	2.65329771	-20	2.78950981
75	2.70037866	-75	2.73662805
200	2.71151712	-800	2.71998270
5,000	2.71801005	$-2,000$	2.71896171
20,000	2.71821387	$-75,000$	2.71829995
745,000	2.71828000	$-170,000$	2.71828982
1,700,000	2.71828103	$-10,000,000$	2.71828196

It should be noted that as m increases, the numbers $(1 + 1/m)^m$ increase. However, $(1 + 1/m)^m$ is always bounded above by 3. In fact, the tables indicate that $(1 + 1/m)^m$ is closing in on a number; namely

$$2.718281 \ldots ,$$

which is customarily denoted by the letter e, in honor of the famous Swiss mathematician Leonhard Euler.* The number e is irrational (the above decimal is nonrepeating), a fact which is too difficult to prove here.

6.2.3
Fact

Each of the limits

$$\lim_{m \to +\infty} \left(1 + \frac{1}{m}\right)^m \quad \text{and} \quad \lim_{m \to -\infty} \left(1 + \frac{1}{m}\right)^m$$

exists and equals the irrational number which is denoted by e ($e \approx 2.718$).

Returning to the differentiability of $f(x) = \log_{10} x$, we have

$$\lim_{h \downarrow 0} \frac{\log_{10}(x + h) - \log_{10} x}{h} = \lim_{m \to +\infty} \frac{1}{x} \log_{10} \left(1 + \frac{1}{m}\right)^m$$

$$= \frac{1}{x} \log_{10} \left[\lim_{m \to +\infty} \left(1 + \frac{1}{m}\right)^m \right]$$

$$= \frac{1}{x} \log_{10} e.$$

* Leonhard Euler (1707–1783) was one of the most prolific writers in the history of mathematics.

Likewise,

$$\lim_{h \uparrow 0} \frac{\log_{10}(x + h) - \log_{10} x}{h} = \lim_{m \to -\infty} \frac{1}{x} \log_{10}\left(1 + \frac{1}{m}\right)^m$$

$$= \frac{1}{x} \log_{10} e.$$

We conclude that

$$\frac{d}{dx} \{\log_{10} x\} = \frac{1}{x} \log_{10} e.$$

EXAMPLE 2   What is $\dfrac{d}{dx} \{(3 + 2 \log_{10} x)^7\}$?

$$\frac{d}{dx} \{(3 + 2 \log_{10} x)^7\} = 7(3 + 2 \log_{10} x)^6 \frac{d}{dx} \{3 + 2 \log_{10} x\}$$

$$= 7(3 + 2 \log_{10} x)^6 \left(\frac{2 \log_{10} e}{x}\right).$$

From the relations

$$u = \log_a x \Leftrightarrow a^u = x$$
$$v = \log_{10} x \Leftrightarrow 10^v = x$$
$$w = \log_{10} a \Leftrightarrow 10^w = a$$

we see that

$$10^{uw} = (10^w)^u = a^u = x = 10^v.$$

Therefore

$$uw = v \qquad \text{or} \qquad u = \frac{v}{w};$$

that is,

$$\log_a x = \frac{\log_{10} x}{\log_{10} a} \qquad (a > 0).$$

It follows that the function L defined by

$$L(x) = \log_a x$$

is also continuous and differentiable for $x > 0$, and, in fact,

$$\frac{d}{dx} \{\log_a x\} = \frac{\log_{10} e}{\log_{10} a}\left(\frac{1}{x}\right) = \frac{1}{x} (\log_a e).$$

EXAMPLE 3   Find $\lim_{x \to 2}[(\log_4 x)^2 + 7x]/(x^2 - 1)$.

Since $\log_4 x$ is continuous at 2, $(\log_4 x)^2$ is continuous at 2 and so

$$\lim_{x \to 2}(\log_4 x)^2 = (\log_4 2)^2 = (\tfrac{1}{2})^2 = \tfrac{1}{4}.$$

It follows that

$$\lim_{x \to 2} \frac{(\log_4 x)^2 + 7x}{x^2 - 1} = \frac{\lim_{x \to 2}[(\log_4 x)^2 + 7x]}{\lim_{x \to 2}(x^2 - 1)}$$

$$= \frac{\frac{1}{4} + 14}{3} = \frac{57}{12}.$$

EXAMPLE 4   Find dy/dx if $y = \log_e x$.
              Now

$$\frac{dy}{dx} = \frac{\log_e e}{x}.$$

But $\log_e e = 1$, thus

$$\frac{dy}{dx} = \frac{1}{x}.$$

EXERCISES

1. Find the following limits.

(a) $\lim_{x \to 4}(\log_2 x)(\log_4 x)$.

(b) $\lim_{x \to 5} \sqrt{\log_{10} 20x}$.

(c) $\lim_{x \to +\infty} \log_{10}[(2x - 3)/(x - 1)]$.

(d) $\lim_{x \to -\infty} \log_{10}[(2x - 3)/(x - 1)]$.

(e) $\lim_{x \to 1} \log_3(2x^2 + 7)$.

2. Evaluate the following limits.

(a) $\lim_{m \to +\infty} (1 + 1/m)^{2m}$.

(b) $\lim_{m \to +\infty} (1 + 1/m)^{20}$.

(c) $\lim_{m \to +\infty} (1 + 1/m)^{-m}$.

(d) $\lim_{m \to +\infty} (1 + 1/(2m))^m$. [*Hint:* Let $n = 2m$.]

(e) $\lim_{m \to +\infty} (1 - 1/m)^m$. [*Hint:* $(1 - 1/m)^m = (1 + 1/(-m))^m$.]

(f) $\lim_{m \to +\infty} (1 + 1/m)^{km}$.

(g) $\lim_{m \to +\infty} (1 + 1/k)^m$ $(k > 0)$.

3. Let $a > 1$. Show that for any number $y > 0$, there exists an $x > 1$ such that $\log_a x = y$. [*Hint:* Choose number n with the property that $\log_a n > y$. Then f defined by $f(x) = \log_a x$ is continuous on $[0, n]$; use the Intermediate Value Theorem for continuous functions.]

4. Show that there exists an x in $[1, 20]$ such that $x = \log_{10} x + 10$. [*Hint:* Apply the Intermediate Value Theorem to the function $f(x) = \log_{10} x + 10 - x$.]

5. Find the derivative of each of the following.

(a) $\log_6 x$.

(b) $\log_6 (2x)$.

(c) $(2 + \log_2 x)^3$.

(d) $\sqrt{(\log_{10} x)^5}$.

(e) $x^2 \log_3 x$.

(f) $(x^2 + 2x - \log_{10} x)/(x + 1)$.

6. Evaluate the following.

(a) $\int [(\log_5 e)/x]\, dx.$

(b) $\int_1^3 [(\log_3 e)/x]\, dx.$

(c) $\int_1^e (1/x)\, dx.$ (See Example 4.)

6.3 NATURAL LOGARITHM FUNCTION

It has been observed that for any rational number $r \neq -1$,

$$\frac{d}{dx}\left\{\frac{x^{r+1}}{r+1}\right\} = x^r$$

or

$$\int x^r\, dr = \frac{x^{r+1}}{r+1} \qquad (r \neq -1).$$

Up to this point, we did not know a function f such that

$$\frac{d}{dx}\{f(x)\} = \frac{1}{x}.$$

But at the end of Section 6.2 we obtained

$$\frac{d}{dx}\{\log_a x\} = \log_a e\, \frac{1}{x}.$$

By choosing $a = e$,

$$\frac{d}{dx}\{\log_e x\} = \frac{1}{x}.$$

Therefore we see that the choice of e for the base of the logarithm function is a natural one in the study of calculus.

**6.3.1
Definition**

The logarithm function to the base e,

$$\log_e : \mathbf{R}^+ \to \mathbf{R},$$

is called the **natural logarithm function**.

For convenience this function will be denoted by log rather than $\log_e$. Thus

$$\log x = y \quad \Leftrightarrow \quad x = e^y$$

and

$$\frac{d}{dx}\{\log x\} = \frac{1}{x}.$$

It is interesting to note a geometric interpretation for the number log x. Recall that for each function f continuous on $[a, b]$ the function G defined by

$$G(x) = \int_a^x f(t)\, dt$$

is an antiderivative of f; that is, $G'(x) = f(x)$. Therefore the function F defined by

$$F(x) = \int_1^x \frac{1}{t}\, dt$$

is differentiable and

$$F'(x) = \frac{1}{x}.$$

By 4.7.4,
$$F(x) - \log x = C, \qquad \text{where } C \text{ is a constant.}$$

Letting $x = 1$,
$$F(1) - \log 1 = C,$$

and so
$$C = 0.$$

Hence
$$\log x = F(x)$$

or

(6.3.2) $$\log x = \int_1^x \frac{1}{t}\, dt.$$

This shows that log x can be interpreted in terms of the area of the region bounded by the curves

$$y = \frac{1}{t}, \qquad t = 1$$

$$y = 0, \qquad t = x,$$

as shown in Figure 6.1. In particular, log 10 is the area of the shaded region

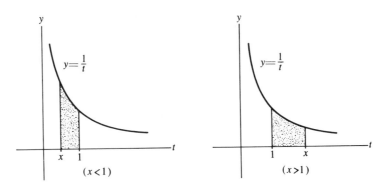

Figure 6.1

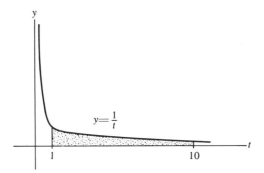

Figure 6.2

in Figure 6.2. By using the method in Section 5.2, we can obtain an approximate value of log 10, namely,

$$\log 10 \approx 2.3026.$$

Let us now sketch the graph of log x. If $f(x) = \log x$, then $f'(x) = 1/x$ and $f''(x) = -1/x^2$. The log function is differentiable (and hence continuous) at each positive number x. This means that its graph is an unbroken and smooth curve.

If $x > 0$, then $f'(x) = 1/x > 0$ and hence log is increasing on $\mathbf{R}^+$, and since $f''(x) = -1/x^2 < 0$, the graph of log is concave down on $\mathbf{R}^+$. Some of the values include

$$f\left(\frac{1}{e}\right) = -1, \quad f(1) = 0, \quad f(e) = 1,$$

and so $(1/e - 1)$, $(1, 0)$, $(e, 1)$ are points on the graph.

Finally, we want to investigate the behavior of log x as $x \to +\infty$ and as $x \downarrow 0$. Let $x = 10^n$; then

$$\log x = \log 10^n = n \log 10.$$

It is clear that

we can get n log 10 as large positively as we like by taking n large enough positively; that is, we can get log x as large positively as we like by taking x large enough positively.

Therefore

$$\lim_{x \to +\infty} \log x = +\infty.$$

Now let $x = 1/y$; then

$$x \downarrow 0 \quad \Leftrightarrow \quad y \to +\infty.$$

Therefore

$$\lim_{x \downarrow 0} \log x = \lim_{y \to +\infty} \log \frac{1}{y} = \lim_{y \to +\infty} (\log 1 - \log y)$$

$$= \lim_{y \to +\infty} (-\log y)$$

$$= -\infty.$$

The **graph of the natural log function** is drawn in Figure 6.3.

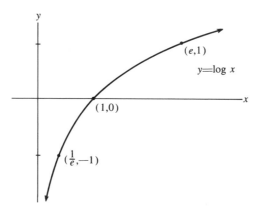

Figure 6.3

Occasionally it is helpful to have a formula for

$$\frac{d}{dx} \{\log f(x)\},$$

where f is a positive and differentiable function. To this end, let

$$u = f(x) \qquad \text{and} \qquad g(u) = \log u.$$

Then $\log f(x) = g(f(x))$ and

$$\frac{d}{dx} \{\log f(x)\} = \frac{d}{dx} \{g(f(x))\}$$

$$= g'(f(x))f'(x)$$

$$= \left(\frac{1}{f(x)}\right) f'(x).$$

6.3.3

Fact If f is a positive and differentiable function, then $\log f$ is differentiable, and, in fact,

$$\frac{d}{dx} \{\log f(x)\} = \frac{f'(x)}{f(x)}.$$

EXAMPLE 1

$$\frac{d}{dx}\{\log(x^2)\} = \frac{\dfrac{d}{dx}\{x^2\}}{x^2} = \frac{2x}{x^2} = \frac{2}{x}.$$

EXAMPLE 2

$$\frac{d}{dx}\{\log(5x^2 + 10x + 2)\} = \frac{\dfrac{d}{dx}\{5x^2 + 10x + 2\}}{5x^2 + 10x + 2}$$

$$= \frac{10x + 10}{5x^2 + 10x + 2}.$$

EXAMPLE 3

$$\frac{d}{dx}\{[\log(5x + 3)]^2\} = 2[\log(5x + 3)]^1 \frac{d}{dx}\{\log(5x + 3)\}$$

$$= 2\log(5x + 3)\frac{\dfrac{d}{dx}\{5x + 3\}}{5x + 3}$$

$$= 2\log(5x + 3)\frac{5}{5x + 3}$$

$$= \frac{10\log(5x + 3)}{5x + 3}.$$

EXAMPLE 4   Determine $\dfrac{d}{dx}\{\log|x|\}$.

Case 1: $x > 0$. Then $\log|x| = \log x$, and so

$$\frac{d}{dx}\{\log|x|\} = \frac{d}{dx}\{\log x\} = \frac{1}{x}.$$

Case 2: $x < 0$. Then $\log|x| = \log(-x)$, and so

$$\frac{d}{dx}\{\log|x|\} = \frac{d}{dx}\{\log(-x)\} = \frac{-1}{-x} = \frac{1}{x}.$$

Hence, for each $x \neq 0$,

$$\frac{d}{dx}\{\log|x|\} = \frac{1}{x}.$$

The preceding example gives us the following.

6.3.4
Fact

$$\int \frac{du}{u} = \log|u| + C.$$

EXAMPLE 5   Evaluate $\int [1/(3x + 2)]\, dx$.
Let $u = 3x + 2$; then $du = 3\, dx$ and

$$\int \frac{dx}{3x + 2} = \int \frac{\frac{1}{3}\, du}{u}$$

$$= \frac{1}{3} \int \frac{du}{u}$$

$$= \frac{1}{3} \log|u| + C$$

$$= \frac{1}{3} \log|3x + 2| + C.$$

EXAMPLE 6   Evaluate $\int_0^1 [x/(x + 1)]\, dx$.
Let $u = x + 1$; then $du = dx$. Note that $u = 2$ if $x = 1$ and $u = 1$ if $x = 0$. Then we have

$$\int_0^1 \frac{x}{x + 1}\, dx = \int_1^2 \frac{u - 1}{u}\, du$$

$$= \int_1^2 \left(1 - \frac{1}{u}\right) du$$

$$= (u - \log u)\big|_1^2$$

$$= (2 - \log 2) - (1 - \log 1)$$

$$= 1 - \log 2.$$

EXAMPLE 7   Evaluate $\int_1^e (\log x/x)\, dx$.
If we let $u = \log x$, then $du = 1/x\, dx$ and

$$\int_1^e \frac{\log x}{x}\, dx = \int_0^1 u\, du$$

$$= \frac{u^2}{2}\bigg|_0^1$$

$$= \frac{1}{2}.$$

EXAMPLE 8   Find $f'(x)$ if

$$f(x) = \sqrt[5]{\frac{(2x + 3)^2(x^2 + 2x - 5)^3}{(x^2 + 1)^7}}.$$

We shall use a method called **logarithmic differentiation**:

$$\log f(x) = \log \left[\frac{(2x + 3)^2(x^2 + 2x - 5)^3}{(x^2 + 1)^7}\right]^{1/5}$$

$$= \frac{1}{5} \log \left[\frac{(2x + 3)^2(x^2 + 2x - 5)^3}{(x^2 + 1)^7}\right]$$

$$= \frac{1}{5} [\log(2x + 3)^2(x^2 + 2x - 5)^3 - \log(x^2 + 1)^7]$$

$$= \frac{1}{5} [\log(2x + 3)^2 + \log(x^2 + 2x - 5)^3 - \log(x^2 + 1)^7]$$

$$= \frac{1}{5} [2 \log(2x + 3) + 3 \log(x^2 + 2x - 5) - 7 \log(x^2 + 1)]$$

$$= \frac{2}{5} \log(2x + 3) + \frac{3}{5} \log(x^2 + 2x - 5) - \frac{7}{5} \log(x^2 + 1).$$

Now if we differentiate both sides, we obtain

$$\frac{f'(x)}{f(x)} = \frac{2}{5}\left(\frac{2}{2x + 3}\right) + \frac{3}{5}\left(\frac{2x + 2}{x^2 + 2x - 5}\right) - \frac{7}{5}\left(\frac{2x}{x^2 + 1}\right).$$

Hence

$$f'(x) = f(x)\left[\frac{4}{5(2x + 3)} + \frac{3(2x + 2)}{5(x^2 + 2x - 5)} - \frac{14x}{5(x^2 + 1)}\right].$$

EXERCISES

1. Find the derivative of each of the following.

(a) $\log 1/x$.

(b) $1/\log x$.

(c) $\log 3x$.

(d) $(\log x)^3$.

(e) $3 \log x$.

(f) $\log|\log x|$.

(g) $\log \sqrt{x}$.

(h) $\log \sqrt{x^2}$.

(i) $\log|x + 1|$.

(j) $\log(3x^2 - 4x + 2)$.

(k) $\log[(x^2 + 1)/(x + 2)]$.

(l) $\sqrt{\log(4 - x^2)}$, $(-2 < x < 2)$.

(m) $x^2 \log(2x)$.

2. Evaluate the following.

(a) $\int 1/(2x)\, dx.$

(b) $\int 2/x\, dx.$

(c) $\int [1/(x + 2)]\, dx.$

(d) $\int [2/(x + 2)]\, dx.$

(e) $\int [(\log x)^2/x]\, dx.$

(f) $\int [(\log x^2)^2/x]\, dx.$

(g) $\int [6/(6x - 1)]\, dx.$

(h) $\int [(x + 1)/(x^2 + 2x - 5)]\, dx.$

3. Evaluate each of the following.

(a) $\int_1^{2e} 1/x\, dx.$

(b) $\int_{e^2}^{e^3} 1/x\, dx.$

(c) $\int_0^1 [(2x + 1)/(x^2 + x + 5)]\, dx.$

(d) $\int_1^e [(\log x)^2/x]\, dx.$

(e) $\int_0^1 [1/(2x + 3)]\, dx.$

(f) $\int_{-1}^0 [1/(5 - 6x)]\, dx.$

(g) $\int_0^1 [1/(2x - 5)^2]\, dx.$

(h) $\int_1^4 [1/\sqrt{x}(\sqrt{x} + 4)]\, dx.$

(i) $\int_1^2 [(2x + 1)/x]\, dx.$

(j) $\int_{-1}^0 [2x/(1 - 2x)]\, dx.$

(k) $\int_1^e \log x\, dx.$ [*Hint*: Let $f(x) = \log x$ and $g'(x) = 1$ in Exercise 10 of Section 5.4.]

(l) $\int_{2e}^{e^2} x \log x\, dx.$ [*Hint*: Let $f(x) = \log x$ and $g'(x) = x$ in Exercise 10 of Section 5.4.]

4. In each of the following, find the slope of the tangent line, the equation of the tangent line, and the equation of the normal line at the specified point.

(a) $y = \log x,\ (e, 1).$

(b) $y = \log 1/x,\ (1/e, 1).$

(c) $y = (\log x)/x,\ (e, 1/e).$

5. In each case sketch the region bounded by the given curves and find its area.

(a) $y = 3/x,\ y = 0,\ x = 2,\ x = 4.$

(b) $y = 1/x,\ y = 2,\ x = 3.$

(c) $y = 4/x,\ y = 5 - x.$

(d) $y = 1/(x - 1),\ y = -1,\ x = -1.$

6. Sketch the graph of each of the following equations.

(a) $y = \log 5x$.   (b) $y = \log x^5$.

(c) $y = \log 1/x$.   (d) $y = \log 1/x^2$.

(e) $y = \log|x|$.   (f) $y = \log x - x$.

(g) $y = x \log x$. (*Note:* $\lim_{x \downarrow 0} x \log x = 0$.)

7. Assume that $y = f(x)$ satisfies the equation $xy^3 + y \log x = 2$. Determine $f'(x)$.

8. Using the method of logarithmic differentiation find $f'(x)$.

(a) $f(x) = \sqrt[4]{(x + 1)^3/x}$.

(b) $f(x) = \{[(x - 1)(2x + 3)(x - 5)]/(2x^4 + 3)\}^3$.

9. (a) Determine constants A and B such that

$$1/(x^2 + 3x + 2) = 1/[(x + 1)(x + 2)] = [A/(x + 1)] + [B/(x + 2)].$$

[*Hint:* Write the right-hand side as a fraction and then equate numerators.]

(b) Evaluate $\int_0^1 [1/(x^2 + 3x + 2)]\, dx$.

10. (a) Determine constants A, B, C such that $1/[(x - 1)^2(2x + 3)] = [A/(x - 1)] + [B/(x - 1)^2] + [C/(2x + 3)]$.

(b) Evaluate $\int_2^3 \{1/[(x - 1)^2(2x + 3)]\}\, dx$.

11. Show that the following are true.

(a) $\dfrac{d}{dx} \{\log(x + \sqrt{x^2 + 1})\} = 1/\sqrt{1 + x^2}$, $(-\infty < x < +\infty)$.

(b) $\dfrac{d}{dx} \{\log(x + \sqrt{x^2 - 1})\} = 1/\sqrt{x^2 - 1}$, $(x > 1)$.

(c) $\dfrac{d}{dx} \{\tfrac{1}{2} \log|(1 + x)/(1 - x)|\} = 1/(1 - x^2)$, $(|x| < 1)$.

12. Evaluate the following.

(a) $\displaystyle\int_0^2 (1/\sqrt{1 + x^2})\, dx$.

(b) $\displaystyle\int_2^3 [1/(1 - x^2)]\, dx$. [*Hint:* See Exercise 11.]

6.4   NATURAL EXPONENTIAL FUNCTION

Our purpose now is to study a function which is closely related to the natural logarithm function. These two functions appear in the solution of many important problems. As we shall see in Section 6.5, the fact that this new function is its own derivative is quite significant. That is, the new function that we shall study in this section will give us a solution to the differential equation

$$\frac{d}{dx}\{fx\} = f(x).$$

The function

$$\exp: \mathbf{R} \to \mathbf{R}^+$$

defined by

$$\exp x = e^x$$

is called the **natural exponential function**.

The fact

(6.4.2) $\exp x = y \quad \Leftrightarrow \quad x = \log y$

indicates the relationship between the log and exp functions. This type of relationship occurs between other functions as well.

More generally, let us consider a function $f: I \to J$, where I and J are intervals, and suppose that f is continuous and strictly increasing on I (see Figure 6.4). We note in this case that for every w in J there is exactly one z in I such that

$$f(z) = w.$$

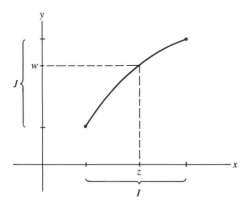

Figure 6.4

Therefore we have a function $g: J \to I$ defined by the relationship

$$g(w) = z \quad \Leftrightarrow \quad w = f(z),$$

where w is in J and z is in I. Such a function g is called the **inverse function of f**. *Therefore, 6.4.2 indicates that* exp *is the inverse of* log.

To determine the graph of an inverse function, let us make some observations. It is easy to see that the point $(1, 2)$ is the mirror image of the point $(2, 1)$ in the line $y = x$. In general, (a, b) is the mirror image of (b, a) in the line $y = x$ (see Figure 6.5). We know that if g is the inverse function of f, then

$$y = g(x) \quad \Leftrightarrow \quad x = f(y).$$

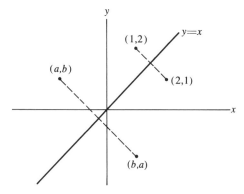

Figure 6.5

Therefore the graph of

$$y = g(x)$$

is the mirror image in the line $y = x$ of the graph of

$$y = f(x),$$

as shown in Figure 6.6.

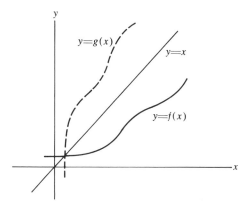

Figure 6.6

To sketch the graph of $y = e^x$, that is, $x = \log y$, we sketch the mirror image of $y = \log x$ in the line $y = x$ (see Figure 6.7). This gives the **graph of the natural exponential function**.

We make some comments about the graph of exp. Reflecting the graph of a strictly increasing, concave down function in the line $y = x$ yields the

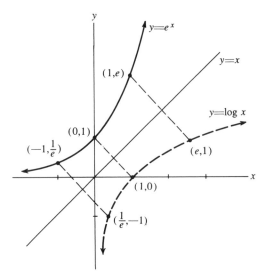

Figure 6.7

graph of a function which is also strictly increasing but concave up. Also, since

$$\lim_{x \to +\infty} \log x = +\infty \qquad \text{it follows that} \qquad \lim_{x \to +\infty} \exp x = +\infty;$$

and

$$\lim_{x \to 0^+} \log x = -\infty \qquad \text{implies that} \qquad \lim_{x \to -\infty} \exp x = 0.$$

Moreover, since the graph of log is unbroken, the graph of exp is unbroken; that is, exp *is continuous on* **R**.

Let us now turn our attention to determining

$$\frac{d}{dx}\{e^x\}.$$

If $y = e^x$, then $x = \log y$; and if $y + k = e^{x+h}$, then $x + h = \log(y + k)$. It follows that

$$h = \log(y + k) - \log y \qquad \text{and} \qquad k = e^{x+h} - e^x.$$

Since exp is strictly increasing,

$$h \neq 0 \Rightarrow k \neq 0,$$

and since exp is continuous,

$$h \to 0 \Rightarrow k \to 0.$$

Therefore, if $h \neq 0$,

$$\frac{e^{x+h} - e^x}{h} = \frac{k}{\log(y + k) - \log y}$$

$$= \frac{1}{[\log(y + k) - \log y]/k}.$$

Now $h \to 0$ implies that $k \to 0$, and since

$$\lim_{k \to 0} \frac{\log(y + k) - \log y}{k} = \frac{1}{y},$$

we have

$$\frac{d}{dx} \{e^x\} = \frac{1}{1/y} = y = e^x.$$

6.4.3
Facts

$$\frac{d}{dx} \{e^x\} = e^x$$

and

$$\int e^u \, du = e^u + C.$$

It is also useful to have a formula for

$$\frac{d}{dx} \{e^{f(x)}\},$$

where f is itself a differentiable function. To this end, let

$$u = f(x) \qquad \text{and} \qquad g(u) = e^u.$$

Then $e^{f(x)} = g(f(x))$ and

$$\frac{d}{dx} \{e^{f(x)}\} = \frac{d}{dx} \{g(f(x))\}$$

$$= g'(f(x))f'(x)$$

$$= e^{f(x)}f'(x).$$

6.4.4
Fact

If f is a differentiable function, then so is exp f, and, in fact,

$$\frac{d}{dx} \{e^{f(x)}\} = f'(x)e^{f(x)}.$$

EXAMPLE 1   Let $y = e^{x^2 + 2x + 4}$; what is dy/dx?

$$\frac{dy}{dx} = (2x + 2)e^{x^2 + 2x + 4}.$$

EXAMPLE 2   What is $f'(x)$ if $f(x) = e^{x^3} \log 2x$?

$$f'(x) = e^{x^3} \frac{d}{dx} \{\log 2x\} + \log 2x \frac{d}{dx} \{e^{x^3}\}$$

$$= e^{x^3} \left(\frac{2}{2x}\right) + (\log 2x)(3x^2 e^{x^3})$$

$$= \frac{e^{x^3}}{x} + 3x^2 (\log 2x) e^{x^3}.$$

EXAMPLE 3   The power rule

$$\frac{d}{dx} \{x^r\} = rx^{r-1}$$

is also valid for irrational numbers r.

To see this, recall that $x = e^{\log x}$, and so

$$x^r = e^{r \log x}.$$

Therefore

$$\frac{d}{dx} \{x^r\} = \frac{d}{dx} \{e^{r \log x}\}$$

$$= \frac{r}{x} e^{r \log x}$$

$$= \frac{r}{x} x^r$$

$$= rx^{r-1}.$$

EXAMPLE 4   Evaluate $\int_1^4 (e^{\sqrt{x}}/\sqrt{x}) \, dx$.

Let $u = \sqrt{x}$; then $du = [1/(2\sqrt{x})] \, dx$. If $x = 1$, then $u = 1$, and if $x = 4$, then $u = 2$.

$$\int_1^4 \frac{e^{\sqrt{x}}}{\sqrt{x}} \, dx = \int_1^2 e^u (2 \, du)$$

$$= 2 \int_1^2 e^u \, du$$

$$= 2e^u \Big|_1^2$$

$$= 2(e^2 - e).$$

EXAMPLE 5   Evaluate $\int [e^x/(1 + e^x)\, dx$.

Let $u = 1 + e^x$; then $du = e^x\, dx$. Therefore we have

$$\int \frac{e^x}{1 + e^x}\, dx = \int \frac{du}{u}$$

$$= \log|u| + C$$

$$= \log(1 + e^x) + C.$$

EXERCISES

1. Find $\dfrac{d}{dx}\{f(x)\}$ given that $f(x)$ is the following.

   (a) $6e^{2x}$.

   (b) $3e^{x^2}$.

   (c) $\log e^x$.

   (d) $e^{\log x}$.

   (e) $\log(1 + e^x)$.

   (f) $(x - 1)/(e^{x+3})$.

   (g) $e^{\sqrt{x}}$.

   (h) $e^{(4x+2)^{5/3}}$.

   (i) $x^2 e^{x^7}$.

   (j) $(2x^2 + 3x - 8)e^{(4x^3 - 9x)}$.

   (k) $e^{1/x^2}/x$.

   (l) $\log x/(e^{x^2} + e^3)$.

2. Assume that $y = f(x)$ satisfies the equation $xy^3 + ye^x = 2$. Determine $f'(x)$.

3. Evaluate the following.

   (a) $\displaystyle\int e^{3x}\, dx$.

   (b) $\displaystyle\int e^{x/2}\, dx$.

   (c) $\displaystyle\int 5e^{6x}\, dx$.

   (d) $\displaystyle\int xe^{x^2}\, dx$.

   (e) $\displaystyle\int x^9 e^{x^{10}}\, dx$.

   (f) $\displaystyle\int (3x^2 + 4x - 7)e^{x^3 + 2x^2 - 7x + 13}\, dx$.

   (g) $\displaystyle\int (1/\sqrt{x}e^{\sqrt{x}})\, dx$.

   (h) $\displaystyle\int [(e^x + e^{-x})/(e^x - e^{-x})]\, dx$.

4. Evaluate the following.

   (a) $\displaystyle\int_0^1 e^{3x}\, dx$.

   (b) $\displaystyle\int_{-5}^2 5e^{2x}\, dx$.

   (c) $\displaystyle\int_{-4}^1 e^{-x}\, dx$.

   (d) $\displaystyle\int_1^2 (e^x + 3)\, dx$.

   (e) $\displaystyle\int_1^2 (e^x - e^{x-1})\, dx$.

   (f) $\displaystyle\int_1^4 (1/\sqrt{x}\, e^{\sqrt{x}})\, dx$.

   (g) $\displaystyle\int_0^1 xe^x\, dx$. [*Hint:* See Exercise 10 of Section 5.4.]

5. Determine the following limits.

(a) $\lim\limits_{x \to +\infty} xe^x$.

(b) $\lim\limits_{x \downarrow 0} e^{1/x}$.

(c) $\lim\limits_{x \uparrow 0} e^{1/x}$.

(d) $\lim\limits_{x \downarrow 0} e^{(1/x)^2}$.

(e) $\lim\limits_{x \uparrow 0} e^{(1/x)^2}$.

(f) $\lim\limits_{x \to +\infty} e^{-x^2}$.

(g) $\lim\limits_{x \to -\infty} e^{-x^2}$.

6. Sketch the graph of the following equations.

(a) $y = 3e^{2x}$.

(b) $y = e^{-4x}$.

(c) $y = e^{1/x}$.

(d) $y = e^{1/x^2}$.

(e) $y = e^{-x^2}$.

7. Find the equation of the tangent line and the equation of the normal line at the given point.

(a) $y = e^{-x}$ at $(-1, e)$.

(b) $y = 3e^{2x}$ at $(1, 3e^2)$.

(c) $y = e^{1/x^2}$ at $(-\sqrt{2}, \sqrt{e})$.

8. In each case, sketch the region bounded by the given curves and find its area.

(a) $y = e^x$, $y = 0$, $x = 4$, $x = 0$.

(b) $y = e^{3x}$, $x = -5$; $y = 1$, $x = -2$.

(c) $y = e^x$, $y = 4$, $x = 0$.

(d) $y = e^x$, $x = 0$; $y = x^2$, $x = 1$.

(e) $y = e^{4x}$, $y = 8x + 1$, $x = -\frac{1}{2}$.

9. Let $s(x) = (e^x - e^{-x})/2$ and $c(x) = (e^x + e^{-x})/2$. Show that the following hold.

(a) $s'(x) = c(x)$.

(b) $c'(x) = s(x)$.

(c) $c^2(x) - s^2(x) = 1$.

(d) $c(x + y) = c(x)c(y) + s(x)s(y)$.

(e) $D\{s(x)/c(x)\} = 1/c^2(x)$.

The functions s, c are called **hyperbolic sine** and **hyperbolic cosine**, respectively.

10. Consider the function $f(x) = e^{|x|}$.

(a) Show that f is differentiable everywhere except zero.

(b) Determine the local extrema of f.

(c) Determine where f is concave up (down).

(d) Sketch the graph of $y = e^{|x|}$.

11. For any $x > 0$, the expression x^x is defined to be $e^{x \log x}$.

(a) If $f(x) = x^x$, find $f'(x)$ directly using the above definition.

(b) Find $f'(x)$ by the method of logarithmic differentiation.

12. Let $(e^y - e^{-y})/2 = x$.

   (a) Solve for y in terms of x. [*Hint:* Multiply both sides by e^y and use the quadratic formula to solve for e^y.]

   (b) Compute dy/dx from your answer in part (a).

   (c) Find dy/dx by implicit differentiation and compare your answer with part (b).

13. Consider the function $f(x) = 1/(2 + e^{1/x})$. Determine the following limits.

   (a) $\lim\limits_{x \to +\infty} f(x)$.

   (b) $\lim\limits_{x \to -\infty} f(x)$.

   (c) $\lim\limits_{x \downarrow 0} f(x)$.

   (d) $\lim\limits_{x \uparrow 0} f(x)$.

   Sketch a rough graph of $y = f(x)$ using this information.

14. Evaluate $\lim\limits_{h \to 0} [(e^h - 1)/h]$. [*Hint:* Consider the derivative of e^x at zero.]

15. (a) Show that for any $x > 0$, $e^x > 1 + x$. [*Hint:* Apply the Mean Value Theorem to $[0, x]$, getting $(e^x - e^0)/(x - 0) = e^{x_0} > 1, 0 < x_0 < x$.]

   (b) Show that for any $x > 0$, $e^x > 1 + x + x^2/2$. [*Hint:* $e^t > 1 + t$, $0 < t \leq x$; integrate and use Exercise 13 of Section 5.4.]

   (c) If n is a positive integer, let the symbol $n!$ (read "n factorial") represent the product of the first n positive integers; that is, $n! = 1 \cdot 2 \cdot 3 \cdots (n - 1) \cdot n$. Show that for any $x > 0$ and any positive integer n, $e^x > 1 + x + x^2/2! + x^3/3! + \cdots + x^n/n!$.

16. Show that for any positive integer n, $\lim\limits_{x \to +\infty} e^x/x^n = +\infty$. [*Hint:* From Exercise 15 we know that for any positive integer n, $e^x > x^{n+1}/(n + 1)!$.]

17. Let $f: \mathbf{R}^+ \to \mathbf{R}^+$ be defined by $f(x) = x^2$. What is the inverse of f?

18. Let $f: \mathbf{R} \to \mathbf{R}$ be defined by $f(x) = (2x - 2)/3$. What is the inverse of f?

19. Let $f: \mathbf{R} \to \mathbf{R}$ be defined by $f(x) = (e^x - e^{-x})/2$. Show that $g: \mathbf{R} \to \mathbf{R}$ defined by $g(x) = \log(x + \sqrt{x^2 + 1})$ is the inverse of f.

20. Let $f: I \to J$ and let $f'(x) > 0$ on I. Show that the inverse $g: J \to I$ has a derivative $g'(w) = 1/(f'(z))$, where $w = f(z)$.

6.5 APPLICATIONS INVOLVING EXPONENTIAL AND LOGARITHM FUNCTIONS

The functions that have just been studied arise naturally in the solution of a wide variety of problems.

   For instance, in the mathematical analysis of a certain problem one may be faced with the equation

(6.5.1) $$f'(t) = kf(t) \qquad (k \text{ constant})$$

and the task of finding a function f which satisfies it. It is easy to verify that if A is any constant,

$$Ae^{kt}$$

is one function which satisfies Equation 6.5.1. One can show that every solution to 6.5.1 has the form Ae^{kt} as follows. Suppose that function f also satisfies 6.5.1. Then

$$\frac{d}{dt}\left\{\frac{f(t)}{Ae^{kt}}\right\} = \frac{Ae^{kt}f'(t) - f(t)kAe^{kt}}{(Ae^{kt})^2} = \frac{f'(t) - kf(t)}{Ae^{kt}} = 0.$$

Due to 4.7.3, it must be that

$$\frac{f(t)}{Ae^{kt}} = \text{constant};$$

that is,

$$f(t) = Ce^{kt}$$

for some constant C.

6.5.2

Fact   Every function f which satisfies the equation

$$f'(t) = kf(t),$$

where k is a constant, has the form

$$Ce^{kt}$$

for some constant C.

EXAMPLE 1   **(growth)** Suppose that three partially filled swimming pools (50 gallons of water in each) are about to be filled completely, each pool having a different water pump.

Pool 1 has a standard pump which operates at a constant rate of 2 gallons per second. We wish to find the number of gallons $w_1(t)$ in pool 1 at time t.

If we assign $t = 0$ to the instant when the pump is turned on, then our given data can be expressed mathematically as

$w_1(0) = 50$ (initially the pool contains 50 gallons)

$w_1'(t) = 2$ (the pump operates at a rate of 2 gallons per second).

The second of these equations, in light of 4.7.4, tells us that

$$w_1(t) = 2t + C$$

for some constant C, and using the first equation,

$$50 = w_1(0) = 2(0) + C,$$

Therefore
$$C = 50.$$
Our desired result is
$$w_1(t) = 2t + 50.$$

This is an example of **linear growth**.

Pool 2 has a more elaborate pump, operating at a rate of $2t$ gallons per second, where t is the number of seconds the pump has been in operation. Again, we seek the number of gallons $w_2(t)$ in pool 2 at time t.

Assigning $t = 0$ to the instant when the pump is turned on, then our given data can be translated as

$w_2(0) = 50$ (initially the pool contains 50 gallons)

$w_2'(t) = 2t$ (the pump operates at a rate of $2t$ gallons per second).

The second equation, because of 4.7.4, gives
$$w_2(t) = t^2 + C$$
for some constant C, and the first equation gives
$$50 = w_2(0) = (0)^2 + C$$
or
$$C = 50.$$
The desired result is
$$w_2(t) = t^2 + 50.$$

This is an example of **quadratic growth**.

Pool 3 has an even more elaborate pump (actually an exotic one), which operates at a rate of $2w_3$ gallons per second, where $w_3(t)$ is the number of gallons in pool 3 at time t. Once more, we wish to determine w_3.

As above, let us assign $t = 0$ to the instant when the pumping is initiated; therefore

$w_3(0) = 50$ (initially the pool contains 50 gallons)

$w_3'(t) = 2w_3$ (the pump operates at a rate of $2w_3$ gallons per second).

The second equation, because of our result in 6.5.2, tells us that
$$w_3(t) = Ce^{2t}$$
for some constant C, and using the first equation,
$$50 = w_3(0) = Ce^0 = C,$$
and so again
$$C = 50.$$
The desired solution is
$$w_3(t) = 50e^{2t}.$$

This is an example of **exponential growth**. The graphs of w_1, w_2, w_3 are drawn in Figure 6.8.

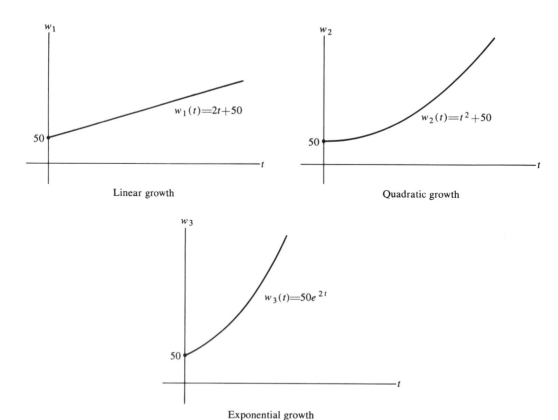

Figure 6.8

EXAMPLE 2 **(exponential growth)** It is observed that the number of cells in a particular cancerous growth increases at a rate proportional to the number of cells present. If this growth at one point had 1000 cells and was found to have 10,000 cells 1 hour later, determine the number of cells at time t.

Let $N(t)$ be the number of cells present at time t, measured in hours. If we assign $t = 0$ to the instant when 1000 cells were present, then our given data become

$N(0) = 1000$ (at a particular instant, 1000 cells present)

$N(1) = 10,000$ (1 hour later, 10,000 cells present)

$N'(t) = kN(t)$ (the rate of change in the number of cells is pro-
                     portional to the number of cells),

where k is some constant.

The third equation, due to 6.5.2, yields

$$N(t) = Ce^{kt}$$

for some constant C. The first two equations may be used to determine constants k and C as follows:

$$1000 = N(0) = Ce^0 = C, \quad \text{and so } C = 1000;$$

and

$$10,000 = N(1) = 1000e^k, \quad \text{hence } 10 = e^k \quad \text{or} \quad k = \log 10.$$

Since $1000e^{kt} = 1000(e^k)^t = 1000(10^t) = 10^{t+3}$, our desired solution is

$$N(t) = 10^{t+3} \quad \text{(see Figure 6.9)}.$$

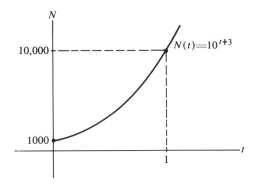

Figure 6.9

EXAMPLE 3   **(exponential decay)** It is known that a radioactive substance diminishes in mass at a rate proportional to the mass present. The **half-life** of such a substance is the number h of years such that if one begins with n grams of the substance, then $n/2$ grams will remain after h years. If one begins with 2 grams of radium, a radioactive substance with a half-life of 1550 years, how much remains after 1000 years?

Let $G(t)$ be the number of grams of radium present at time t, measured in years. Assigning $t = 0$ to the instant when 2 grams are present, we have

$G(0) = 2$ (initially, 2 grams are present)

$G(1550) = 1$ (half-life of radium is 1550 years)

$G'(t) = kG(t)$ (the rate of change in mass is proportional to mass present),

where k is some constant.

Due to the third equation and 6.5.2, we have

$$G(t) = Ce^{kt}$$

for some constant C. Now, to determine constants k and C from the first two equations:

$$2 = G(0) = Ce^0 = C, \quad \text{thus } C = 2;$$

and

$$1 = G(1550) = 2e^{k\,1550}, \quad \text{hence } \tfrac{1}{2} = e^{k\,1550}$$

or

$$k = \frac{\log \tfrac{1}{2}}{1550}.$$

Since

$$2e^{kt} = 2(e^{k\,1550})^{t/1550}$$
$$= 2(\tfrac{1}{2})^{t/1550}$$
$$= 2(2^{-t/1550})$$
$$= 2^{(1-t/1550)},$$

we have

$$G(t) = 2^{(1-t/1550)} \quad \text{(see Figure 6.10)}.$$

In particular, when $t = 1000$,

$$G(1000) = 2^{1-1000/1550} = 2^{550/1550} = 2^{11/31},$$

and so our desired result is

$$2^{11/31} \text{ grams.}$$

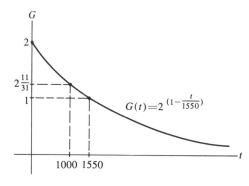

Figure 6.10

EXAMPLE 4   **(exponential decay) Newton's law of cooling** states that a body immersed in cooler surroundings will cool at a rate proportional to the difference in temperature between the body and its surroundings (assuming that the temperature of the surroundings remains fixed).

In a fluid whose temperature remains fixed at 20°C, a body of 100°C is immersed; it cools to 90°C in 1 minute. How long will it take to cool to 50°C?

Let $T(t)$ be the temperature of the immersed body at time t. If we assign $t = 0$ to the instant of immersion and measure time in minutes, then our given data become

$T(0) = 100$ (at immersion, body temperature is 100°C)

$T(1) = 90$ (1 minute after immersion, body temperature is 90°C)

$T'(t) = k(T - 20)$ (rate of cooling is proportional to the difference between T and 20°C),

where k is some constant.

Now let $f(t) = T(t) - 20$. Then

$$f'(t) = T'(t),$$

and so we can rewrite the third equation above as

$$f'(t) = kf(t).$$

By 6.5.2

$$f(t) = Ce^{kt}$$

for some constant C, and so

$$T(t) = Ce^{kt} + 20.$$

Again, let us use the first two equations to determine constants k and C:

$$100 = T(0) = C + 20, \quad \text{and so } C = 80;$$

and

$$90 = T(1) = 80e^k + 20$$

$$70 = 80e^k$$

$$\tfrac{7}{8} = e^k$$

or

$$k = \log \tfrac{7}{8}.$$

Hence our general solution is

$$T(t) = 80e^{(\log 7/8)t} + 20$$

$$= 80(\tfrac{7}{8})^t + 20 \quad \text{(see Figure 6.11)}.$$

Since we seek the time it takes for the body to cool to 50°C, then all we need do is solve for t in the equation

$$50 = 80(\tfrac{7}{8})^t + 20.$$

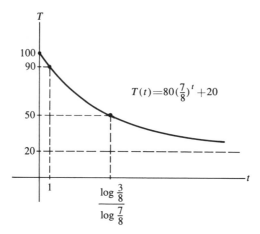

Figure 6.11

This can be done as follows:

$$30 = 80(\tfrac{7}{8})^t$$

$$\tfrac{3}{8} = (\tfrac{7}{8})^t$$

$$\log \tfrac{3}{8} = (\log \tfrac{7}{8})t$$

$$t = \frac{\log \tfrac{3}{8}}{\log \tfrac{7}{8}}.$$

Suppose that we are confronted with the equation

(6.5.3) $$f'(t) = \frac{k}{t} \qquad (k \text{ constant})$$

and wish to find a function f which satisfies it. Since we know how to differentiate logarithm functions, a short computation will show that

$$k \log|At|,$$

where A is a constant, satisfies this equation. To see that every solution to 6.5.3 has this form, let function f satisfy 6.5.3. Then by 4.7.4

$$f(t) = k \log|At| + B$$

for some constant B. Now from the graph of log in Section 6.3 we see that for any number B there is a positive number E such that

$$\log E = B.$$

Hence

$$f(t) = k \log|At| + \log E$$
$$= \log|At|^k + \log E$$
$$= \log E|At|^k$$
$$= \log|E^{1/k}At|^k$$
$$= k \log|(E^{1/k}A)t|.$$

6.5.4

Fact   Every function f which satisfies the equation

$$f'(t) = \frac{k}{t},$$

where k is a constant, has the form

$$k \log|Ct|$$

for some constant C.

EXAMPLE 5   (**logarithmic growth**) One theory in psychology hypothesizes that the change in sensation with respect to stimulus is inversely proportional to the stimulus. If a subject records a sensation of 1 prior to the application of the stimulus (say, stimulus = 1) and records a sensation of 1.3 when the stimulus is 2, determine the sensation for any given stimulus between 1 and 5. (For simplicity here we shall avoid mentioning units of measurement for stimulus and sensation. The meaning and measurement of sensation are difficult problems.)

Let $S(t)$ be the sensation recorded for stimulus t. Then we can translate the given data as follows:

$S(1) = 1$ (sensation 1 when stimulus 1)

$S(2) = 1.3$ (sensation 1.3 when stimulus 2)

$S'(t) = \dfrac{k}{t}$ (the rate of change in sensation is inversely proportional to stimulus),

where k is some constant. The third equation, because of 6.5.4, tells us that

$$S(t) = k \log|Ct|$$

for some constant C. As before, we can determine constants k and C by using the first two equations:

$$1 = S(1) = k \log|C|$$

$$1.3 = S(2) = k \log|2C|;$$

thus

$$1.3 = k \log|2C|$$

$$= k \log 2 + k \log|C|$$

$$= k \log 2 + 1$$

or

$$k = \frac{0.3}{\log 2}.$$

Also,

$$1 = \frac{0.3}{\log 2} \log|C|$$

$$\log|C| = \frac{\log 2}{0.3}$$

$$|C| = e^{(\log 2)/0.3}$$

$$= e^{\log(2^{10/3})}$$

$$= 2^{10/3}.$$

Our desired solution for any t in $[1, 5]$ is

$$S(t) = \frac{0.3}{\log 2} \log(2^{10/3} t) \qquad \text{(see Figure 6.12).}$$

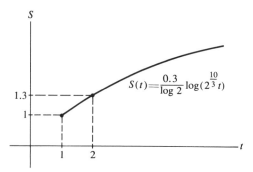

Figure 6.12

EXAMPLE 6   (**compound interest**) A man wishes to open a savings account at one of two local banks. Bank A offers $5\frac{1}{4}\%$ interest compounded yearly, and Bank B offers $5\frac{1}{5}\%$ interest compounded quarterly (that is, four times a year). Which bank should he choose?

If he were to invest P dollars in Bank A, at the end of one year his account would show

$$P + 0.0525P = 1.0525P \text{ dollars.}$$

But if he were to invest these P dollars in Bank B instead, we can compute his account after one year in the following manner. At the end of the first quarter, his account has

$$P + \frac{0.052}{4} P = \left(1 + \frac{0.052}{4}\right) P$$

$$= 1.013P \text{ dollars;}$$

at the end of the second quarter,

$$1.013P + \frac{0.052}{4}(1.013P) = \left(1 + \frac{0.052}{4}\right)(1.013P)$$

$$= (1.013)^2 P \text{ dollars;}$$

at the end of the third quarter,

$$(1.013)^2 P + \frac{0.052}{4} [(1.013)^2 P] = \left(1 + \frac{0.052}{4}\right) [(1.013)^2 P]$$

$$= (1.013)^3 P \text{ dollars};$$

and at the end of the year (fourth quarter),

$$(1.013)^3 P + \frac{0.052}{4} [(1.013)^3 P] = \left(1 + \frac{0.052}{4}\right) [(1.013)^3 P]$$

$$= (1.013)^4 P.$$

Since

$$(1.013)^4 = e^{4 \log(1.013)} \approx 1.05302,$$

then his account in Bank B at the end of one year is

$$1.05302 P.$$

Therefore the man should open his account in Bank B.

EXAMPLE 7   **(compound interest)** In a particular neighborhood, competition between banks results in more and more frequent compounding at 5%. Is it likely that this more frequent compounding will lead to bankruptcy?

Let us first consider the situation of a bank which gives $r\%$ interest compounded n times a year (n is some positive integer). If P dollars are invested, then at the end of the first period, the account would have

$$P + \frac{r}{n} P = \left(1 + \frac{r}{n}\right) P \text{ dollars};$$

at the end of the second period,

$$\left(1 + \frac{r}{n}\right) P + \frac{r}{n} \left[\left(1 + \frac{r}{n}\right) P\right] = \left(1 + \frac{r}{n}\right)\left(1 + \frac{r}{n}\right) P$$

$$= \left(1 + \frac{r}{n}\right)^2 P \text{ dollars};$$

at the end of the third period,

$$\left(1 + \frac{r}{n}\right)^2 P + \frac{r}{n} \left[\left(1 + \frac{r}{n}\right)^2 P\right] = \left(1 + \frac{r}{n}\right)\left(1 + \frac{r}{n}\right)^2 P$$

$$= \left(1 + \frac{r}{n}\right)^3 P \text{ dollars} \ldots, \text{ and}$$

at the end of the year (nth period),

$$\left(1 + \frac{r}{n}\right)^{n-1} P + \frac{r}{n} \left[\left(1 + \frac{r}{n}\right)^{n-1} P\right] = \left(1 + \frac{r}{n}\right)\left(1 + \frac{r}{n}\right)^{n-1} P$$

$$= \left(1 + \frac{r}{n}\right)^n P \text{ dollars}.$$

If we are interested in the situation where n grows larger, we have (keeping in mind that P and r are fixed)

$$\lim_{n \to +\infty} \left[P\left(1 + \frac{r}{n}\right)^n \right] = P \lim_{n \to +\infty} \left(1 + \frac{r}{n}\right)^n$$

$$= P \lim_{n \to +\infty} \left[\left(1 + \frac{1}{n/r}\right)^{n/r} \right]^r$$

$$= P \lim_{m \to +\infty} \left[\left(1 + \frac{1}{m}\right)^m \right]^r \qquad \left(m = \frac{n}{r} \right)$$

$$= P \left[\lim_{m \to +\infty} \left(1 + \frac{1}{m}\right)^m \right]^r$$

$$= Pe^r.$$

Thus, when interest is compounded continuously (that is, at each instant), the account at the end of one year would be

$$Pe^r.$$

In our particular case, $r = 0.05$ and

$$Pe^{0.05} \approx 1.05127P.$$

Observe that this amount is less than the amount which would result if interest were compounded annually at $5\frac{1}{5}\%$, namely, $1.052P$. Therefore it does not seem likely that this increased frequency in compounding will endanger the banks.

Now generalizing upon our work in Example 7, we can say that P dollars, when compounded n times a year at a rate of $r\%$ will become

$$P\left(1 + \frac{r}{n}\right)^n \text{ dollars}$$

at the end of one year. Hence at the end of 2 years P dollars becomes

$$\left[P\left(1 + \frac{r}{n}\right)^n \right]\left(1 + \frac{r}{n}\right)^n = P\left(1 + \frac{r}{n}\right)^{2n};$$

at the end of 3 years it becomes

$$\left[P\left(1 + \frac{r}{n}\right)^{2n} \right]\left(1 + \frac{r}{n}\right)^n = P\left(1 + \frac{r}{n}\right)^{3n};$$

and, more generally, at the end of T years P dollars becomes

$$P\left(1 + \frac{r}{n}\right)^{Tn} \text{ dollars}.$$

EXAMPLE 8   A father wishes to give his son (age 6) $10,000 on the son's twenty-first birthday. Assuming that the father can invest in a bank which pays 5% compounded quarterly, how much must the father invest now to fulfill his wish?

The problem is to determine the number P of dollars such that

$$P\left(1 + \frac{0.05}{4}\right)^{15(4)} = 10,000.$$

Solving this equation for P, we have

$$P = \frac{10,000}{(1.0125)^{60}}$$

$$= 10,000e^{-60 \log(1.0125)}$$

$$\approx \$4745.68.$$

In Example 7 we saw that P dollars, when compounded continuously at $r\%$, will become

$$Pe^r \text{ dollars}$$

at the end of 1 year. Thus at the end of 2 years P dollars becomes

$$(Pe^r)e^r = Pe^{2r};$$

at the end of three years it becomes

$$(Pe^{2r})e^r = Pe^{3r};$$

and, in general, at the end of T years P dollars becomes

$$Pe^{Tr} \text{ dollars.}$$

Again, a question of importance is: *What amount P put in the bank today will yield A dollars T years from now?* That is, given A, T, r, for what P will

$$Pe^{Tr} = A?$$

Of course, solving this equation for P, we get

$$P = \frac{A}{e^{Tr}},$$

or

(6.5.5) $$P = Ae^{-Tr},$$

which is called the **present value of A dollars T years in the future**.

Now suppose that profit from a certain business venture is spread over several years, say T years, and for any t in $[0, T]$

$$P(t)$$

gives the profit earned at time t. Then, in light of 6.5.5,

$$P(t)e^{-tr}$$

is the present value of the profit earned at time t. Moreover, recalling 5.9.1, if P is a continuous function on $[0, T]$, then so is $P(t)e^{-tr}$ and

(6.5.6) $$\int_0^T P(t)e^{-tr}\, dt$$

is the **present value of profit earned over the next T years**.

EXAMPLE 9

(present value) After heavy capital expenditures in opening a chain of restaurants, an entrepreneur expects his profits to rise steadily over the next 5 years. In fact, he estimates that for any t in $[0, 5]$

$$P(t) = 1.5^t$$

will be the profit (in thousands of dollars) earned at time t (in years). Then, using 6.5.6,

$$\int_0^5 1.5^t e^{-tr}\, dt$$

is the present value V of his profit earned over the next 5 years. (Note that r is some fixed rate of interest.) Finally,

$$\begin{aligned}
V &= \int_0^5 1.5^t e^{-tr}\, dt \\
&= \int_0^5 e^{t\log 1.5} e^{-tr}\, dt \\
&= \int_0^5 e^{t(\log 1.5 - r)}\, dt \\
&= \left. \frac{e^{t(\log 1.5 - r)}}{(\log 1.5 - r)} \right|_0^5 \\
&= \frac{e^{5(\log 1.5 - r)} - 1}{(\log 1.5 - r)} \\
&= \frac{(1.5)^5 e^{-5r} - 1}{(\log 1.5 - r)}.
\end{aligned}$$

EXERCISES

1. In each case, determine the mass function m in terms of time t, graph it, and describe the growth (or decay) in mass as exponential, linear, quadratic, or cubic.

   (a) $m'(t) = -\frac{3}{2}$, $m(0) = 0$. (b) $m'(t) = t$, $m(0) = 2$.

   (c) $m'(t) = -3t$, $m(0) = 2$. (d) $m'(t) = 4t^2$, $m(0) = 0$.

   (e) $m'(t) = 5m$, $m(0) = 1$. (f) $m'(t) = -5m$, $m(0) = 1$.

2. It is observed that a certain body is attracting mass at a rate proportional to the mass present. If the body weighed 100 grams at an initial instant and 200 grams 2 hours later, what was the weight of the body 1 hour after the initial instant?

3. If 1 gram of a radioactive substance decreases to 0.8 gram after 10 years, what is its half-life?

4. The length L of a given metal rod depends on the temperature T of its surrounding medium. As T changes, it is observed that L changes at a rate proportional to $L - 90$ (and so the longer the rod, the greater the rate of expansion). If the rod is 100 inches in length at one instant and 101 inches 30 minutes later, find its length at any time t.

5. The population of a particular city is known to be growing at a rate proportional to the population. Twenty years ago its population was 100,000 and today it is 150,000. What will the population be 20 years from now?

6. An object, initially at rest, falls vertically under the influence of gravity. If the air resistance at time t is $(\frac{1}{96})v$, where $v = v(t)$ is the velocity of the object at time t, determine $v(t)$ for any $t \geq 0$.

7. A brine solution of 50 gallons contains 150 pounds of dissolved salt. Fresh water enters at a rate of 3 gallons per minute and the solution is drawn off at the same rate. Assuming that the solution is thoroughly stirred, determine the time at which the solution contains 100 pounds of dissolved salt. [*Hint:* Let $x(t)$ = number of pounds of salt dissolved in the solution at time t, and show that $(dx/dt) = -3x/50$.]

8. A tank contains 50 gallons of pure water. At a certain instant $t = 0$ a brine solution containing 2 pounds of dissolved salt per gallon flows into the tank at a rate of 3 gallons per minute. The mixture is kept uniform by stirring, and the well-stirred mixture is simultaneously drawn off at the same rate. Determine how much salt is in the tank at any time $t \geq 0$. [*Hint:* Let $x(t)$ = number of pounds of salt dissolved in the solution at time t, and show that $(dx/dt) = 6 - 3x/50$.]

9. In Example 4 of the text, suppose that the cooling rate were proportional to half the difference in temperature between the body and its surroundings. Assuming the other data given in that example, how long will it take for the object to cool to 50°C?

10. Gas is being compressed into a rubberized container (so it can stretch), and it is observed that the volume of the container is increasing at a rate inversely proportional to time. If the volume of the container is 100 cubic centimeters at one instant and 200 cubic centimeters 30 seconds later, find the volume of the container at any time $t \geq 0$.

11. What will be the interest accumulated by $1000 if it is invested for 3 years in a bank which compounds as follows.

   (a) Annually at $5\frac{1}{4}\%$.　　　　　　(b) Monthly at $5\frac{1}{8}\%$.
   (c) Weekly at $5\frac{1}{8}\%$.　　　　　　(d) Continuously at 5%.
   [*Note:* $(1.0525)^3 \approx 1.16591$, $(1,205,125/1,200,000)^{36} \approx 1.16582$,
   　　$(5,205,125/5,200,000)^{156} \approx 1.16611$, and $e^{0.15} \approx 1.16183$.]

12. (a) What amount, when invested now at $5\frac{1}{2}\%$ interest compounded semi-annually, will yield \$100,000 in 25 years? [*Note:* $(1.0275)^{50} \approx 3.88232.$]

   (b) What amount, when invested now at $5\frac{1}{2}\%$ interest compounded continuously, will yield \$100,000 in 25 years? (*Note:* $e^{1.375} \approx 3.95508.$)

13. A businessman invests his profits in a bank which compounds continuously at 5% interest. If the profit (in hundreds of dollars) at time t is estimated by the businessman as $p(t) = 2$, where t (in years) is in $[0, 3]$, then determine the present value of the profit earned over the next 3 years.

14. If the profit earned (in thousands of dollars) at time t is given by $p(t) = 0.9^t$, where t (in years) is in $[0, 10]$, then determine the present value of the profit earned over the next 10 years.

Chapter 7

FUNCTIONS OF
TWO VARIABLES

Thus far we have studied exclusively the calculus of functions of one variable. However, in many situations it is necessary to analyze a function of two or more variables. For instance, in geometry, the volume of a right circular cylinder is a function of the radius of the base and the height. In physics, force is a function of the mass and the acceleration. In economics, a manufacturer's profit is a function of several variables including material cost and labor cost.

The purpose of this chapter is to provide an introduction to the calculus of such functions. After defining a function of two variables, we shall discuss limits, continuity, and differentiability (Sections 7.2 and 7.3). In Section 7.4 we shall develop techniques for handling certain maximization and minimization problems.

7.1 COORDINATES IN THREE-DIMENSIONAL
SPACE

We shall begin our treatment of functions of two variables by discussing analytic geometry in three-dimensional space.

To place coordinates on points in three-dimensional space, we use three mutually perpendicular lines which intersect at a point O, called the **origin**. Taking the origin O as the zero point on each of the three lines, we can set up a coordinate system (as in Section I.3) on each of the three lines. We label these lines as the x-axis, y-axis, and z-axis, and call them the **coordinate axes** (see Figure 7.1).

The plane containing the x-axis and the y-axis is called the x-y plane; other planes containing two of the coordinate axes are referred to in the obvious way.

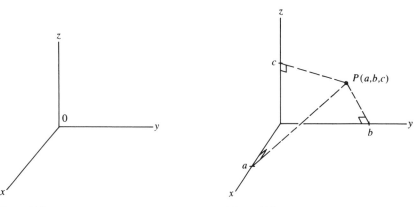

Figure 7.1   Figure 7.2

Now to label a point P in 3-space, we first draw a line through P and the x-axis and perpendicular to the x-axis. If the number a is the coordinate of the intersection of this line with the x-axis, then number a is called the **x-coordinate of P**. Similarly, by drawing lines through P and perpendicular to the y- and z-axes, one determines numbers (say b and c) called the **y-coordinate of P** and the **z-coordinate of P**, respectively. Thus we label P using these coordinates,

$$(a,\ b,\ c),$$

and sometimes write

$$P(a,\ b,\ c) \qquad \text{(see Figure 7.2)}.$$

Observe that each point on the x-axis has coordinates of the form $(a, 0, 0)$, each point on the y-axis has coordinates of the form $(0, b, 0)$, and each point on the z-axis has coordinates of the form $(0, 0, c)$.

On the other hand, if one is given a triple of numbers

$$(a,\ b,\ c),$$

then one can find the corresponding point in 3-space as follows: First, locate the point

$$(a,\ b,\ 0)$$

in the x-y plane;

then move upward a distance c (if $c \geq 0$) or move downward a distance $|-c|$ (if $c < 0$).

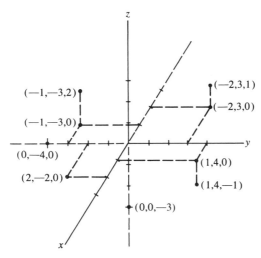

Figure 7.3

EXAMPLE 1   Several points are plotted in Figure 7.3.

With the introduction of coordinates in 3-space, it is possible to determine a formula for the distance $d(P, Q)$ between two points P and Q.

Let P have coordinates (x_1, y_1, z_1) and let Q have coordinates (x_2, y_2, z_2). Consider the rectangular parallelepiped with opposite vertices P and Q, as in Figure 7.4. Note that P and P_1 lie in the plane $z = z_1$, and so by the theorem of Pythagoras,

$$d^2(P, P_1) = (x_1 - x_2)^2 + (y_1 - y_2)^2.$$

Again, using the theorem of Pythagoras, we have

$$d^2(P, Q) = d^2(P, P_1) + d^2(P_1, Q)$$
$$= (x_1 - x_2)^2 + (y_1 - y_2)^2 + (z_1 - z_2)^2.$$

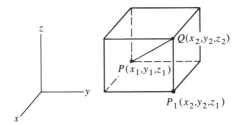

Figure 7.4

7.1.1
Distance
Between Two
Points

Let $P(x_1, y_1, z_1)$ and $Q(x_2, y_2, z_2)$ be two points in three-dimensional space. Then

$$d(P, Q) = \sqrt{(x_1 - x_2)^2 + (y_1 - y_2)^2 + (z_1 - z_2)^2}$$

is the distance between P and Q.

EXAMPLE 2   Find the distance between the points $P(4, 3, -1)$ and $Q(-2, 6, 2)$:

$$\begin{aligned} d(P, Q) &= \sqrt{(4 - (-2))^2 + (3 - 6)^2 + (-1 - 2)^2} \\ &= \sqrt{6^2 + 3^2 + 3^2} \\ &= \sqrt{54}. \end{aligned}$$

EXAMPLE 3   Do the three points $P(-1, 0, 2)$, $Q(2, 1, 0)$, and $R(5, 2, -2)$ lie on a line?

$$d(P, Q) = \sqrt{(-1 - 2)^2 + (0 - 1)^2 + (2 - 0)^2} = \sqrt{14}$$

$$d(Q, R) = \sqrt{(2 - 5)^2 + (1 - 2)^2 + (0 - (-2))^2} = \sqrt{14}$$

$$\begin{aligned} d(P, R) = \sqrt{(-1 - 5)^2 + (0 - 2)^2 + (2 - (-2))^2} &= \sqrt{56} \\ &= \sqrt{(4)(14)} \\ &= 2\sqrt{14}. \end{aligned}$$

Since

$$d(P, R) = d(P, Q) + d(Q, R),$$

we have

$$P, Q, R \text{ lie on a line.}$$

Suppose that we have an equation involving the variables x, y, z. The collection of points in 3-space whose coordinates satisfy the given equation is called the **graph of the equation**.

EXAMPLE 4   The graphs of the equations $x = 0$, $y = 0$, $z = 0$ are, respectively, the y-z plane, the x-z plane, and the x-y plane (see Figure 7.5).

EXAMPLE 5   The graph of the equation $y = 3$ is a plane parallel to the x-z plane and 3 units to the right, as shown in Figure 7.6.

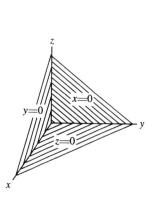

Figure 7.5

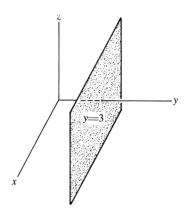

Figure 7.6

EXAMPLE 6   What is the graph of $x + y + 2z = 1$?

The graph of $x + y + 2z = 1$ in the plane $z = 0$ is the line

$$x + y = 1;$$

in the plane $y = 0$, it is the line

$$x + 2z = 1;$$

and in the plane $x = 0$, it is the line

$$y + 2z = 1.$$

The plane determined by these three lines is the graph of the given equation (see Figure 7.7).

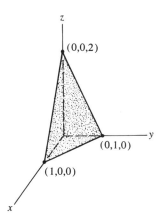

Figure 7.7

EXAMPLE 7   What is the graph of $x^2 + y^2 + z^2 + 2x - 4y + 4 = 0$?

Completing the square, one obtains

$$(x^2 + 2x + 1) + (y^2 - 4y + 4) + z^2 = 1$$
$$(x + 1)^2 + (y - 2)^2 + z^2 = 1$$
$$\sqrt{(x + 1)^2 + (y - 2)^2 + z^2} = 1.$$

Each point on the graph is the distance one from $(-1, 2, 0)$. Thus it is a sphere with center $(-1, 2, 0)$ and radius 1, as drawn in Figure 7.8.

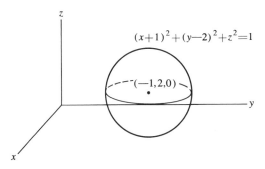

Figure 7.8

EXAMPLE 8   Sketch the graph of $x^2 + y^2 = 9$.

The graph of $x^2 + y^2 = 9$ in the x-y plane is a circle. Any line L through this circle and perpendicular to the x-y plane consists of points whose co-ordinates satisfy the given equation. The collection of all such lines is the cylinder shown in Figure 7.9; it is the graph of $x^2 + y^2 = 9$ in 3-space.

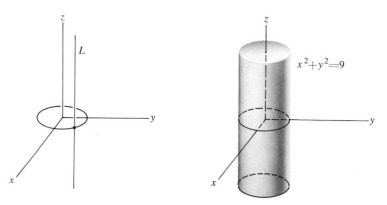

Figure 7.9

EXAMPLE 9   Sketch the graph of $z = x^2 + y^2$.

Since $x^2 + y^2 \geq 0$, the graph lies above the x-y plane. Moreover, for any positive number c, the intersection of the graph of

$$z = x^2 + y^2$$

with the horizontal plane

$$z = c$$

is the circle

$$x^2 + y^2 = c.$$

The graph of the given equation is drawn in Figure 7.10.

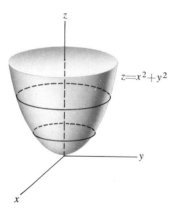

Figure 7.10

EXERCISES

1. Plot the following points in three-dimensional space: $(0, 0, 1)$, $(\frac{1}{2}, 0, -4)$, $(-4, -5, 2)$, $(1, 1, 0)$, $(0, -3, \frac{5}{3})$, $(3, -\frac{1}{2}, \frac{3}{4})$, $(-2, 3, 1)$, $(1, 2, -1)$, and $(-5, -2, -\frac{9}{10})$.

2. Find the distance between the two given points.

   (a) $(0, 0, 0)$, $(2, -2, 1)$.   (b) $(3, 4, 5)$, $(-1, 2, 5)$.

   (c) $(-2, 0, 4)$, $(1, 1, 3)$.   (d) $(4, 2, 1)$, $(4, 2, -3)$.

   (e) $(6, -3, 1)$, $(2, 0, -4)$.

3. Determine whether or not the three given points lie on a straight line.

   (a) $(-1, 3, 2)$, $(1, 2, 6)$, $(5, 0, 14)$.

   (b) $(0, 0, 0)$, $(4, -2, 2)$, $(-2, -1, 3)$.

   (c) $(0, -4, 5)$, $(2, -8, 7)$, $(-2, 0, 3)$.

   (d) $(-1, -1, -1)$, $(2, 3, 1)$, $(4, -1, 2)$.

4. Consider the set of equations $x = 2 + t$, $y = -1 + 3t$, and $z = 4 - 2t$, where t is a number.

   (a) Choose any three distinct numbers t_1, t_2, t_3 and plot in three-dimensional space the points $(2 + t_1, -1 + 3t_1, 4 - 2t_1)$, $(2 + t_2, -1 + 3t_2, 4 - 2t_2)$, and $(2 + t_3, -1 + 3t_3, 4 - 2t_3)$.

   (b) Verify that the points in part (a) lie on a straight line.

5. Graph each of the following equations in three-dimensional space. (*Note:* In each case the graph is a plane; find the points of intersection with the coordinate axes.)

   (a) $y = 2$.   (b) $x = -3$.   (c) $z = e$.

   (d) $x + y = 1$.   (e) $y + z = 1$.   (f) $x + z = 1$.

   (g) $x + y + z = 1$.   (h) $x - 2y + 3z = 2$.   (i) $3x + y - 4z = 4$.

6. Graph each of the following equations in three-dimensional space. (*Note:* In each case the graph is a sphere; find the center and radius.)

   (a) $x^2 + y^2 + z^2 = 1$.

   (b) $x^2 + y^2 + z^2 = 9$.

   (c) $x^2 + (y - 1)^2 + (z + 3)^3 = 4$.

   (d) $(x + 2)^2 + (y + 1)^2 + (z - 2)^2 = 16$.

   (e) $x^2 + y^2 + z^2 + 4y - 2z + 4 = 0$.

   (f) $x^2 + y^2 + z^2 - 2x + 4y - 6z + 12 = 0$.

7. Graph each of the following equations in three-dimensional space.

   (a) $x^2 + z^2 = 1$.   (b) $y = x^2$.   (c) $y = x^3$.

   (d) $y = x^2 + z^2$.   (e) $z^2 = x^2 + y^2$.   (f) $z = \log y$.

7.2 FUNCTIONS OF TWO VARIABLES, LIMITS AND CONTINUITY

In Chapter 2 we studied certain correspondences between sets of real numbers called functions. For example, the volume V of a sphere is a function of its radius, given by

$$V(r) = \tfrac{4}{3}\pi r^2.$$

Here V is a function of r. However, the volume of a right circular cylinder depends on both the radius r of the base and the height h; that is, the volume V depends on the pair of real numbers r, h:

$$V(r, h) = \pi r^2 h.$$

Therefore we say that V is a function of two variables r and h.

7.2.1
Function of
Two Variables

Let A be a set of points in the plane and B a set of real numbers.

A **function f from A to B** is a correspondence which associates with each point (x, y) in A one and only one number z in B.

In this case we use the notation

$$f: A \rightarrow B$$

and say that f is defined on A—called **domain of f**. We also write

$$z = f(x, y),$$

read "z equals f at (x, y)." The number $f(x, y)$ is called the **value of f at (x, y)**. In the following examples the symbol $\mathbf{R}^2$ represents the collection of all pairs of numbers.

EXAMPLE 1   Let $f: \mathbf{R}^2 \rightarrow \mathbf{R}$ be defined by

$$f(x, y) = \sqrt{x^2 + y^2}.$$

Note that the value of f at (x, y) represents the distance from $(0, 0)$ to (x, y).

EXAMPLE 2   Let $V: \mathbf{R}^2 \rightarrow \mathbf{R}$ be defined by

$$V(r, h) = \tfrac{1}{3}\pi r^2 h.$$

Then the value of V at (r, h) represents the volume of a right circular cone with radius r and height h.

When the domain of a given function f is not specified, it is taken to be the largest set of points (x, y) for which $f(x, y)$ is defined.

EXAMPLE 3   What is the domain of the function f defined by

$$f(x, y) = \frac{x}{x + y} \, ?$$

Since $x/(x + y)$ is a real number provided that $x + y \neq 0$, the domain of f is the entire plane with the exception of the points on the line $y = -x$.

The graph of the equation

$$z = f(x, y)$$

is called the **graph of the function f**. Just as the graph of a function of one variable is often a curve in the plane, the graph of a function of two variables is often a surface in 3-space.

EXAMPLE 4   Sketch the graph of the function f defined by

$$f(x, y) = x^2 + y^2.$$

The graph of the function f is the graph of the equation

$$z = x^2 + y^2.$$

This surface was sketched in Example 9 of the preceding section.

We shall now turn our attention to the concept of the limit of a function of two variables. We ask a question analogous to one we asked for functions of one variable: What happens to the values $f(x, y)$ as (x, y) approaches the point (a, b)?

Recall that for a function of one variable there are just two directions of approaching a point c. However, for a point (a, b) in the plane, there are infinitely many directions of approaching the point (see Figure 7.11).

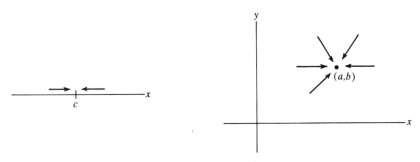

Figure 7.11

We say that

$$\lim_{(x,y)\to(a,b)} f(x, y) = L$$

if we can get $f(x, y)$ as close to L as we like by choosing any (x, y) within a certain distance of (a, b), $(x, y) \neq (a, b)$.

In 7.2.2 we are assuming that f is defined at each point in some sufficiently small disk with center (a, b), although not necessarily at (a, b) itself. In addition, it should be noted that "any (x, y) within a certain distance of (a, b), $(x, y) \neq (a, b)$" means "any (x, y) in a disk with its center (a, b) deleted."

EXAMPLE 5   Investigate $\lim\limits_{(x,y)\to(0,0)} f(x, y)$, where

$$f(x, y) = x^2 + y^2.$$

Observe that the value of f at (x, y) represents the square of the distance from $(0, 0)$ to (x, y). Thus we can get $f(x, y)$ as close to 0 as we like by choosing any (x, y) within a certain distance of $(0, 0)$. Therefore

$$\lim_{(x,y)\to(0,0)} (x^2 + y^2) = 0.$$

EXAMPLE 6   Investigate $\lim\limits_{(x,y)\to(0,0)} f(x, y)$, where

$$f(x, y) = xy.$$

Since

$$|x| \leq \sqrt{x^2 + y^2} \qquad \text{and} \qquad |y| \leq \sqrt{x^2 + y^2},$$

we see that

$$0 \leq |xy| \leq x^2 + y^2.$$

Since we can get $x^2 + y^2$ as close to 0 as we like, we can get xy as close to 0 as we like by choosing any (x, y) within a certain distance of $(0, 0)$, $(x, y) \neq (0, 0)$. Therefore

$$\lim_{(x,y)\to(0,0)} xy = 0.$$

EXAMPLE 7   Investigate

$$\lim_{(x,y)\to(0,0)} \frac{xy}{\sqrt{x^2 + y^2}}.$$

Since $\lim\limits_{(x,y)\to(0,0)} xy = 0$ and $\lim\limits_{(x,y)\to(0,0)} \sqrt{x^2 + y^2} = 0$, the result is not obvious. However, since

$$|xy| \le x^2 + y^2,$$

we have

$$0 \le \frac{|xy|}{\sqrt{x^2 + y^2}} \le \sqrt{x^2 + y^2}.$$

Since we can get $\sqrt{x^2 + y^2}$ as close to 0 as we like, we can get

$$\frac{xy}{\sqrt{x^2 + y^2}}$$

as close to 0 as we like by choosing any (x, y) within a certain distance of $(0, 0)$, $(x, y) \ne (0, 0)$. Thus

$$\lim_{(x,y)\to(0,0)} \frac{xy}{\sqrt{x^2 + y^2}} = 0.$$

EXAMPLE 8   Investigate $\lim\limits_{(x,y)\to(0,0)} f(x, y)$, where

$$f(x, y) = \frac{x}{x + y}.$$

Note that

$$f(0, y) = 0 \qquad \text{and} \qquad f(x, 0) = 1;$$

that is, the values of f on the x-axis are all 1 and the values of f on the y-axis are all 0. Therefore in every disk with its center $(0, 0)$ deleted, there are values of the function which are 0 and 1, which are not close. Therefore

$$\lim_{(x,y)\to(0,0)} \frac{x}{x + y} \qquad \text{does not exist.}$$

EXAMPLE 9   Investigate $\lim\limits_{(x,y)\to(0,0)} f(x, y)$, where

$$f(x, y) = \frac{xy}{x^2 + y^2}.$$

Note that

$$f(0, y) = 0 \qquad \text{and} \qquad f(x, 0) = 0;$$

hence the values are the same on the axes. However, on the line $y = x$, we see that

$$f(x, x) = \frac{x^2}{x^2 + x^2} = \frac{1}{2}.$$

Therefore in every disk with center $(0, 0)$ deleted, there are values of the function which are 0 and $\frac{1}{2}$, which are not close. Thus

$$\lim_{(x,y)\to(0,0)} \frac{xy}{x^2 + y^2} \qquad \text{does not exist.}$$

EXAMPLE 10   Investigate $\lim\limits_{(x,y)\to(1,2)} (x^2 - 2xy)$.

It is reasonably clear that we can get

$$x^2$$

as close to 1 as we like, and we can get

$$2xy$$

as close to 4 as we like by choosing any (x, y) within a certain distance of $(1, 2)$, $(x, y) \neq (1, 2)$. Hence we can get

$$x^2 - 2xy$$

as close to

$$1 - 4 = -3$$

as we like by choosing any (x, y) within a certain distance of $(1, 2)$, $(x, y) \neq (1, 2)$. Therefore

$$\lim_{(x,y)\to(1,2)} (x^2 - 2xy) = -3.$$

The concept of continuity for functions of two variables is similar to that for functions of one variable.

7.2.3
Continuity of
Functions of
Two Variables

Function f is **continuous at point (a, b)** if

1. f is defined at (a, b),

2. $\lim\limits_{(x,y)\to(a,b)} f(x, y)$ exists,

3. $\lim\limits_{(x,y)\to(a,b)} f(x, y) = f(a, b)$.

EXAMPLE 11   The polynomial in two variables

$$P(x, y) = a_0 x^n + a_1 x^{n-1} y + a_2 x^{n-2} y^2 + \cdots + a_{n-1} xy^{n-1} + a_n y^n$$

is continuous at each point (a, b):

$$\lim_{(x,y)\to(a,b)} P(x, y) = a_0 a^n + a_1 a^{n-1} b + a_2 a^{n-2} b^2 + \cdots$$
$$+ a_{n-1} ab^{n-1} + a_n b^n.$$

EXAMPLE 12   Let

$$f(x, y) = \begin{cases} \dfrac{xy}{\sqrt{x^2 + y^2}}, & (x, y) \neq (0, 0), \\ 0, & (x, y) = (0, 0). \end{cases}$$

In Example 7, we observed that

$$\lim_{(x,y)\to(0,0)} \frac{xy}{\sqrt{x^2 + y^2}} = 0.$$

Since $f(0, 0) = 0$, f is continuous at $(0, 0)$.

EXAMPLE 13 Let

$$f(x, y) = \begin{cases} \dfrac{xy}{x^2 + y^2}, & (x, y) \neq (0, 0), \\ 0, & (x, y) = (0, 0). \end{cases}$$

In Example 9, we observed that

$$\lim_{(x,y)\to(0,0)} \frac{xy}{x^2 + y^2} \quad \text{does not exist.}$$

Therefore, f is not continuous at $(0, 0)$.

If $(a, b) \neq (0, 0)$, it is easy to see that

$$\lim_{(x,y)\to(a,b)} \frac{xy}{x^2 + y^2} = \frac{ab}{a^2 + b^2}.$$

Thus f is continuous everywhere except at $(0, 0)$.

We say that f is **continuous on a set S** if f is continuous at each point of S. If (a, b) is on the boundary of the set, then by continuity of f at (a, b), we mean that $f(x, y)$ can be made as close to $f(a, b)$ as we like by choosing any (x, y) in S within a certain distance of (a, b).

Recall that a continuous function of one variable on $[a, b]$ has a graph which is an unbroken curve. Similarly, *a continuous function of two variables on S has a graph which is an unbroken surface.*

Theorem 3.3.4(1) has a generalization to functions of several variables.

7.2.4

Maximum-Minimum Theorem for Functions of Two Variables

Let R be the rectangle given by

$$a \leq x \leq b$$
$$c \leq y \leq d.$$

If f is continuous on R, then there exist points (α, β) and (γ, δ) in R such that

$$f(\alpha, \beta) \leq f(x, y) \leq f(\gamma, \delta)$$

for each (x, y) in R.

$f(\alpha, \beta)$ is called the **minimum value of f on R** and $f(\gamma, \delta)$ is called the **maximum value of f on R.**

EXAMPLE 14   Let $f(x, y) = \sqrt{x^2 + y^2}$. On the rectangle R given by

$$-1 \le x \le 1$$
$$-1 \le y \le 2,$$

we see that

$$f(0, 0) = 0 \text{ is the minimum value of } f \text{ on } R$$

and that

$$f(1, 2) = f(-1, 2) = \sqrt{5} \text{ is the maximum value of } f \text{ on } R.$$

EXERCISES

1. In each of the following, determine the domain of f and graph it.

    (a) $f(x, y) = 5$.
               (b) $f(x, y) = 3x - 2y + 4$.
    (c) $f(x, y) = \sqrt{9 - x^2 - y^2}$.
       (d) $f(x, y) = x^2$.

2. In each of the following, determine the domain of f.

    (a) $f(x, y) = 2x^2 - y$.
    (b) $f(x, y) = 1/(2x^2 - y)$.
    (c) $f(x, y) = 3/[(x - 1)^2 + (y + 2)^2]$.
    (d) $f(x, y) = 2xy/(|x| + |y|)$.
    (e) $f(x, y) = \log xy$.
    (f) $f(x, y) = (e^{2x - 3y})/xy$.

3. Investigate each of the following limits.

    (a) $\displaystyle\lim_{(x,y)\to(0,0)} x^2 - y^2$.
       (b) $\displaystyle\lim_{(x,y)\to(-5,4)} \sqrt{(x + 5)^2 + (y - 4)^2}$.

    (c) $\displaystyle\lim_{(x,y)\to(0,0)} 3e^{x+y}$.
       (d) $\displaystyle\lim_{(x,y)\to(2,-1)} \log(2x + y^2)$.

    (e) $\displaystyle\lim_{(x,y)\to(0,0)} x/y$.
       (f) $\displaystyle\lim_{(x,y)\to(0,0)} [x^2/(x^2 + y^2)]$.

    (g) $\displaystyle\lim_{(x,y)\to(0,0)} [(x + y)/(x^2 + y)]$.
    (h) $\displaystyle\lim_{(x,y)\to(0,0)} [x^2y^2/(x^2 + y^2)]$.

    (i) $\displaystyle\lim_{(x,y)\to(0,0)} [xy^2/(x^2 + y^4)]$.

4. Given

$$f(x, y) = \begin{cases} \dfrac{x^2}{x^2 + y^2}, & (x, y) \ne (0, 0), \\ 0, & (x, y) = (0, 0), \end{cases}$$

at what points of the plane is f continuous?

5. Can either of the following functions be defined at $(0, 0)$ so as to make the resulting function continuous at $(0, 0)$?

(a) $f(x, y) = (x + y)/(x^2 + y)$.

(b) $f(x, y) = 3 + [x^2y^2/(x^2 + y^2)]$.

6. Find the maximum and minimum values of the given function f in the given rectangle R.

(a) $f(x, y) = xy$, R: $\begin{cases} 0 \le x \le 1, \\ 0 \le y \le 1. \end{cases}$

(b) $f(x, y) = xy$, R: $\begin{cases} -1 \le x \le 1, \\ -1 \le y \le 1. \end{cases}$

(c) $f(x, y) = x^2 + y$, R: $\begin{cases} 2 \le x \le 4, \\ -1 \le y \le 3. \end{cases}$

7.3   PARTIAL DIFFERENTIATION

As for functions of one variable, the derivatives of functions of two variables involve rates of change of the function. For a function of two variables, it is useful to consider the rate of change of the function with respect to one variable while holding the other variable constant.

If we differentiate the function $f(x, y)$ with respect to x, assuming that y is a constant, we obtain what is called the **partial derivative of f with respect to x**, denoted by

$$f_x(x, y).$$

Assuming that x is constant and differentiating with respect to y yields the **partial derivative of f with respect to y**, denoted by

$$f_y(x, y).$$

In terms of limits, we have the following.

**7.3.1
Partial
Derivatives of
Functions of
Two Variables**

$$f_x(x, y) = \lim_{h \to 0} \frac{f(x + h, y) - f(x, y)}{h}$$

$$f_y(x, y) = \lim_{k \to 0} \frac{f(x, y + k) - f(x, y)}{k}.$$

Recall that for functions of one variable, if

$$y = f(x),$$

then

$$f'(x) = \frac{dy}{dx} = \frac{d}{dx}\{f(x)\}.$$

Similar notation exists for partial derivatives. If $z = f(x, y)$,

$$f_x(x, y) = \frac{\partial z}{\partial x} = \frac{\partial}{\partial x}\{f(x, y)\}$$

and

$$f_y(x, y) = \frac{\partial z}{\partial y} = \frac{\partial}{\partial y}\{f(x, y)\}.$$

EXAMPLE 1   Let $f(x, y) = x^3 + 7x^2y^4 + 6y^3$.

Treating y as a constant, we have

$$f_x(x, y) = 3x^2 + 14xy^4.$$

With x held constant, we obtain

$$f_y(x, y) = (7x^2)(4y^3) + 18y^2$$
$$= 28x^2y^3 + 18y^2.$$

EXAMPLE 2   Let $f(x, y) = \log(7x^2 + 2y^2)$.

Then

$$f_x(x, y) = \frac{1}{7x^2 + 2y^2}\frac{\partial}{\partial x}\{7x^2 + 2y^2\}$$

$$= \frac{14x}{7x^2 + 2y^2}$$

and

$$f_y(x, y) = \frac{1}{7x^2 + 2y^2}\frac{\partial}{\partial y}\{7x^2 + 2y^2\}$$

$$= \frac{4y}{7x^2 + 2y^2}.$$

EXAMPLE 3   Let

$$f(x, y) = \begin{cases} \dfrac{xy}{x^2 + y^2}, & (x, y) \neq (0, 0), \\ 0, & (x, y) = (0, 0). \end{cases}$$

If $(x, y) \neq (0, 0)$, then

$$f_x(x, y) = \frac{(x^2 + y^2)y - xy(2x)}{(x^2 + y^2)^2}$$

$$= \frac{y^3 - x^2y}{(x^2 + y^2)^2}.$$

Likewise, one obtains

$$f_y(x, y) = \frac{x^3 - y^2 x}{(x^2 + y^2)^2}.$$

For $(x, y) = (0, 0)$,

$$f_x(0, 0) = \lim_{h \to 0} \frac{f(0 + h, 0) - f(0, 0)}{h}.$$

Since $f(h, 0) = 0$ and $f(0, 0) = 0$,

$$f_x(0, 0) = \lim_{h \to 0} \frac{0}{h} = 0.$$

Similarly, one can verify that

$$f_y(0, 0) = 0.$$

EXAMPLE 4 Given $f(x, y) = |y|$ and the point $(1, 0)$, let us investigate

$$\lim_{h \to 0} \frac{f(1 + h, 0) - f(1, 0)}{h} \quad \text{and} \quad \lim_{k \to 0} \frac{f(1, 0 + k) - f(1, 0)}{k}.$$

Now

$$\lim_{h \to 0} \frac{f(1 + h, 0) - f(1, 0)}{h} = \lim_{h \to 0} \frac{0 - 0}{h} = 0.$$

Therefore

$$f_x(1, 0) = 0.$$

On the other hand,

$$\lim_{k \to 0} \frac{f(1, 0 + k) - f(1, 0)}{k} = \lim_{k \to 0} \frac{|k|}{k}.$$

The fact that

$$\lim_{k \downarrow 0} \frac{|k|}{k} = 1 \quad \text{and} \quad \lim_{k \uparrow 0} \frac{|k|}{k} = -1$$

implies that

$$f_y(1, 0) \quad \text{does not exist.}$$

It is important to note the geometric interpretation of

$$f_x(a, b).$$

Consider the intersection of the surface

$$z = f(x, y)$$

and the plane

$$y = b;$$

it is the curve

$$z = f(x, b),$$

designated by C in Figure 7.12. Then $f_x(a, b)$ is the slope of the tangent line to curve C,

$$z = f(x, b),$$

at the point $P(a, b, f(a, b))$. There is a similar interpretation for

$$f_y(a, b).$$

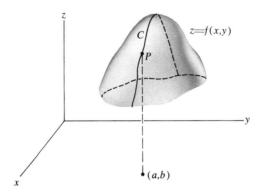

Figure 7.12

Now consider the surface

$$z = f(x, y)$$

and a point $P(a, b, f(a, b))$ on it. If tangent lines to *all* curves lying on the surface passing through the point P lie in one plane, then this plane T is called the **tangent plane** to the surface at P (see Figure 7.13).

Recall that for a function f of one variable differentiability at a point x_0 means the existence of a tangent line to the curve $y = f(x)$ at $(x_0, f(x_0))$. Likewise for functions of two variables we say that f is **differentiable at** (a, b)

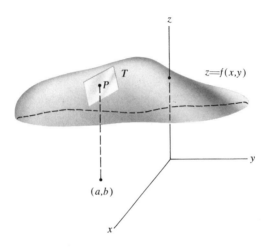

Figure 7.13

provided that there is a tangent plane to the surface $z = f(x, y)$ at $(a, b, f(a, b))$.

As before, it follows that differentiability at a point implies continuity there. Moreover, if f is differentiable at each point of a set S, then the graph of the surface

$$z = f(x, y)$$

on S is not only unbroken but also smooth. The following examples illustrate these ideas.

EXAMPLE 5   Let $z = f(x, y) = x^2 + y^2$. The entire graph of f is smooth and unbroken. In particular, the plane $z = 0$ is the tangent plane to the surface at the point $(0, 0, 0)$ (see Figure 7.14).

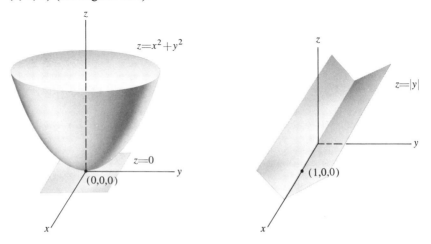

Figure 7.14   Figure 7.15

EXAMPLE 6   The graph of $z = f(x, y) = |y|$ is given in Figure 7.15. It is unbroken, but it has an edge (and so is *not* smooth) at points on the x-axis. In particular, there is *no* tangent plane at the point $(1, 0, 0)$. (See Example 4 of this section.)

It turns out that differentiability of f at point (a, b) implies the existence of the partial derivatives $f_x(a, b)$ and $f_y(a, b)$. However, the converse of this statement is false. (See Example 13 of Section 7.2 and Example 3 of this section.)

The process of partial differentiation may be repeated. The function

$$f_x(x, y)$$

may be differentiated with respect to x and with respect to y, yielding

$$f_{xx}(x, y) \quad \text{and} \quad f_{xy}(x, y),$$

respectively. The derivatives

$$f_{yx}(x, y) \quad \text{and} \quad f_{yy}(x, y)$$

have similar meanings. These four derivatives are called the **second partial derivatives**. Other notations are

$$f_{xx} = \frac{\partial}{\partial x}\left\{\frac{\partial f}{\partial x}\right\} = \frac{\partial^2 f}{\partial x^2}$$

$$f_{xy} = \frac{\partial}{\partial y}\left\{\frac{\partial f}{\partial x}\right\} = \frac{\partial^2 f}{\partial y\, \partial x}$$

$$f_{yx} = \frac{\partial}{\partial x}\left\{\frac{\partial f}{\partial y}\right\} = \frac{\partial^2 f}{\partial x\, \partial y}$$

and

$$f_{yy} = \frac{\partial}{\partial y}\left\{\frac{\partial f}{\partial y}\right\} = \frac{\partial^2 f}{\partial y^2}.$$

EXAMPLE 7   Let $f(x, y) = x^3 + 7x^2 y^4 + 6y^3$. Then

$$f_x(x, y) = 3x^2 + 14xy^4$$

$$f_{xx}(x, y) = 6x + 14y^4$$

$$f_{xy}(x, y) = 56xy^3$$

$$f_y(x, y) = 28x^2 y^3 + 18y^2$$

$$f_{yx}(x, y) = 56xy^3$$

$$f_{yy}(x, y) = 84x^2 y^2 + 36y.$$

In Example 7, note that the mixed partials were equal; that is,

(7.3.2) $$f_{xy}(x, y) = f_{yx}(x, y).$$

More generally, Equation 7.3.2 holds whenever the mixed partials of f are continuous at (x, y).

EXERCISES

1. Find $\partial z/\partial x$, $\partial z/\partial y$ in each of the following cases.

(a) $z = x^3 y + 7xy^2 + 2$.   (b) $z = (2x - y)/(x^2 + y^3)$.

(c) $z = 2x\sqrt{6xy} + 8x^4$.   (d) $z = \log(x^2 + y^2)$.

(e) $z = e^{x+y}$.   (f) $z = \log(1 + e^{xy})$.

2. Find $\partial^2 z/\partial x^2$, $\partial^2 z/\partial y^2$, $\partial^2 z/\partial x\, \partial y$, $\partial^2 z/\partial y\, \partial x$ for each part in Exercise 1.

3. Show that $(\partial^2 f/\partial x^2) + (\partial^2 f/\partial y^2) = 0$ in each of the following cases.

(a) $f(x, y) = x^2 - y^2$.   (b) $f(x, y) = 2xy$.

(c) $f(x, y) = x/(x^2 + y^2)$.   (d) $f(x, y) = \log\sqrt{x^2 + y^2}$.

4. In each of the following, assume that $z = f(x, y)$ satisfies the given equation. Find $\partial f/\partial x$ and $\partial f/\partial y$.

   (a) $x^2 + y^2 + z^2 = 16$.   (b) $x^2 - y^2 + z^2 = 1$.

5. (a) Let C be the curve of intersection of the surfaces $z = x^2 + y^2$ and $y = 1$. Find the slope of C at the point $(2, 1, 5)$.

   (b) Let C be the curve of intersection of the surfaces $z = \sqrt{9 - x^2 - y^2}$ and $x = 2$. Find the slope of C at the point $(2, -1, 2)$.

7.4 MAXIMA AND MINIMA

In Section 4.6 we used the derivative to help solve maximum-minimum problems which involve functions of one variable. Here we shall employ partial derivatives to deal with problems of maximization and minimization of functions of two variables.

7.4.1
Definitions

1. f has a **local maximum value at** (x_0, y_0) if

$$f(x, y) \le f(x_0, y_0)$$

for all (x, y) in some disk with center (x_0, y_0).

2. f has a **local minimum value at** (x_0, y_0) if

$$f(x, y) \ge f(x_0, y_0)$$

for all (x, y) in some disk with center (x_0, y_0).

If either f has a local maximum value at (x_0, y_0) or f has a local minimum value at (x_0, y_0), we say that f has a **local extreme value at** (x_0, y_0).

EXAMPLE 1   Let $f(x, y) = 4 - x^2 - y^2$.
Since

$$f(x, y) = 4 - (x^2 + y^2) \le 4 = f(0, 0),$$

f has a local maximum at $(0, 0)$.

In 7.4.1, let

$$x = x_0 + h \quad \text{and} \quad y = y_0 + k.$$

Then

f has a local maximum at (x_0, y_0)

if

$$f(x_0 + h, y_0 + k) \le f(x_0, y_0)$$

or

$$f(x_0 + h, y_0 + k) - f(x_0, y_0) \le 0$$

for all (h, k) in some disk with center $(0, 0)$.
Likewise,

$$f \text{ has a local minimum at } (x_0, y_0)$$

if

$$f(x_0 + h, y_0 + k) - f(x_0, y_0) \geq 0$$

for all (h, k) in some disk with center $(0, 0)$.

Recall that for functions of one variable the existence of a derivative at a local extreme value implies that the derivative is zero there. Analogously, we have the following.

7.4.2
Theorem   If f has a local extreme value at (x_0, y_0) and

$$f_x(x_0, y_0), f_y(x_0, y_0) \text{ exist,}$$

then

$$f_x(x_0, y_0) = 0 \quad \text{and} \quad f_y(x_0, y_0) = 0.$$

Proof: We shall prove the case when f has a local maximum value at (x_0, y_0). The proof is similar when f has a local minimum at (x_0, y_0).

Consider the curve on the surface passing through (x_0, y_0) and with the equation

$$z = f(x, y_0).$$

It has a local maximum at x_0. By Theorem 4.6.2,

$$f_x(x_0, y_0) = 0.$$

Likewise, considering the curve $z = f(x_0, y)$, one can show that

$$f_y(x_0, y_0) = 0.$$

To summarize, for f to have a local extreme value at (x_0, y_0) it is necessary that either

$$f_x(x_0, y_0) \quad \text{or} \quad f_y(x_0, y_0) \quad \text{fail to exist}$$

or

$$f_x(x_0, y_0) = 0 \quad \text{and} \quad f_y(x_0, y_0) = 0.$$

Points (x_0, y_0) satisfying either of the above conditions are called **critical points of f**.

EXAMPLE 2   Let $f(x, y) = 2y - x^2 - y^2$. Then

$$f_1(x, y) = -2x$$

$$f_2(x, y) = 2 - 2y,$$

and solving

$$f_1(x, y) = -2x = 0$$

$$f_2(x, y) = 2 - 2y = 0$$

gives $(0, 1)$ as the only critical point. Does f have a local extreme value at $(0, 1)$?

$$f(0 + h, 1 + k) - f(0, 1) = 2(1 + k) - h^2 - (1 + k)^2 - 1$$
$$= -(h^2 + k^2) < 0.$$

Therefore, f has a local maximum at $(0, 1)$.

It should be noted that since $(0, 1)$ is the only critical point and f has partial derivatives at each point, $f(0, 1) = 1$ must be the maximum value of the function.

EXAMPLE 3   Let $f(x, y) = x^2 + xy + y^2$. Then

$$f_1(x, y) = 2x + y$$
$$f_2(x, y) = x + 2y.$$

The system

$$2x + y = 0$$
$$x + 2y = 0$$

has the solution $(0, 0)$; that is,

$(0, 0)$ is the only critical point.

Does f have a local extreme value at $(0, 0)$?

$$f(0 + h, 0 + k) - f(0, 0) = h^2 + hk + k^2$$
$$= (h^2 + hk + \tfrac{1}{4}k^2) + k^2 - \tfrac{1}{4}k^2$$
$$= (h + \tfrac{1}{2})^2 + \tfrac{3}{4}k^2 > 0.$$

Therefore, f has a local minimum at $(0, 0)$.

EXAMPLE 4   Let $f(x, y) = x^2 - y^2$. Then

$$f_1(x, y) = 2x$$
$$f_2(x, y) = -2y,$$

and so

$$f_1(0, 0) = 0 \quad \text{and} \quad f_2(0, 0) = 0.$$

Does f have a local extreme value at $(0, 0)$?

$$f(0 + h, 0 + k) - f(0, 0) = h^2 - k^2 \quad \begin{cases} h^2 - k^2 < 0 \text{ if } |h| < |k|, \\ h^2 - k^2 > 0 \text{ if } |h| > |k|. \end{cases}$$

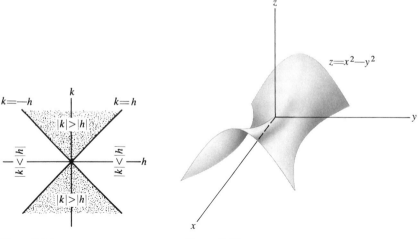

Figure 7.16   Figure 7.17

See Figure 7.16. Hence, $f(0 + h, 0 + k) - f(0, 0)$ does not have the same sign in any disk with center $(0, 0)$. Therefore, f has neither a local maximum nor a local minimum at $(0, 0)$. Such a critical point is called a **saddle point**. The graph of f is given in Figure 7.17.

For functions of one variable we had the first derivative test and the second derivative test for determining the nature of a local extreme value. There is a test for functions of two variables which we shall state without proof.

7.4.3
Derivative Test   Suppose that

$$\frac{\partial f}{\partial x} = \frac{\partial f}{\partial y} = 0 \qquad \text{at } (x_0, y_0).$$

Let

$$D = \left(\frac{\partial^2 f}{\partial x \, \partial y}\right)^2 - \left(\frac{\partial^2 f}{\partial x^2}\right)\left(\frac{\partial^2 f}{\partial y^2}\right)$$

evaluated at (x_0, y_0).

1. If $D > 0$, then f has a saddle point at (x_0, y_0).
2. If $D < 0$ and $\partial^2 f/\partial x^2 < 0$ at (x_0, y_0), then f has a local maximum at (x_0, y_0).
3. If $D < 0$ and $\partial^2 f/\partial x^2 > 0$ at (x_0, y_0), then f has a local minimum at (x_0, y_0).
4. If $D = 0$, then no conclusion can be drawn.

EXAMPLE 5   Determine the critical points of $f(x, y) = x^2 - 9y + y^3$ and classify them.

First $\dfrac{\partial f}{\partial x} = 2x$ and $\dfrac{\partial f}{\partial y} = -9 + 3y^2$.

Solving
$$2x = 0 \quad \text{and} \quad -9 + 3y^2 = 0$$
yields
$$(0, \sqrt{3}) \quad \text{and} \quad (0, -\sqrt{3})$$

as the only critical points. Now
$$\frac{\partial^2 f}{\partial x^2} = 2, \qquad \frac{\partial^2 f}{\partial y^2} = 6y, \qquad \frac{\partial^2 f}{\partial y\,\partial x} = 0.$$

Hence, at $(0, \sqrt{3})$,
$$\left(\frac{\partial^2 f}{\partial y\,\partial x}\right)^2 - \left(\frac{\partial^2 f}{\partial x^2}\right)\left(\frac{\partial^2 f}{\partial y^2}\right) = 0 - (2)(6\sqrt{3}) < 0$$

and $\partial^2 f/\partial x^2 > 0$. Therefore, f has a local minimum at $(0, \sqrt{3})$. However, at $(0, -\sqrt{3})$,
$$\left(\frac{\partial^2 f}{\partial y\,\partial x}\right)^2 - \left(\frac{\partial^2 f}{\partial x^2}\right)\left(\frac{\partial^2 f}{\partial y^2}\right) = 0 - (2)(-6\sqrt{3}) > 0.$$

Thus, f has a saddle point at $(0, -\sqrt{3})$.

EXAMPLE 6   Show that the rectangular box of given volume with minimum surface area is a cube.

Let x, y, z be the dimensions of the rectangular box. Then the volume V is given by
$$V = xyz$$
and the surface area is given by
$$S = 2xy + 2xz + 2yz.$$
Since $z = V/xy$,
$$S = S(x, y) = 2xy + \frac{2V}{y} + \frac{2V}{x}.$$

We seek the point at which S has a minimum value. Now
$$\frac{\partial S}{\partial x} = 2y - \frac{2V}{x^2} \quad \text{and} \quad \frac{\partial S}{\partial y} = 2x - \frac{2V}{y^2}.$$

Observe that $\partial S/\partial x$ and $\partial S/\partial y$ do not exist at $(0, 0)$. Since the dimensions of the box are positive, the critical point $(0, 0)$ yields no solution to the problem. The solution of the system
$$2y - \frac{2V}{x^2} = 0 \quad \text{and} \quad 2x - \frac{2V}{y^2} = 0$$

is $(\sqrt[3]{V}, \sqrt[3]{V})$.

Let us compute

$$\left(\frac{\partial^2 S}{\partial y\,\partial x}\right)^2 - \frac{\partial^2 S}{\partial x^2}\left(\frac{\partial^2 S}{\partial y^2}\right)$$

at $(\sqrt[3]{V}, \sqrt[3]{V})$:

$$\frac{\partial^2 S}{\partial x^2} = +\frac{4V}{x^3}, \qquad \text{and at } (\sqrt[3]{V}, \sqrt[3]{V})\ \frac{\partial^2 S}{\partial x^2} = 4;$$

$$\frac{\partial^2 S}{\partial y^2} = \frac{4V}{y^3}, \qquad \text{and at } (\sqrt[3]{V}, \sqrt[3]{V})\ \frac{\partial^2 S}{\partial y^2} = 4;$$

$$\frac{\partial^2 S}{\partial y\,\partial x} = 2.$$

Therefore, at $(\sqrt[3]{V}, \sqrt[3]{V})$,

$$\left(\frac{\partial^2 S}{\partial y\,\partial x}\right)^2 - \left(\frac{\partial^2 S}{\partial x^2}\right)\left(\frac{\partial^2 S}{\partial y^2}\right) = 4 - (4)(4) = -12 < 0.$$

Moreover,

$$\frac{\partial^2 S}{\partial x^2} > 0 \qquad \text{at } (\sqrt[3]{V}, \sqrt[3]{V}),$$

and so S has a local minimum at $(\sqrt[3]{V}, \sqrt[3]{V})$. Since it is the only critical point besides $(0, 0)$, $S(\sqrt[3]{V}, \sqrt[3]{V})$ is the minimum value of S.

Finally, $V = xyz$, $x = \sqrt[3]{V}$, $y = \sqrt[3]{V}$ imply that

$$z = \sqrt[3]{V};$$

thus the box is a cube.

EXAMPLE 7   Find the point on the plane $2x - y + z = 2$ that is closest to the origin.
The distance from $(0, 0, 0)$ to (x, y, z) is

$$\sqrt{x^2 + y^2 + z^2}.$$

Since $z = 2 - 2x + y$, we wish to minimize

$$\sqrt{x^2 + y^2 + (2 - 2x + y)^2}.$$

To make the computation simpler, let us minimize the square of the distance; that is, we seek the point where

$$f(x, y) = x^2 + y^2 + (2 - 2x + y)^2$$

is a minimum.

$$f(x, y) = 5x^2 + 2y^2 - 8x - 4xy + 4y + 4$$
$$f_x(x, y) = 10x - 8 - 4y$$
$$f_y(x, y) = 4y - 4x + 4.$$

The solution to the system

$$10x - 4y = 8 \quad \text{and} \quad -4x + 4y = -4$$

is $x = \frac{2}{3}$, $y = -\frac{1}{3}$ (for which $z = \frac{1}{3}$).

Since there is no point on the plane which is farthest from the origin, the desired point is $(\frac{2}{3}, -\frac{1}{3}, \frac{1}{3})$.

EXAMPLE 8   A manufacturer produces x items of a product A and y items of a product B. His cost and revenue are given by

$$C(x, y) = x^2 + 200x + \tfrac{1}{10}y^2 - 100y + xy$$

and

$$R(x, y) = 1800x + 780y - \tfrac{1}{3}x^2 - y^2 + xy + 5000.$$

What level of production for items A and B will give maximum profit?

Since the profit P is given by

$$P(x, y) = R(x, y) - C(x, y),$$

we shall determine x and y such that

$$P(x, y) = -\tfrac{4}{3}x^2 - \tfrac{11}{10}y^2 + 1600x + 880y + 5000$$

is a minimum. First

$$\frac{\partial P}{\partial x} = -\frac{8}{3}x + 1600$$

$$\frac{\partial P}{\partial y} = -\frac{11}{5}y + 880.$$

Solving

$$-\tfrac{8}{3}x + 1600 = 0 \quad \text{and} \quad -\tfrac{11}{5}y + 880 = 0$$

yields

$$(600, 400)$$

as the only critical point.

Since

$$\frac{\partial^2 P}{\partial x^2} = -\frac{8}{3} < 0$$

and

$$\left(\frac{\partial^2 P}{\partial y \, \partial x}\right)^2 - \left(\frac{\partial^2 P}{\partial x^2}\right)\left(\frac{\partial^2 P}{\partial y^2}\right) = 0 - \left(-\frac{8}{3}\right)\left(-\frac{11}{5}\right) < 0,$$

P has its maximum value at $(600, 400)$.

EXAMPLE 9   Find the maximum and minimum values of $f(x, y) = 2y - x^2 - y^2$ on the rectangle $-1 \le x < 1$, $-1 \le y \le 2$.

The maximum and minimum values occur either at points on the boundary or at points in the interior of the rectangle. If these values are assumed at an interior point, then f has a local extreme value there.

From Example 2, we observed that the interior point

$$(0, 1)$$

is a critical point of f. Therefore

$$f(0, 1) = 1$$

is a candidate for maximum value. The other possibilities occur on the boundary. Let us look at the functional values on each side of the rectangle (see Figure 7.18):

$$\phi_1(x) = f(x, -1) = -x^2 - 3, \qquad -1 \le x \le 1,$$
$$\phi_2(y) = f(1, y) = -1 + 2y - y^2, \qquad -1 \le y \le 2,$$
$$\phi_3(x) = f(x, 2) = -x^2, \qquad -1 \le x \le 1,$$
$$\phi_4(y) = f(-1, y) = -1 + 2y - y^2, \qquad -1 \le y \le 2.$$

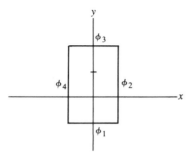

Figure 7.18

We seek the maximum and minimum values of each of these functions on the given intervals.

Since $\phi_1'(x) = -2x$ and $\phi_1'(0) = 0$, $\phi_1(0) = -3$ is the maximum value of ϕ_1, and $\phi_1(1) = \phi_1(-1) = -4$ is the minimum value of ϕ_1. Now $\phi_2'(y) = 2 - 2y$ and $\phi_2'(1) = 0$; therefore, $\phi_2(1) = 0$ is the maximum value of ϕ_2. Since $\phi_2(-1) = -4$ and $\phi_2(2) = -1$, we see that $\phi_2(-1) = -4$ is the minimum value of ϕ_2. It is easy to see that $\phi_3(0) = 0$ is the maximum value of ϕ_3 and that $\phi_3(-1) = \phi_3(1) = -1$ is the minimum value of ϕ_3. Finally, $\phi_4(y) = \phi_2(y)$.

Let us summarize: The only candidates for the maximum and minimum values of f are

$$f(0, 1) = 1$$
$$f(0, -1) = -3$$
$$f(1, -1) = -4$$
$$f(1, 1) = 0$$
$$f(0, 2) = 0$$
$$f(1, 2) = -1.$$

Hence 1 is the maximum value and -4 is the minimum value of f on the given rectangle.

EXERCISES

1. Determine the critical points of each of the following and classify them.

   (a) $f(x, y) = x^2 + xy + y^2 - 6x + 2$.

   (b) $f(x, y) = -x^2 + xy - y^2 + 4x + 2y$.

   (c) $f(x, y) = 2x + 2y - x^2 + y^2 + 5$.

   (d) $f(x, y) = x^3 + y^3 - 3xy$.

   (e) $f(x, y) = xy + 8/y$.

   (f) $f(x, y) = |x| + |y|$.

2. Determine the smallest value of $x^2 + xz + 2y^2$ subject to the restriction that $2x + y - z = 3$.

3. Find the point on the plane $x + 2y + z = 6$ which is closest to the origin.

4. A rectangular box is to have surface area of 100. Find the dimensions which will maximize the volume.

5. A rectangular box with no top is to have surface area of 100. Find the dimensions which will maximize the volume.

6. An aquarium with a volume of 18 cubic feet is to be constructed in the form of a rectangular box with no top. If the base costs $3 per square foot, the front face $2 per square foot, and the side and back faces $1 per square foot, find the dimensions of the aquarium for which the cost will be least.

TABLE OF COMMON LOGARITHMS

n	$\log_{10} n$	$\log_{10} n.1$	$\log_{10} n.2$	$\log_{10} n.3$	$\log_{10} n.4$	$\log_{10} n.5$	$\log_{10} n.6$	$\log_{10} n.7$	$\log_{10} n.8$	$\log_{10} n.9$
10	1.0000	1.0043	1.0086	1.0128	1.0170	1.0212	1.0253	1.0294	1.0334	1.0374
11	1.0414	1.0453	1.0492	1.0531	1.0569	1.0607	1.0645	1.0682	1.0719	1.0755
12	1.0792	1.0828	1.0864	1.0899	1.0934	1.0969	1.1004	1.1038	1.1072	1.1106
13	1.1139	1.1173	1.1206	1.1239	1.1271	1.1303	1.1335	1.1367	1.1399	1.1430
14	1.1461	1.1492	1.1523	1.1553	1.1584	1.1614	1.1644	1.1673	1.1703	1.1732
15	1.1761	1.1790	1.1818	1.1847	1.1875	1.1903	1.1931	1.1959	1.1987	1.2014
16	1.2041	1.2068	1.2095	1.2122	1.2148	1.2175	1.2201	1.2227	1.2253	1.2279
17	1.2304	1.2330	1.2355	1.2380	1.2405	1.2430	1.2455	1.2480	1.2504	1.2529
18	1.2553	1.2577	1.2601	1.2625	1.2648	1.2672	1.2695	1.2718	1.2742	1.2765
19	1.2788	1.2810	1.2833	1.2856	1.2878	1.2900	1.2923	1.2945	1.2967	1.2989
20	1.3010	1.3032	1.3054	1.3075	1.3096	1.3118	1.3139	1.3160	1.3181	1.3201
21	1.3222	1.3243	1.3263	1.3284	1.3304	1.3324	1.3345	1.3365	1.3385	1.3404
22	1.3424	1.3444	1.3464	1.3483	1.3502	1.3522	1.3541	1.3560	1.3579	1.3598
23	1.3617	1.3636	1.3655	1.3674	1.3692	1.3711	1.3729	1.3747	1.3766	1.3784
24	1.3802	1.3820	1.3838	1.3856	1.3874	1.3892	1.3909	1.3927	1.3945	1.3962
25	1.3979	1.3997	1.4014	1.4031	1.4048	1.4065	1.4082	1.4099	1.4116	1.4133
26	1.4150	1.4166	1.4183	1.4200	1.4216	1.4232	1.4249	1.4265	1.4281	1.4298
27	1.4314	1.4330	1.4346	1.4362	1.4378	1.4393	1.4409	1.4425	1.4440	1.4456
28	1.4472	1.4487	1.4502	1.4518	1.4533	1.4548	1.4564	1.4579	1.4594	1.4609
29	1.4624	1.4639	1.4654	1.4669	1.4683	1.4698	1.4713	1.4728	1.4742	1.4757
30	1.4771	1.4786	1.4800	1.4814	1.4829	1.4843	1.4857	1.4871	1.4886	1.4900
31	1.4914	1.4928	1.4942	1.4955	1.4969	1.4983	1.4997	1.5011	1.5024	1.5038
32	1.5051	1.5065	1.5079	1.5092	1.5105	1.5119	1.5132	1.5145	1.5159	1.5172
33	1.5185	1.5198	1.5211	1.5224	1.5237	1.5250	1.5263	1.5276	1.5289	1.5302
34	1.5315	1.5328	1.5340	1.5353	1.5366	1.5378	1.5391	1.5403	1.5416	1.5428
35	1.5441	1.5453	1.5465	1.5478	1.5490	1.5502	1.5514	1.5527	1.5539	1.5551
36	1.5563	1.5575	1.5587	1.5599	1.5611	1.5623	1.5635	1.5647	1.5658	1.5670
37	1.5682	1.5694	1.5705	1.5717	1.5729	1.5740	1.5752	1.5763	1.5775	1.5786
38	1.5798	1.5809	1.5821	1.5832	1.5843	1.5855	1.5866	1.5877	1.5888	1.5899
39	1.5911	1.5922	1.5933	1.5944	1.5955	1.5966	1.5977	1.5988	1.5999	1.6010
40	1.6021	1.6031	1.6042	1.6053	1.6064	1.6075	1.6085	1.6096	1.6107	1.6117
41	1.6128	1.6138	1.6149	1.6160	1.6170	1.6180	1.6191	1.6201	1.6212	1.6222
42	1.6232	1.6243	1.6253	1.6263	1.6274	1.6284	1.6294	1.6304	1.6314	1.6325
43	1.6335	1.6345	1.6355	1.6365	1.6375	1.6385	1.6395	1.6405	1.6415	1.6425
44	1.6435	1.6444	1.6454	1.6464	1.6474	1.6484	1.6493	1.6503	1.6513	1.6522
45	1.6532	1.6542	1.6551	1.6561	1.6571	1.6580	1.6590	1.6599	1.6609	1.6618
46	1.6628	1.6637	1.6646	1.6656	1.6665	1.6675	1.6684	1.6693	1.6702	1.6712
47	1.6721	1.6730	1.6739	1.6749	1.6758	1.6767	1.6776	1.6785	1.6794	1.6803
48	1.6812	1.6821	1.6830	1.6839	1.6848	1.6857	1.6866	1.6875	1.6884	1.6893
49	1.6902	1.6911	1.6920	1.6928	1.6937	1.6946	1.6955	1.6964	1.6972	1.6981
50	1.6990	1.6998	1.7007	1.7016	1.7024	1.7033	1.7042	1.7050	1.7059	1.7067
51	1.7076	1.7084	1.7093	1.7101	1.7110	1.7118	1.7126	1.7135	1.7143	1.7152
52	1.7160	1.7168	1.7177	1.7185	1.7193	1.7202	1.7210	1.7218	1.7226	1.7235
53	1.7243	1.7251	1.7259	1.7267	1.7275	1.7284	1.7292	1.7300	1.7308	1.7316
54	1.7324	1.7332	1.7340	1.7348	1.7356	1.7364	1.7372	1.7380	1.7388	1.7396

TABLE OF COMMON LOGARITHMS

n	$\log_{10} n$	$\log_{10} n.1$	$\log_{10} n.2$	$\log_{10} n.3$	$\log_{10} n.4$	$\log_{10} n.5$	$\log_{10} n.6$	$\log_{10} n.7$	$\log_{10} n.8$	$\log_{10} n.9$
55	1.7404	1.7412	1.7419	1.7427	1.7435	1.7443	1.7451	1.7459	1.7466	1.7474
56	1.7482	1.7490	1.7497	1.7505	1.7513	1.7520	1.7528	1.7536	1.7543	1.7551
57	1.7559	1.7566	1.7574	1.7582	1.7589	1.7597	1.7604	1.7612	1.7619	1.7627
58	1.7634	1.7642	1.7649	1.7657	1.7664	1.7672	1.7679	1.7686	1.7694	1.7701
59	1.7709	1.7716	1.7723	1.7731	1.7738	1.7745	1.7752	1.7760	1.7767	1.7774
60	1.7782	1.7789	1.7796	1.7803	1.7810	1.7818	1.7825	1.7832	1.7839	1.7846
61	1.7853	1.7860	1.7868	1.7875	1.7882	1.7889	1.7896	1.7903	1.7910	1.7917
62	1.7924	1.7931	1.7938	1.7945	1.7952	1.7959	1.7966	1.7973	1.7980	1.7987
63	1.7993	1.8000	1.8007	1.8014	1.8021	1.8028	1.8035	1.8041	1.8048	1.8055
64	1.8062	1.8069	1.8075	1.8082	1.8089	1.8096	1.8102	1.8109	1.8116	1.8122
65	1.8129	1.8136	1.8142	1.8149	1.8156	1.8162	1.8169	1.8176	1.8182	1.8189
66	1.8195	1.8202	1.8209	1.8215	1.8222	1.8228	1.8235	1.8241	1.8248	1.8254
67	1.8261	1.8267	1.8274	1.8280	1.8287	1.8293	1.8299	1.8306	1.8312	1.8319
68	1.8325	1.8331	1.8338	1.8344	1.8351	1.8357	1.8363	1.8370	1.8376	1.8382
69	1.8388	1.8395	1.8401	1.8407	1.8414	1.8420	1.8426	1.8432	1.8439	1.8445
70	1.8451	1.8457	1.8463	1.8470	1.8476	1.8482	1.8488	1.8494	1.8500	1.8506
71	1.8513	1.8519	1.8525	1.8531	1.8537	1.8543	1.8549	1.8555	1.8561	1.8567
72	1.8573	1.8579	1.8585	1.8591	1.8597	1.8603	1.8609	1.8615	1.8621	1.8627
73	1.8633	1.8639	1.8645	1.8651	1.8657	1.8663	1.8669	1.8675	1.8681	1.8686
74	1.8692	1.8698	1.8704	1.8710	1.8716	1.8722	1.8727	1.8733	1.8739	1.8745
75	1.8751	1.8756	1.8762	1.8768	1.8774	1.8779	1.8785	1.8791	1.8797	1.8802
76	1.8808	1.8814	1.8820	1.8825	1.8831	1.8837	1.8842	1.8848	1.8854	1.8859
77	1.8865	1.8871	1.8876	1.8882	1.8887	1.8893	1.8899	1.8904	1.8910	1.8915
78	1.8921	1.8927	1.8932	1.8938	1.8943	1.8949	1.8954	1.8960	1.8965	1.8971
79	1.8976	1.8982	1.8987	1.8993	1.8998	1.9004	1.9009	1.9015	1.9020	1.9025
80	1.9031	1.9036	1.9042	1.9047	1.9053	1.9058	1.9063	1.9069	1.9074	1.9079
81	1.9085	1.9090	1.9096	1.9101	1.9106	1.9112	1.9117	1.9122	1.9128	1.9133
82	1.9138	1.9143	1.9149	1.9154	1.9159	1.9165	1.9170	1.9175	1.9180	1.9186
83	1.9191	1.9196	1.9201	1.9206	1.9212	1.9217	1.9222	1.9227	1.9232	1.9238
84	1.9243	1.9248	1.9253	1.9258	1.9263	1.9269	1.9274	1.9279	1.9284	1.9289
85	1.9294	1.9299	1.9304	1.9309	1.9315	1.9320	1.9325	1.9330	1.9335	1.9340
86	1.9345	1.9350	1.9355	1.9360	1.9365	1.9370	1.9375	1.9380	1.9385	1.9390
87	1.9395	1.9400	1.9405	1.9410	1.9415	1.9420	1.9425	1.9430	1.9435	1.9440
88	1.9445	1.9450	1.9455	1.9460	1.9465	1.9469	1.9474	1.9479	1.9484	1.9489
89	1.9494	1.9499	1.9504	1.9509	1.9513	1.9518	1.9523	1.9528	1.9533	1.9538
90	1.9542	1.9547	1.9552	1.9557	1.9562	1.9566	1.9571	1.9576	1.9581	1.9586
91	1.9590	1.9595	1.9600	1.9605	1.9609	1.9614	1.9619	1.9624	1.9628	1.9633
92	1.9638	1.9643	1.9647	1.9652	1.9657	1.9661	1.9666	1.9671	1.9675	1.9680
93	1.9685	1.9689	1.9694	1.9699	1.9703	1.9708	1.9713	1.9717	1.9722	1.9727
94	1.9731	1.9736	1.9741	1.9745	1.9750	1.9754	1.9759	1.9763	1.9768	1.9773
95	1.9777	1.9782	1.9786	1.9791	1.9795	1.9800	1.9805	1.9809	1.9814	1.9818
96	1.9823	1.9827	1.9832	1.9836	1.9841	1.9845	1.9850	1.9854	1.9859	1.9863
97	1.9868	1.9872	1.9877	1.9881	1.9886	1.9890	1.9894	1.9899	1.9903	1.9908
98	1.9912	1.9917	1.9921	1.9926	1.9930	1.9934	1.9939	1.9943	1.9948	1.9952
99	1.9956	1.9961	1.9965	1.9969	1.9974	1.9978	1.9983	1.9987	1.9991	1.9996

ANSWERS TO SELECTED EXERCISES

SECTION I.2

1. (a) True   (f) False
     (c) False   (h) False
     (d) True

2. (a) False   (f) False
     (c) True   (h) True
     (d) True

3. (a) False   (f) False
     (c) False   (h) False
     (d) True

4. (a) True
     (d) False

5. (b) True
     (d) True
     (e) False

7. (b) True
     (f) False
     (g) True

10. (c) False   (f) True
      (e) False   (g) True

SECTION I.3

2. (a) True   (i) True
     (d) False   (l) False

6. $-\frac{1}{2} = -0.50\bar{0}\ldots$
     $\frac{13}{3} = 4.33\bar{3}\ldots$
     $\frac{1}{7} = .142857\overline{142857}\ldots$

8. (a) $2.2358\ldots$
     (b) $\frac{22}{10}, \frac{223}{100}, \frac{2235}{1000}, \frac{22.358}{10.000}, \ldots$

10. $3.1462\ldots$

3. (a) 4   (c) No
     (b) -13 or -14

7. (a) $\frac{2}{3}$ (e) $\frac{1}{9999}$
     (c) $\frac{28}{99}$ (g) 1

11. (a) $8.82497\ldots$
      (b) $7.82497\ldots$

14. (b) $2\sqrt{5}$ (e) $\frac{9}{5}$ **15.** (b) rational
    (d) $\frac{2}{7}$ (f) $4\sqrt{3}$ (d) irrational
16. (b) False   (e) False
    (c) True

SECTION I.4

1. (b) $-\frac{5}{3}$ (e) $1, -1$ (f) $-3, -2$
2. (b) $-\frac{1}{3}$ ·   (e) 2 (f) never zero
3. (a) $0, \frac{3}{4}$ (c) $-4, 5$
4. (a) $x > -\frac{5}{2}$ (g) $x \geq 5$ or $x \leq 1$
    (c) $x \geq \frac{3}{2}$ (i) $x < -5$ or $-2 < x < 1$
    (d) $x > -\frac{4}{5}$ (j) $x > 2$ or $x < -2$
6. (a) zero when $x = 0$; not defined when $x = 1$; positive when $x > 1$ or
        $x < 0$; negative when $0 < x < 1$
    (c) zero when $x = 1$; not defined when $x = 0$, $x = 2$; positive when $x > 2$
        or $0 < x < 1$; negative when $1 < x < 2$ or $x < 0$
    (g) zero when $x = 1$; always defined; positive when $x > 1$; negative when
        $x < 1$
7. $r = 2$, $s = 3$, $t = -1$

SECTION I.5

1. (a) 5 (c) 1 (g) $1/\sqrt{3}$
3. (b) no solution   (c) 3 or -3
4. (b) $-3 < x < 2$ (d) $-3 < x < 2$, $x \neq -\frac{1}{2}$
5. (b) $x \leq -3$ or $x \geq 3$ (d) $x \leq -\frac{7}{2}$ or $x \geq -\frac{3}{2}$
7. (a) $\frac{3}{2}$ (c) $\frac{5}{6}$
16. (a) $|3x|$ (c) $|x/x + 1|$

SECTION 1.1

3. (a) $5\sqrt{2}$ (d) $\sqrt{2}$ (g) $\sqrt{61}$
4. (a) $(-\frac{3}{2}, \frac{7}{2})$ (c) $(1, 3)$ (e) $(\frac{11}{4}, \frac{1}{4})$
5. (a) Yes   (d) No
6. (a) isosceles triangle   (c) collinear
7. (a) $(4, 4)$, square of side $\sqrt{5}$ (b) $(\frac{7}{2}, \frac{5}{2})$

SECTION 1.2

1. (a) $\frac{1}{5}$ (c) $\frac{21}{2}$ (e) 0 (d) $-\frac{3}{4}$ (f) no slope
2. (b) right triangle   (d) collinear
5. (a) parallel   (b) intersect but not perpendicular   (d) perpendicular
6. (a) $(1, -1)$ or $(-1, 1)$

SECTION 1.3

1. (b) $x - y - 2 = 0$ (d) $x = -2$ (f) $\sqrt{22}y = x$
2. (a) $y = 3x$ (c) $3y = 2x - 10$ (f) $2y = 2 - x$
3. (a) $y = 2$, $x = 1$ (c) $y = 1$, $x = 0$
5. (b) $y = -2$ (d) $y = -\sqrt{2}x + \frac{1}{3}$
6. (a) $m = 2$, $b = \frac{2}{3}$ (d) $m = -\frac{1}{4}$, $b = \frac{3}{4}$
8. (a) $x/4 + y/5 = 1$ (c) $4x - 15y = 3$
9. (c) $a = 3$, $b = \frac{3}{4}$ (f) both intercepts 0
10. (b) $x + y = 5$
12. (b) $2y = -7x + 3$ (d) $y = 5$ (g) $2y = x + 11$
13. (b) $7y = 6x - 8$, $5y = -4x + 15$
18. 3
19. (b) $(\frac{7}{6}, 1)$ (f) $(-\frac{39}{29}, -\frac{7}{29})$

SECTION 1.4

1. (a) $x^2 + y^2 = 9$ (c) $(x + \frac{7}{3})^2 + (y - 4)^2 = \frac{1}{4}$
2. (c) center $(2, 3)$, radius $\sqrt{2}$
   (f) center $(1, -3)$, radius 1
4. (c) one point $(-3, -7)$ (d) no graph
5. (c) $(x + 2)^2 = -4(y + 1)$
6. (a) $(x - 2)^2 = -4(y + 2)$ (b) $(y + 4)^2 = -8(x + 1)$
9. $(1 + \sqrt{3}, 1 + \sqrt{3})$, $(1 - \sqrt{3}, 1 - \sqrt{3})$
10. (b) $-\sqrt{2} \leq k \leq \sqrt{2}$
12. (a) $(0, 0)$ and $(1, 1)$
13. (c) $-\frac{3}{2}$

SECTION 1.5

5. (a) $(x - y)(x + y)(x^2 + y^2)$ (b) $(x + y)^2$ (e) $(x + y)^3$

SECTION 2.1

1. (a) $f(6) = 45$, $g(-1) = -\frac{1}{5}$, $h(4) = 60$, $k(\frac{1}{2}) = \frac{1}{4}$
   (c) $f(g(x)) = 14x/(x^2 + 9) + 3$
   (d) $(7x - 14)/(x - 2)$
2. Some possibilities are:
   (b) $(1, -2)$, $(0, -5)$, $(-1, -8)$, $(\frac{1}{3}, -4)$
   (f) $(0, 0)$, $(1, 2)$, $(-1, 2)$, $(-3, 6)$
   (k) $(1, 1)$, $(2, 1)$, $(-6, 1)$, $(-\frac{7}{2}, 1)$
5. (b) R (c) R except $3, -3$ (d) $-5 \leq x \leq 5$ (i) R
6. Three possibilities are:
   $$f(x) = \sqrt{36 - x^2}$$
   $$g(x) = -\sqrt{36 - x^2}$$
   $$h(x) = \begin{cases} \sqrt{36 - x^2}, & -6 \leq x \leq 2 \\ -\sqrt{36 - x^2}, & 2 < x \leq 6 \end{cases}$$
8. (b) perimeter $= 4s$, area $= s^2$

10. $15,000 + 600x$ (x months)

15. $8[x] + 8$

16. (b) yes (c) no (d) yes

19. (a) $|x|$ is even; x^3 is odd; $x^2 + x$ is neither even nor odd

SECTION 2.2

1. (a) $f(g(x)) = (2x + 1)^5$, $g(f(x)) = 2x^5 + 1$
  (d) $f(g(x)) = x/(x + 1)$, $g(f(x)) = x + 1$

3. (c) $f(x) = x^{2/3}$, $g(x) = 7x + 2$
  (f) $f(x) = 3/x$, $g(x) = \sqrt{2x}$

5. $A = \pi t/4$

SECTION 3.1

1. (a) (1) $1, 1, .5, 1.5, .1, 1.9$
    (2) $4, 2, x \uparrow 2, 4$

2. (b) (1) $.5, 8.5, .1, 8.1, .08, 8.08$
    (2) $16, 8, x \downarrow 8, 16$

3. (a) (1) $2, 45, .5, 46.5, .03, 46.97$
    (2) $7, 47, x \uparrow 47, 7$

5. $.5, 5, x \to 5, .5$

6. $-4.875, .5, x \to .5, -4.875$

8. (a) no (b) no (c) yes, 0

SECTION 3.2

1. (a) 12 (c) 18 (e) 0 (g) $-\frac{39}{8}$ (j) 6 (m) 0

2. (b) $\frac{3}{4}$ (d) -2 (e) 0 (f) 4 (h) $\frac{1}{4}$

3. (c) $2x_0$ (e) $12x_0^2$ (g) $-1/x_0^2$

5. (a) 0 (c) 8 (e) -12 (f) 3

6. (a) does not exist (b) 1 (f) -3.03 (i) 0

SECTION 3.3

5. (a) no discontinuities (g) 1
  (b) 1 (i) each positive integer
  (d) 7 (l) no discontinuities

6. (a) all points (b) all points except 0 (c) all points except $\frac{1}{2}$

7. In reference to exercise 5:
  (b) jump (d) removable (g) neither jump nor removable

12. maximum .98, minimum $-.9$

SECTION 3.4

1. (b) $+\infty, +\infty$ (c) $-\infty, +\infty$

2. (a) $+\infty, -\infty$ (b) $-\infty, -\infty$

3. (a) 0   (b) $+\infty$ (c) $+\infty$
4. (a) 0   (b) $\frac{3}{2}$ (c) $-\infty$ (d) -5
5. (a) (1) $+\infty$, $-\infty$
    (2) 1, 1
7. (a) $+\infty$ (d) $+\infty$ (g) 0   (i) 0
8. (a) $-\infty$ (b) $\frac{2}{3}$

SECTION 4.2

4. (a) 0   (h) $2/x^{3/2}$
    (c) $(\frac{2}{13})x$ (j) $(-\frac{40}{7})x^{-17/7}$
    (e) $3/(2\sqrt{x})$
5. (a) $6x^2$, 0, 96   (c) 12, 12, 12
6. (b) $-2 + 14x$ (e) $1/(2\sqrt{x+1})$
    (d) $-2x/(x^2 + 1)^2$ (g) $-1/x^2$
7. (a) 1, -1, nondifferentiable
    (c) 0, 0, differentiable
    (d) 0, $\frac{1}{1000}$, nondifferentiable
9. (a) 0   **10.** (d) yes   (e) yes
11. (d) no   (e) no

SECTION 4.3

1. (a) -2 (c) $\frac{1}{6}$ (d) $-\frac{1}{48}$
2. (a) $2x + y + 1 = 0$ (c) $6y - x = 9$ (d) $x + 48y = 32$
4. (a) continuous everywhere, not differentiable at 7
    (c) continuous and differentiable everywhere
    (e) continuous and differentiable everywhere
    (f) discontinuous at -1 and not differentiable at -1, 1
5. (a) no   (d) no   (c) yes
6. $+\infty$, $+\infty$ **7.** $+\infty$ **8.** no

SECTION 4.4

1. (a) $15x^2 - 12x + 2$ (h) $18x^2 + 38x + 2$
    (d) $1 + 1/x^2$ (k) $(-7x^2 + 2x + 28)/(x^2 + 4)^2$
    (e) $-1/(3x^{2/3}) - 1/x^{3/2}$ (m) $(1 - x)/[2\sqrt{x}(x + 1)^2]$
    (f) $8x - 4$
2. (a) $30x - 12$ (f) 8
    (d) $-2/x^3$ (h) $36x + 38$
    (e) $3/(2x^{5/2}) + 2/(9x^{5/3})$
3. (b) $y - 8x + 3 = 0$, $x + 8y - 41 = 0$
    (e) $3y - x = 7$, $y + 3x = 29$
4. (a) $2f(x)f'(x)$ (b) $-g'(x)/[g(x)]^2$
    (d) $f'(x)g(x)h(x) + f(x)g'(x)\,h(x) + f(x)g(x)h'(x)$
5. (a) $2[f(x)]^2, f(x)$ (b) $[f(x)]^2 - [g(x)]^2$

SECTION 4.5

1. $-30x(1 - 5x^2)^2$
2. (d) $-30(6x - 5)^{-6}$ (g) $x/\sqrt{x^2 + 5}$
   (h) $[3x(1 - x)/\sqrt{3x^2 + 1}] - \sqrt{3x^2 + 1}$
   (k) $30[(3x - 2)^5 + 8](3x - 2)^4$
3. (e) $-6(x - 1/x)^2/x^3 + 6(1 + 1/x^2)^2(x - 1/x)$
5. (b) x^2/y^2 (c) $(3x^2 + 2y)/(2y - 2x)$
6. (b) $3y - 2x + 5 = 0$
7. (b) $2f'(2x)$ (c) $f'(x)/(2\sqrt{f(x)})$
   (e) $-6[f(-2x)]^2f'(-2x)$

SECTION 4.6

1. (a) max. 8, min. -22 (f) max. 3, min. 0
   (c) max. 9, min. 8   (j) max. $\frac{1}{2}$, min. $-\frac{1}{2}$
3. 12, 12   **4.** no solution   **5.** $\sqrt{3}/2$
7. length $4R/\sqrt{5}$, height $R/\sqrt{5}$ **9.** radius $(\frac{2}{3})r$, height $(\frac{1}{3})h$
13. 25, 25   **15.** $\sqrt[3]{4}, \sqrt[3]{4}, \sqrt[3]{4}$
17. $P = \$137.50$ will minimize cost

SECTION 4.7

1. (b) $\frac{3}{2}$ (c) $\frac{9}{4}$
2. (a) $-\frac{1}{2}$ (c) 1
3. no
6. (a) $x < 0, 0 < x < 1, 1 < x < 2, x > 2$

SECTION 4.8

1. (a) always decreasing, concave up if $x > 0$, concave down if $x < 0$, no points of inflection
   (d) increasing if $-3 < x < 0$, decreasing if $0 < x < 3$, always concave down, no points of inflection
   (h) neither increasing nor decreasing on any interval, neither concave up nor concave down on any interval, no points of inflection
   (j) decreasing if $x < 0$, increasing if $x > 0$, always concave up, no points of inflection
2. (c) increasing if $x > \frac{1}{2}$, decreasing if $x < \frac{1}{2}$, always concave up, no points of inflection
   (d) increasing if $x > 1$ or $x < -1$, decreasing if $-1 < x < 1$, concave up if $x > 0$, concave down if $x < 0$, $(0, 1)$ is a point of inflection.
   (g) always decreasing, concave up if $x > 1$, concave down if $x < 1$, no points of inflection
3. (a) local max. at 0   (g) no local extreme points
   (c) local min. at 0   (h) local min. at 0 and local max. at 1

4. (a) local min. at 4   (c) local max. at 3, local min. at -3

    (g) no local extreme points   (i) local max. at $\frac{1}{256}$

7. f increasing when $\phi(x) \geq 0$ and decreasing when $\phi(x) \leq 0$, g decreasing

8. g increasing

SECTION 4.10

1. (b) 12   (d) 18   (f) 24

2. (b) $v(t) = 16 - 8t$, $a(t) = -8$

    (e) $v(t) = 1 - 9/t^2$, $a(t) = 18/t^3$

3. (a) 3   (b) $\frac{1}{2}$

5. 216π **7.** $32/(9\pi h^2)$ **9.** $10\sqrt{61}$

11. max. height 1600 at $t = 10$; returns to earth after 20 seconds with a velocity of -320

13. (a) $x'(t) = 2$, $y'(t) = 2t$ (b) $4y = x^2 - 4$

SECTION 4.11

1. $15 + .010x$ **3.** $165 - .02x$

4. (a) $-750 + 150x - .015x^2$ (b) 5000

5. \$1.33   **6.** 1000   **7.** 600

9. $(3 - x^2)/(3 + x^2)^2$ **11.** $(5 - x/30)^2(5 - 4x/30)$

SECTION 4.12

1. (c) $(\frac{3}{2})x^{4/3}$ (e) $x^2 - 7x$

    (f) $(\frac{6}{5})x^5 - (\frac{3}{4})x^4 + (\frac{2}{3})x^3 - (\frac{1}{2})x^2 + x$

    (h) $(\frac{5}{7})x^{7/5} - (\frac{1}{2})x^2$ (i) $(\frac{1}{18})(6x - 3)^3$

    (k) $(\frac{1}{15})(x^3 + 3x - 6)^5$ (m) $(-\frac{4}{15})(1 - 5x)^{3/4}$

    (o) $(\frac{5}{11})x^{11/5}$ (q) $(\frac{24}{5})x^{5/6} + 4x^{3/4}$

    (r) $1/[2(1 + x)^2] - 1/(1 + x)$

4. $(\frac{2}{3})x^3 - 2x^{3/2} - 431$ **6.** $(\frac{1}{3})t^3 - (\frac{3}{2})t^2 + 2t$

10. (a) $f(x)g(x)$ (b) $f(x)/g(x)$

11. $2500 - 20x + .04x^2 + .002x^3$

SECTION 5.2

1. $S_4(l) = \frac{11}{2}$, $S_4(r) = \frac{13}{2}$

3. (a) $S_5(l) = 60$, $S_5(r) = 110$ (b) $S_6(m) = 82.51$

6. (b) $\displaystyle\sum_{k=1}^{5} 2k$ (c) $\displaystyle\sum_{k=3}^{7} k$

7. (a) 60   (b) 112   (c) 220

SECTION 5.3

1. (a) 2   (c) $\frac{3}{2}$

2. (b) $-\frac{3}{2}$

4. (a) $p = 2\sqrt{2}$, $q = 2\sqrt{82}$ (b) $p = \frac{1}{2}$, $q = 1$
5. 3   **11.** no

SECTION 5.4

1. (a) $-\frac{5}{2}$ (d) $\frac{19}{3}$ (h) 4   (j) $(\frac{1}{3})(13^{3/2} - 27)$
2. (b) $\frac{19}{15}$ (d) $\frac{5}{6}$
3. (a) $x^2/\sqrt{1 + x^2}$ (c) $1 + 2x \int_0^x dt/(1 + t^2)$
7. (b) no   **8.** no; $f(x) \neq f(x) - f(a)$
9. (b) $f(b)g(b) - f(a)g(a)$ **15.** (b) (ii) $-6xf(3x^2)$

SECTION 5.5

1. (a) $(\frac{2}{3})[7^{3/2} - 8]$ (c) 0
    (b) $(\frac{1}{9})[(\frac{1}{17})^9 - (\frac{1}{21})^9]$ (d) $\frac{26}{3}$
    (h) $(\frac{2}{7})2^{7/2} - (\frac{8}{5})2^{5/2} + (\frac{8}{3})2^{3/2} - \frac{142}{105}$

SECTION 5.6

1. (a) $\frac{81}{2}$ (d) $\frac{5}{3}$ (g) $\frac{513}{9}$
2. (a) $\frac{32}{3}$ (d) $\frac{1}{3}$ (e) 2   (h) $\frac{1}{4}$ (i) 18   (m) $\frac{1}{4}$ (o) $\frac{1}{6}$
4. $\pi a^2/4$ **5.** πab **8.** (c) no
9. (c) yes, 3   **10.** (c) yes, 6   **11.** (c) no

SECTION 5.7

1. (a) $V_x = 32\pi/5$, $V_y = 8\pi$ (f) $V_x = 65536\pi/7$, $V_y = 16384\pi$
    (e) $V_x = 16\pi/3$, $V_y = 16\pi/3$ (h) $V_x = (\frac{32}{3})\pi$, $V_y = 8\pi/3$
3. (c) 18π **5.** $(\frac{665}{6})\pi$ **7.** $\frac{535}{2}$ **8.** (c) yes, 2π

SECTION 5.8

1. $\frac{27}{2}$ **3.** (a) $\frac{1}{8}$ (b) $\frac{1}{2}$ (c) 2
6. 700   **8.** $512k\pi$ **9.** $384k\pi$

SECTION 5.9

1. 304   **3.** 1066   **5.** $14,700
7. $47.20   **13.** $1,388.89

SECTION 5.10

1. $3x - x^2 + (x^3/3) + C$ **4.** $[(3x + 7)^3/9] + C$
7. $(-\frac{2}{3})\sqrt{1 - x^3} + C$
11. $(\frac{2}{7})(2 + x)^{7/2} - (\frac{8}{5})(2 + x)^{5/2} + (\frac{8}{3})(2 + x)^{3/2} + C$
13. (a) $y = 5x + C$ (c) $y = (\frac{2}{3})x^{3/2} - 2x^{1/2} + C$
15. (a) $y^2 = 2x + C$ (c) $1/y = (1/x) + C$ (d) $\sqrt{y} = (x^2/4) + C$

SECTION 6.1

1. (b) 81   (d) $\frac{1}{9}$ (g) $\frac{1}{2}$ (k) -8 (m) b^4 (o) 2
2. $a^5 = AB$, $a^9 = B^3$, $a^{1/2} = A^{1/4}$, $a^{-2} = 1/A$
3. (b) -2 (d) $\frac{1}{2}$ (e) 0   (f) no solution
5. (a) 3   (c) -2 (f) $\frac{1}{2}$ (h) $\frac{2}{3}$ (k) -4
6. (a) $2A$ (c) $B + C$ (f) $2A - C$ (i) $2A + 2B + C$
7. (b) 27   (d) $\pm 10\sqrt{2}$ (e) ± 2

SECTION 6.2

1. (b) $\sqrt{2}$ (c) $\log_{10} 2$ (e) 2
2. (a) e^2 (b) 1   (c) $1/e$ (g) $+\infty$
5. (b) $(\log_6 e)/x$ (c) $[3(\log_2 e)(2 + \log_2 x)^2]/x$
    (e) $2x \log_3 x + x(\log_3 e)$
6. (a) $\log_5 x + C$ (c) 1

SECTION 6.3

1. (a) $-1/x$ (e) $3/x$ (i) $1/(x + 1)$ (m) $x + 2x \log 2x$
2. (a) $(\frac{1}{2}) \log |x| + C$ (f) $(\frac{4}{3})(\log x)^3 + C$
    (d) $2 \log |x + 2| + C$ (h) $(\frac{1}{2}) \log |x^2 + 2x - 5| + C$
3. (a) $1 + \log 2$ (i) $2 + \log 2$
    (d) $\frac{1}{3}$ (l) $(\frac{3}{4}) e^4 - 2e^2 \log 2 - e^2$
    (f) $(\log 11)/6 - (\log 5)/6$
4. (c) tangent line $y = 1/e$, normal line $x = e$
5. (a) $3 \log 2$ (c) $\frac{15}{2} - 8 \log 2$
7. $f'(x) = (-xy^3 - y)/(3x^2 y^2 + x \log x)$
8. (a) $(\frac{1}{4}) f(x)[(2x - 1)/x(x + 1)]$ **9.** (a) $A = 1$, $B = -1$
10. (a) $A = -\frac{2}{25}$, $B = \frac{1}{5}$, $C = \frac{4}{25}$ **12.** (a) $\log (2 + \sqrt{5})$

SECTION 6.4

1. (b) $6xe^{x^2}$ (g) $e^{\sqrt{x}}/(2\sqrt{x})$
    (d) 1   (i) $(7x^8 + 2x)e^{x^7}$
    (e) $e^x/(1 + e^x)$ (k) $(-e^{1/x^2}/x^2)[(2/x^2) + 1]$
2. $(-y^3 - ye^x)/(3xy^2 + e^x)$
3. (a) $(e^{3x}/3) + C$ (g) $(-2/e^{\sqrt{x}}) + C$
    (c) $(\frac{5}{6})e^{6x} + C$ (h) $\log |e^x - e^{-x}| + C$
    (d) $(\frac{1}{2})e^{x^2} + C$
4. (a) $(\frac{1}{3})e^3 - \frac{1}{3}$ (f) $(2/e) - (2/e^2)$
    (c) $e^4 - (1/e)$ (g) 1
5. (a) $+\infty$ (b) $+\infty$ (c) 0   (f) 0   (g) 0
7. (b) $y - 6e^2 x + 3e^2 = 0$, $x + 6e^2 y - 18e^4 - 1 = 0$
8. (b) $3 - (\frac{1}{3})e^{-6} + (\frac{1}{3})e^{-15}$ (d) $e - (\frac{4}{3})$
11. $(1 + \log x)x^x$

13. (a) $\frac{1}{3}$ (b) $\frac{1}{3}$ (c) 0   (d) $\frac{1}{2}$

14. 1   **17.** $g(x) = \sqrt{x}$ **18.** $g(x) = (3x + 2)/2$

SECTION 6.5

1. (a) $m(t) = -(\frac{3}{2})t$, linear decay
(c) $m(t) = -(\frac{3}{2})t^2 + 2$, quadratic decay
(d) $m(t) = (\frac{4}{3})t^3$, cubic growth
(e) $m(t) = e^{5t}$, exponential growth

3. $10 \log (\frac{1}{2})/\log (\frac{4}{5})$ **5.** 225,000

7. $(-\frac{50}{3}) \log (\frac{2}{3})$ **9.** $\log (\frac{3}{8})/\log (\frac{7}{8})$

11. (a) \$1,165.91   (c) \$1,166.11
(b) \$1,165.82   (d) \$1,161.83

13. $40[1 - e^{-.15}]$

SECTION 7.1

2. (b) $2\sqrt{5}$ (d) 4

3. (a) yes   (b) no   (d) no

5. (c) The plane is parallel to the x-y plane and passes through point $(0, 0, e)$.
(h) $(2, 0, 0)$, $(0, -1, 0)$, $(0, 0, \frac{2}{3})$ lie on the plane

6. (e) $x^2 + (y + 2)^2 + (z - 1)^2 = 1$

SECTION 7.2

1. (b) $\mathbf{R}^2$ (c) $x^2 + y^2 \le 9$

2. (b) $\mathbf{R}^2$ except $y = 2x^2$ (d) $\mathbf{R}^2$ except $(0, 0)$ (e) $xy > 0$

3. (a) 0   (c) 3   (e) does not exist   (h) 0

4. all points except $(0, 0)$

5. (a) no   (b) yes, $f(0, 0) = 3$

6. (a) 1, 0   (c) 19, 3

SECTION 7.3

1. (a) $\dfrac{\partial z}{\partial x} = 3x^2y + 7y^2$, $\dfrac{\partial z}{\partial y} = x^3 + 14xy$

(d) $\dfrac{\partial z}{\partial x} = \dfrac{2x}{x^2 + y^2}$, $\dfrac{\partial z}{\partial y} = \dfrac{2y}{x^2 + y^2}$

(e) $\dfrac{\partial z}{\partial x} = e^{x+y}$, $\dfrac{\partial z}{\partial y} = e^{x+y}$

2. (a) $\dfrac{\partial^2 z}{\partial x^2} = 6xy$; $\dfrac{\partial^2 z}{\partial y^2} = 14x$; $\dfrac{\partial^2 z}{\partial y \partial x} = 3x^2 + 14y$; $\dfrac{\partial^2 z}{\partial x \partial y} = 3x^2 + 14y$

(e) $\dfrac{\partial^2 z}{\partial x^2} = e^{x+y}$, $\dfrac{\partial^2 z}{\partial y^2} = e^{x+y}$

4. (a) $\dfrac{\partial f}{\partial x} = -x/z$, $\dfrac{\partial f}{\partial y} = -y/z$

5. (a) 4 (b) $\frac{1}{2}$

SECTION 7.4

1. (a) local min. at $(4, -2)$
(c) saddle point at $(1, -1)$
(d) $(0, 0)$ saddle point; $(1, 1)$ local min.
(e) none
3. $(1, 2, 1)$
6. length 2, depth 3, height 3

INDEX

A

Absolute value, 24
Acceleration:
  average, 205
  instantaneous, 205
Analytic geometry, 29
Antiderivative, 215
Approximation:
  of area, 228
  by tangent line, 163
Archimedes, 1, 237
Area, 1, 266
  of a plane region, 224
Asymptote:
  horizontal, 132
  vertical, 132

B

Barrow, Isaac, 2
Briggs, Henry, 316

C

Chain rule, 168, 174
Circle, 57

Closed interval, 19
Collinearity, 34
Common logarithms, 316
Completion of squares, 58
Compound interest, 348, 349
Concave down, 194
Concave up, 194
Concavity test, 194
Continuity, 115, 366
  on $[a, b]$, 120
  from the left, 119
  at a point, 116
  from the right, 119
  on a set, 367
Converse, 3
Coordinates:
  on a line, 10
  in the plane, 30
  rectangular, 30
  in three-dimensional space, 355
Coordinatized line, 10
Critical points, 376

D

Decimal, 10